The Colors of Mice

This book is gratefully dedicated to Elizabeth S. (Tibby) Russell

After spending trillions of dollars to disassemble nature in the last century, we are just now acknowledging that we have no clue how to continue. . . . Reductionism was the driving force behind much of the 20th century's scientific research. To comprehend nature, it tells us, we first must decipher its components. The assumption is that once we understand the parts it will be easy to grasp the whole. . . . Therefore, for decades we have been forced to see the world through its constituents. We have been trained to study atoms and strings to understand the universe, molecules to comprehend life, individual genes to understand complex human behavior, prophets to see the origins of fads and religions. Now we are close to knowing just about everything there is to know about the pieces, but we are as far as we have ever been from understanding nature as a whole. . . . We have learned that nature is not a well-designed puzzle with only one way to put it back together. In complex systems, the components can fit in so many different ways that it would take billions of years for us to try them all. Yet nature assembles the pieces with the grace and precision honed over millions of years.

Albert-Laszlo Barabasi
Linked, the New Science of Networks.

No clue? Mathematicians and physicists may have no clue, but pigmentation geneticists pioneered, at the very beginning of the last century, the use of mouse models, and these little nodes in the internet of life continue to link our reductionist results with the network(s) that is / are biological reality.

White spotted mouse, genotype $Mitf^{mi}/Mitf^{mi\text{-}sp}$. Photograph by Dr Melanie Ihrig

The Colors of Mice
A Model Genetic Network

M. Lynn Lamoreux
Bryan, Texas, USA

Véronique Delmas
Developmental Genetics of Melanocytes,
Institut Curie, Orsay, France

Lionel Larue
Developmental Genetics of Melanocytes,
Institut Curie, Orsay, France

Dorothy C. Bennett
Molecular and Metabolic Signalling Centre,
St. George's, University of London, UK

WILEY-BLACKWELL
A John Wiley & Sons, Ltd., Publication

Wiley-Blackwell is an imprint of John Wiley & Sons, formed by the merger of Wiley's global Scientific, Technical and Medical business with Blackwell Publishing.

Registered office: John Wiley & Sons Ltd, The Atrium, Southern Gate, Chichester, West Sussex, PO19 8SQ, UK

Other Editorial Offices:
9600 Garsington Road, Oxford, OX4 2DQ, UK
111 River Street, Hoboken, NJ 07030-5774, USA

For details of our global editorial offices, for customer services and for information about how to apply for permission to reuse the copyright material in this book please see our website at www.wiley.com/wiley-blackwell

Library of Congress Cataloging-in-Publication Data
The colors of mice : a model genetic network / M. Lynn Lamoreux ... [et al.].
p. ; cm.
Includes bibliographical references and index.
ISBN 978-1-4051-7954-6 (hardback : alk. paper) 1. Mice–Color–Genetics. 2. Mice as laboratory animals. I. Lamoreux, M. Lynn.
[DNLM: 1. Mice–genetics. 2. Gene Regulatory Networks–physiology. 3. Models, Animal. 4. Models, Genetic. 5. Pigmentation–genetics. 6. Pigments, Biological–genetics. QY 60.R6 C719 2010]

QH470.M52C65 2010
599.35′3138–dc22

2009045866

A catalogue record for this book is available from the British Library.

Set in 10/12.5pt Dante MT by Graphicraft Limited, Hong Kong
Printed and bound in Singapore

1 2010

Contents

Preface

Our understanding of genetic control over living processes underwent a revolution with the hypothesis that one gene encodes one enzyme that performs one function in the cell or in the body. We now know, of course, that life is not so simple and that the functioning of the whole cell, organ, or organism is not an additive phenomenon. To function properly, life relies upon sophisticated interactions among and between molecules, organelles and cells, organs and organisms. At the base of all these levels of biological function is the genetic code. Between that ladder of nucleotides and the resulting phenotype, the complexity of the interacting functions of life remains to be fully explained. Thus, at the same time that the reductionist approach is highly rewarding and is necessary to our understanding of the questions we need to answer, it also imposes conceptual and methodological limits to understanding of biological function. It is too simplistic to represent the interactive reality.

Sequencing, mutagenesis, mapping, transfection experiments, and other yet more elegant methods give us tantalizing and often unexpected information about the regulation of cell function. Such work *in vitro* is more and more clearly defining the questions that must be answered to explain, for example, the functions of the pigment cell. *In vitro* we demonstrate how the biochemical functions, and their interactions, can or could occur, but such studies must be balanced by phenotype-driven genetics if we are to finally determine how they do occur *in vivo*. For example, with regard to gene regulation in any specific instance, is it mediated via:

- availability of substrate (turning genes on and off)?
- relative quantities of interacting protein complexes; competition for active sites?
- compartmentalization in space: transport?

Studies *in vitro* define these questions and many others, and as our technology becomes increasingly elegant, the questions become more focused and sophisticated. However, the final answers will require the whole living environment with all its unknown interacting processes intact; that is, animal models. In this respect, the contribution of laboratory mice with coat-color mutations was recognized from the first days of genetic analysis at the turn of the last century, and has only increased over time: dramatically so in the last decade.

But then another problem must be addressed. We cannot answer sophisticated questions in the absence of strictly defined controls. Animals may differ, one from another, at a multitude of unknown genetic loci that generate an even greater array of unknown biochemical interactions. To define the unexplored territories lying between the code of life imprinted in the DNA and the maintenance of life itself, we require genetically defined animal models that carry defined mutations in defined combinations, segregating against an invariable genetic background, so that differences at each specific gene locus can be evaluated and their functions and interactions compared alone and in combination with other changes at other loci, against uniform and parallel controls, *in vivo*. In the case of pigment mutants, such resources were importantly developed at The Jackson Laboratory (Green 1966; Lauber 1971; Morse 1978) with the development of the inbred mouse. Inbred mice, defined as a group of mice that have been brother/sister mated for 20 generations or more, are now available at several stock centers (see Chapter 7), as are many other specialized mouse models. They should be utilized. We will save ourselves much unnecessary future work if we use appropriate controls within the experiments that we design today. Appropriate controls include biological as well as environmental and experimental ones.

The basis of science is an experimental design that can be repeated and tested in other laboratories. To this end it is essential that the experimental animal model be described in detail when reporting work that includes the use of animals, tissues, or cells. The background genome of these biological materials, though often omitted from the Materials and methods section of published reports, may be as important as the specific gene locus and allele under study.

Our purpose is to present an overview of the processes of pigmentation and the ways in which pigment loci regulate these processes at all the levels of organization of the living organism. We would prefer to cover the field as fully as Silvers, in 1979, was able to do. However, the area of pigment genetics has grown so rapidly since then that we must for the most part limit ourselves to the current state of understanding, pointing primarily to reviews and recent papers that will direct the reader toward their foundations. In so doing, we will unfortunately omit some references to important work that occurred after Silvers was published. Those authors were our mentors, and we do not intend to slight their contribution. The most detailed source of information is the Mouse Genome Informatics database (MGI; www.informatics.jax.org/).

In the Appendix to Chapter 1 we present an index of the mutant loci that are discussed in the text, as well as Tables A1.1 and A1.2, which list of all the loci (cloned and uncloned) that have been reported to affect pigmentation. These lists will continue to be updated and will be available through the International Federation of Pigment Cell Societies (IFPCS) Color Genes initiative, maintained at the website of the European Society for Pigment Cell Research (ESPCR), www.espcr.org/micemut/ (Montoliu et al. 2009).

Acknowledgments

MLL wishes to acknowledge all the people who made possible the maintenance and preservation of the mice that are our subject. These people are far too numerous to mention, but include the following, in no particular sequence: the Department of Veterinary Pathobiology at Texas A&M University, Ann Kier, and especially Jim Womack, Rick Ermel, Melanie Ihrig, Elizabeth Browder, and all the capable people who so competently care for the welfare of the mice at TAMU and before; Aaron Lerner, John Pawelek, Seth Orlow, Ray Boissy, Tom Mayer, Barry Whitney, Mary Lyon, Tony Searle, Bruce Cattanach, Peter Glenister; Matsumoto-sensei, Ozato-sensei, Mizoguchi-sensei, Kawa-san, Wakamatsu-san, and so many people who made my time in Japan productive and enjoyable; Roy Brantley, Chloé King, William Richards and his family, Jo Ann Sette and Julie Wright. I thank Nancy Jenkins for saying the right word at exactly the right moment. Thanks also to members of the pigment cell societies for encouragement, support, and discussions; MMRRC, MGI, and JAX, and the people at these associations who were responsible for preservation of MLL's mutant mouse colony.

We thank St George's, University of London, The Wellcome Trust, Daniel Louvard, Director of the Research Division, Institut Curie, and the E. de Rothschild and Y. Mayent Foundation of the Institut Curie for their crucial support.

We are also indebted to: our intrepid and patient publishing team, Pernille Hammelsoe, Karen Chambers, Kelvin Matthews, Nik Prowse and Angela Cottingham; John Wiley & Sons; our expert chapter reviewers, José Carlos García-Borrón, Vincent Hearing and Bernhard Wehrle-Haller, for their valuable comments, and to Lluís Montoliu and Sheila Schmutz for helpful input and photographs.

And all the mice and other pets.

Statement regarding the use of pictures

Chestnut Oriental shorthair is the breed of this cat. His color phenotype is nonagouti brown (known by fanciers as chestnut or chocolate). His color genotype is *a/a* $Tyrp1^b/Tyrp1^b$.

No, this is not a picture of a mouse! It is a 'chestnut Oriental shorthair' breed of cat, of the genotype a/a $Tyrp1^b/Tyrp1^b$; that is, to a pigmentation geneticist, *Nonagouti Brown*.

The pictures in this book serve a somewhat different purpose than pictures in scientific papers. The pictures in this book, for the most part, are not meant to illustrate some specific reductionist detail of science expounded in the text, but rather to introduce the reader to the broad array of phenotypic consequences of change in gene functions.

Both scientists and animal breeders (from their different perspectives) tend to think of the phenotype as something like a patchwork representation of the genes (factors) that cause them. In fact, the genes are at the most basic level of of the pyramid of functions, while the phenotype is nearly the highest level of organization of the individual organism, with the highest perhaps representing the survival value of pigmentation within the gene pool.

The phenotype is one of the tools we use to understand gene function. To use that tool effectively we need a good understanding of the phenotype in its many manifestations. In research papers, illustrations of phenotypes, if they are present at all, represent the difference between the experimental animal and its otherwise similar control. This is good. However, it doesn't provide a sense of the broad range of phenotypes in real life. In the text of this book you will read about the science of pigmentation genetics; in the pictures you will see it in as many of its varied manifestations as possible. The pigmentary phenotype of an organism is its appearance; there is no such thing as 'no phenotype.'

Statement regarding nomenclature

One problem in a book about phenotypes and genotypes is to differentiate between the use of a word to indicate the gene (*Albino*) or the appearance (phenotype) of the animal that is mutant at that locus (albino). To be as clear as possible in this regard, we have attempted to use standard nomenclature throughout, that is:

- for mouse genes and genotypes: italics, with initial capital for gene symbols and allele symbols (*Lethal spotting*, Tyr^{c});
- for human genotypes: italics, with full capitals for gene symbols ($PAX3^{S84F}$, *EDN3*);
- for phenotypes in any species: ordinary text (albino, megacolon);
- for protein abbreviations in any species: roman capitals (MITF, TYRP1).

We will use the mouse standard when referring to other nonhuman species.

The most official statement of the rules of nomenclature for mice is available at:

www.informatics.jax.org/mgihome/nomen/gene.shtml#gaas.

In the older genetic terminology, mentioned in parts of this book, gene alleles were given upper- and lower-case initials which indicated respectively that they were dominant (their effect is detectable when the other chromosome carries a normal wild-type copy, e.g. E^{so}) or recessive (no effect detectable when the other copy is wild-type, e.g. *p*). This nomenclature is now retained as the superscript part of an allele symbol ($Mc1r^{E\text{-}so}$; $Oca2^{p}$).

Another convention to note is that a dash, as in $A^{y}/-$, means 'any allele'.

Part I

Introduction to the Pigmentary System

Figure 1.1 Mini Mintzmeat, a brown (chestnut, chocolate) tortoiseshell cat with her kittens. The cat's genotype is *a*/*a*, *b*/*b*, O^{+}/O (or *a*/*a*, $Tyrp1^{b}/Tyrp1^{b}$, O^{+}/O). The kitten at the front is *a*/*a*, *b*/*b*, $Mlph^{ln}/Mlph^{ln}$, O^{+}/O^{+} (or *a*/*a*, $Tyrp1^{b}/Tyrp1^{b}$, $Mlph^{ln}/Mlph^{ln}$, O^{+}/O^{+}). The technical name for the phenotype is nonagouti, brown, leaden (or melanophilin-deficient). The fancier's name for the phenotype is often lavender (brown, dilute). The kitten behind is *a*/*a* *b*/*b*, O^{+}/Y (or *a*/*a*, $Tyrp1^{b}/Tyrp1^{b}$, O^{+}/Y); the technical name is nonagouti, brown and the fancier's name for the phenotype is often chocolate or chestnut.

1

Introduction to the Pigmentary System

1.1 Introduction

The colors of mice, horses, dogs, rabbits, humans, and other animals – yellow, black, brown, cream, orange, chocolate, blue, beige, white – result from mutations in genes that influence some of the most basic genetically controlled processes of life. These include specification of cell lineage, migration and homing of specialized cells, tissues, and organs, interactions of cells with surrounding tissues and their differentiation, and maintenance throughout life. Because defects in pigment cells are usually not lethal to the organism, pigmentation is an unusually accessible system in which to study these basic processes of vertebrate life.

In addition to these useful qualities of the pigmentary system, the availability of inbred mice with genetic defects in pigmentation that can be studied using modern techniques of genetics and cell biology – and comparative genetics at the level of sequenced genomes – has resulted in an unprecedented explosion of information about pigmentation.

The pigmentary system of the inbred laboratory mouse is a uniquely useful model of phenotype-based genetics, both basic and applied (Lauber 1971; Morse 1978; Barsh 2007). With the inbred mouse, the sequenced genome, modern technologies, and the pigmentary model we have the necessary tools to unravel the complex web of interacting genic functions that lead from a change in a gene through the multiple communicating processes that result in a specific phenotypic outcome.

Study of pigmentary genetics began, at the turn of the last century, with yellow mice and albino mice (Cuénot 1902; Castle & Allen 1903; Little 1913a; Wright 1917). A few books serve as landmarks of the progress of our studies of the genetic control over murine and comparative phenotypes; prominent among them are Green's (1966) *Biology of the Laboratory Mouse*, Searle's (1968) *Comparative Coat Color Genetics in Mammals*, Silvers' (1979) *The Coat Colors of Mice*, and Nordlund et al.'s (2006) *The Pigmentary System*. See also Foster (1965). More recent technological advances have given us the tools to take a closer look at the interacting processes that generate specific pigmentary phenotypes of the mouse, our most valuable genetically controlled animal model (Fig. 1.2).

The Colors of Mice: A Model Genetic Network, 1st edition. By M. Lynn Lamoreux, Véronique Delmas, Lionel Larue, and Dorothy C. Bennett. Published 2010 by Blackwell Publishing Ltd.

Figure 1.2 White spotted mouse, *red-eyed-white*, genotype $Mitf^{rw}/Mitf^{rw}$.

1.2 Colors of vertebrate animals

The function of melanocytes is to deposit melanin pigment within their cell-specific organelles (melanosomes). In the skin and/or growing hairs (feathers, scales) these melanosomes are often delivered to neighboring keratinocytes. Pigment cells of the eye arise *in situ* and retain their melanosomes; pigment cells of other organs may also retain their melanosomes. The *colors* of the native pigment – but not all the phenotypes of mice and other mammals – are red/yellow pheomelanin and black or brown eumelanin. Yellow, black, and chocolate brown are the three basic colors of pigment produced in mammalian melanocytes (see Chapter 4 on melanogenesis and Chapter 5 on pigment-type switching). Additionally, some colors are affected by the blood in the tissues, as for example in the comb and wattles of chickens, and other colors are influenced primarily by the structures of skin and hair (Quevedo & Holstein 2006), but most color variables, especially in animals that are covered with hair or feathers, are created by genetically controlled modification of the amount and distribution of the three basic colors of melanin pigment and of the melanocytes themselves (Searle 1968; Silvers 1979).

Pheomelanin is yellow/red melanin pigment found, for example in the hairs of Golden Retriever and Irish Setter breeds of dogs; yellow mice, rabbits and guinea pigs, orange or red cats, sorrel or chestnut horses, 'red' chickens (Fig. 1.3), 'brown' cattle (Fig. 1.4), and the bright red hair of humans (Little 1957; Searle 1968; Robinson 1971; Bowling & Ruvinsky 2000; Ostrander et al. 2005; Pontius et al. 2007; Schmutz & Berryere 2007; Barsh 2007; see Chapter 5). Eumelanic pigment is of two varieties: black eumelanin, as in the hairs of black Labrador Retriever dogs, and brown eumelanin, as in brown (chocolate) Labrador Retrievers, as well as chestnut Oriental Shorthair cats (see the picture under Statement regarding the use of pictures at the front of the book), brown mice (Fig. 1.5), and others.

Some of the phenotypes that we think of as 'brown' actually are not brown at the level of pigment color, and this is also true of 'red' or 'yellow.' It's not always easy to tell the difference visually, and

Figure 1.3 Pheomelanic chicken, genotype unknown.

Figure 1.4 Pheomelanic calf/cow, *Mc1r*^{e}/*Mc1r*^{e} (original name *e*/*e*); genotype name *Recessive yellow*. The common name is 'red' or 'brown'; however, brown is technically incorrect as *Brown* is an allele of the black/brown *Tyrp1* locus (Chapter 4).

Figure 1.5 Eumelanic mice, one brown, one black; *Tyrp1^b/Tyrp1^b* (former name *b/b*), phenotype brown; *Tyrp1^B/Tyrp1^B* (formerly *B/B*), phenotype black.

Figure 1.6 Agouti black mouse and nonagouti black mouse; *A/A Tyrp1^B/Tyrp1^B* (former name *A/A B/B*); *a/a Tyrp1^B/Tyrp1^B* (former name *a/a B/B*). The hairs of the agouti mouse are banded with yellow pigment and nonyellow pigment as determined by the *Agouti* locus. The color of the nonyellow pigment is black, as determined at the black/brown (*Tyrp1*) locus.

the common names for color phenotypes are not based on science. Sometimes the word brown is used to describe animals that are pheomelanic, like the cow and calf in Figure 1.4, that may be called brown or red; sometimes the word brown is used instead of the more accurate term agouti for phenotypes such as the mouse in Figure 1.6 and the cat in Figure 1.7. Agouti coloration – such as

Figure 1.7 An *Agouti* (*A*/*A*) cat of the Abyssinian breed. The name of a breed of animal does not relate to the genetics of its pigmentation.

the wild type in the mouse, rabbit, cat, guinea pig, and squirrel (Fig. 2.5), actually most mammals – consists of patterns (bands) of eumelanin and pheomelanin within individual hairs. These are produced by pigment-type switching, with switching between the deposition of eumelanic and pheomelanic pigment, into the growing hair at different times in its creation, as discussed in Chapter 5. We will reserve the word 'brown' for phenotypes that are based on brown eumelanin.

Mammals may produce eumelanin or pheomelanin in the same or different melanocytes. This choice is under direct genetic control. In mice, pheomelanin may be produced in the skin of tail and ears; otherwise pheomelanin is not normally produced outside the hair follicles. Other mammals, however, do produce pheomelanin in the skin; for example, humans, rabbits, and guinea pigs.

Color of pigment is not the only cause of variability in phenotypes. Banding patterns of hairs have already been described. Pheomelanin and eumelanin may also be distributed in specific patterns over the body surface. Normal black (or brown) and yellow color patterns such as those of tigers and ocelots, black (or brown)-and-tan patterns of mice or dogs (Fig. 1.8), brindle dogs (Fig. 1.9) and cows (Fig. 5.11), and tabby cats, as well as the agouti pattern of hair banding (Fig. 5.8), are exquisitely controlled and are caused by pigment-type switching as discussed in Chapter 5.

Figure 1.8 Black-and-tan dog with white spotting; a^t/a^t *Tyrp1*^{B}/*Tyrp1*^{B} (former name a^t/a^t, *B*/*B*). The specific white-spotting genotype is unknown, but it does not relate to the *Agouti* gene locus or the *Tyrp1* locus. This is another example of the patterning function of the *Agouti* locus. The *Agouti* locus regulates patterns of yellow pheomelanin and nonyellow eumelanin on individual hairs, as in Figures 1.6 and 1.7, or patterns of pheomelanin and eumelanin over the surface of the body. The so-called tan portion of the pattern is actually pheomelanic (yellow) pigment. The black color of the eumelanic portion is determined by the *Black*/*brown* locus (Kerns et al. 2004).

Pigment-type switching describes the ability of pigment cells to switch between the production of eumelanin and pheomelanin, under the control of the *Agouti* (*A*) and / or *Extension* (*Mc1r*) loci and modifying genes. The *Agouti* locus regulates the pattern of distribution of eumelanin and pheomelanin on each hair and over the surface of the body. So, agouti is not so much a color as a pattern of pheomelanogenesis that may or may not occur. That the pigment switch is between eumelanin and pheomelanin is illustrated, for example, in the Doberman Pinscher dog (or the similarly pigmented dog in Fig. 1.8). These may display yellow and black pigment patterns, or yellow and brown patterns in chocolate brown (incorrectly named 'red') Doberman Pinscher dogs.

Black and brown combinations do not occur. The type of eumelanin pigment, black versus brown, is genetically determined at birth, by the genotype at the *B* (*Tyrp1*, black / brown) locus, and there is no mechanism for switching between them. Color switching is exclusively a phenomenon involving eumelanin and pheomelanin.

In the mouse and other mammals, the three basic colors of melanin pigment, and their patterns of distribution over the body or within individual hairs, are regulated by the *Agouti* (*A*), *Extension* (*Recessive yellow*, *E*, or *Mc1r*), and black / brown (*B* or *Tyrp1*) loci (see Chapter 4).

Two basic types of pigment pattern exist that involve the death of melanocytes or their failure to differentiate or survive. White spotting is defined as the congenital absence of pigment cells from portions of the body or from the entire body, and we will use the term progressive graying to

Figure 1.9 Franny and Bitsy, a white-spotted dog of the Border Collie breed and brindle-colored Catahoula Hound. Both dogs have white spotting of unknown genotype that is not associated with the other color genotypes. Both dogs are *Black* at the *Black/brown* locus; therefore, eumelanic areas are black. The brindle phenotype is associated with mutation at a third locus that regulates the eumelanin/pheomelanin alternative in dogs (the *K*, *Canine β-defensin 103* locus).

describe ongoing loss of melanocytes in mice. Everyone has seen progressive graying with age, caused by loss or death of pigment cells from the hair follicles. Premature graying or patterned graying may be termed vitiligo or hypopigmentation; these are discussed in Chapter 3 (see also Fig. 1.10). Here we will use vitiligo and avoid the term hypopigmentation because it is not specific to an etiology.

Examples of white spotting include Paint horses, Hereford cattle, white forelock in humans, and the white areas of white-spotted cats, dogs, and mice, many of which are pictured throughout this book, especially in Chapter 3. The terms partial albinism and hypopigmentation are sometimes substituted for white spotting, especially in older publications. However, these are not appropriate to specifically describe white spotting, because hypopigmentation is a term also applied to albinism, which is a fundamentally different process. White-spotting and progressive-graying phenotypes result from pigment cell death or failure to survive, whereas albinism is the failure of deposition of pigment in living pigment cells, as discussed in Chapter 4.

The final step in cutaneous pigmentation occurs when the pigment cell transfers its melanosomes into keratinocytes of the adjacent skin, or the keratinocytes of a growing hair or feather as discussed in Chapter 4. The distribution of melanosomes in hairs is precise, and is an

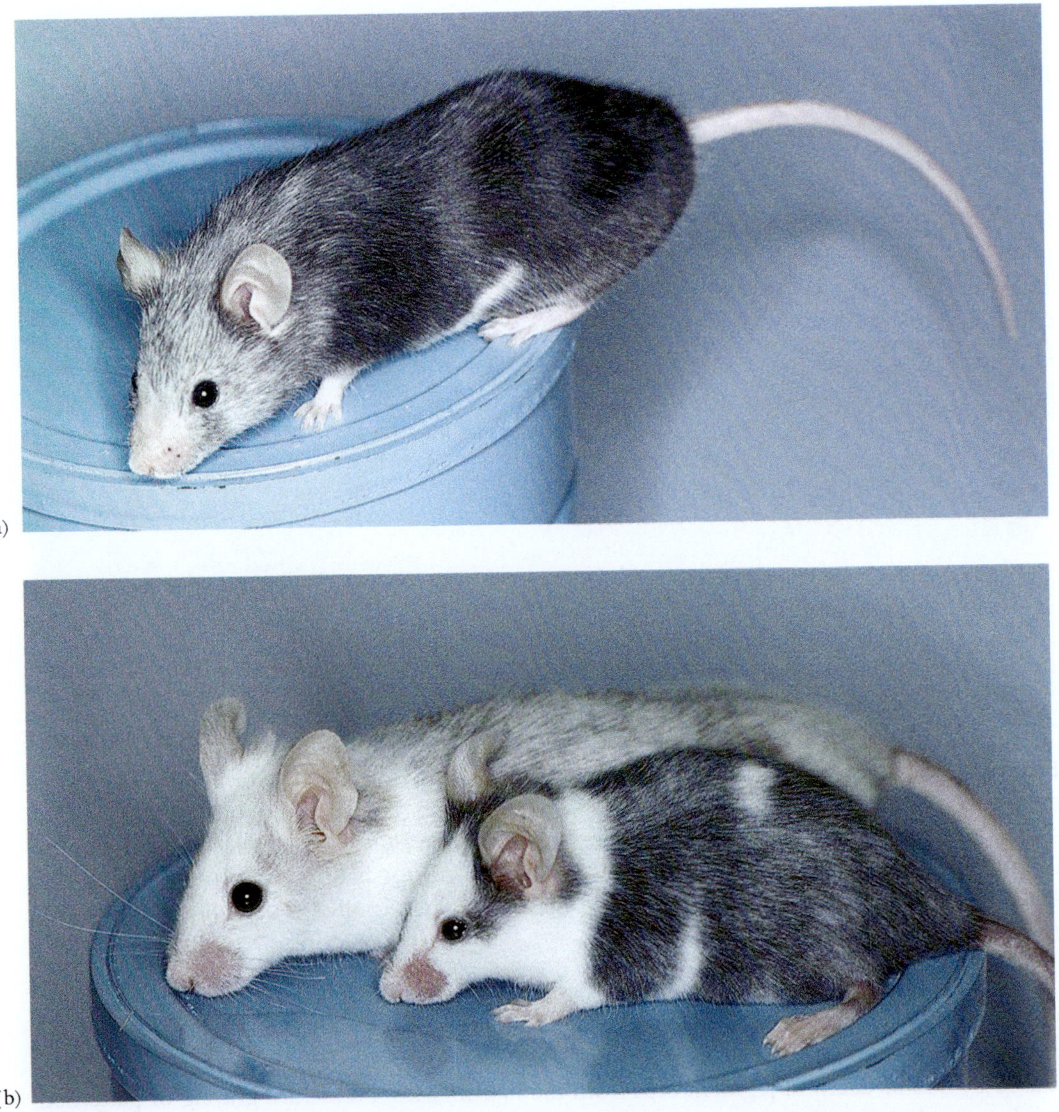

(a)

(b)

Figure 1.10 Vitiligo; (a) *Mitf*^{vit}/*Mitf*$^{mi\text{-}rw}$; (b) *Mitf*$^{mi\text{-}vit}$/*Mitf*$^{mi\text{-}or}$. The *Vitiligo* mouse has both white spotting and vitiligo. (a) This mouse is in the process of molting to a lighter color. (b) The older mouse is shown at the rear.

important component of how we perceive the color. When this pattern of distribution is modified by mutation, if the change causes the fur coat or feathers to reflect light differently to our eyes, the animal may appear to be quite a different color. In mammals, the most common example of this type of genetically controlled color difference is found in the 'dilute' phenotypes that are variously referred to as gray or blue, lavender, or cream (there are yet more names for this phenotype in various species) colors of cats, dogs, and other mammals (see Fig. 1.11). These may be caused by mutation at several different loci. We discuss melanosomal transfer briefly below in section 1.8 and in more detail in Chapter 4.

Figure 1.11 Mini Mizzle, lavender color, oriental shorthair. Oriental shorthair is the name of a breed of cat, and is therefore not a color. 'Lavender' is a cat fancier's term for 'nonagouti, brown, dilute' Unfortunately, 'dilute' is the cat breeder's term for the mouse gene known as *Leaden* (*Melanophilin*). The mouse *Dilute* gene identifies the *Myo5a* locus. The genotype of this cat is a/a $Tyrp1^{b}/Tyrp1^{b}$ $Mlph^{ln}/Mlph^{ln}$, which cat fanciers would refer to as *a/a b/b d/d*; however, the correct original nomenclature is *a/a b/b ln/ln*.

To summarize, the complexity of pigmentation phenotypes results primarily from four basic processes:

1 development of the melanocyte (Chapter 3);
2 differentiation of the melanocyte as it generates melanosomes (Chapter 4);
3 regulation of the type of melanin that is deposited upon the melanosome (pheomelanin, black eumelanin, or brown eumelanin) (Chapter 5);
4 transport of the melanosome within the pigment cell and transfer to the neighboring keratinocytes (Chapter 4).

These four basic functions are regulated by over 300 gene loci in the laboratory mouse (see the Appendix to this chapter). In the mouse, any of these genes can be evaluated using natural or

stem cells in this context) migrate from the bulge area into the hair bulb, around the dermal hair papilla, where they proliferate, differentiate, and initiate melanogenesis.

1.6 Melanosome biogenesis and translocation

The epidermal melanocyte (Fig. 1.13) is dendritic and contains an abundant endoplasmic reticulum and active Golgi region where melanosomal proteins are processed before being carried to their specific locations in stage I, II, III, and the mature stage IV melanosomes, as described in Chapter 4 (Fig. 4.8). Correct processing of the melanosomal proteins and their appropriate transport to the developing melanosome are necessary for normal pigmentation. Pigmentary defects associated with processing and transport are discussed in Chapter 4.

The melanosome is a member of the family of lysosome-related organelles that also includes platelet dense granules, lamellar bodies of type II alveolar epithelial cells, and lytic granules of cytotoxic T lymphocytes and natural killer cells. Premelanosomes appear to derive as 'saccules' that bud from the smooth endoplasmic reticulum (Orlow 1995). In eumelanosomes an internal matrix condenses from amorphous filamentous material to an organized internal structure. This material is largely the protein product of the *Silver* gene locus, as discussed in Chapter 4 (Boissy et al. 2006).

When assembled into a stage II melanosome (lacking pigment) the structure appears in section as highly organized spirals that apparently represent three-dimensional sheets that are rolled or

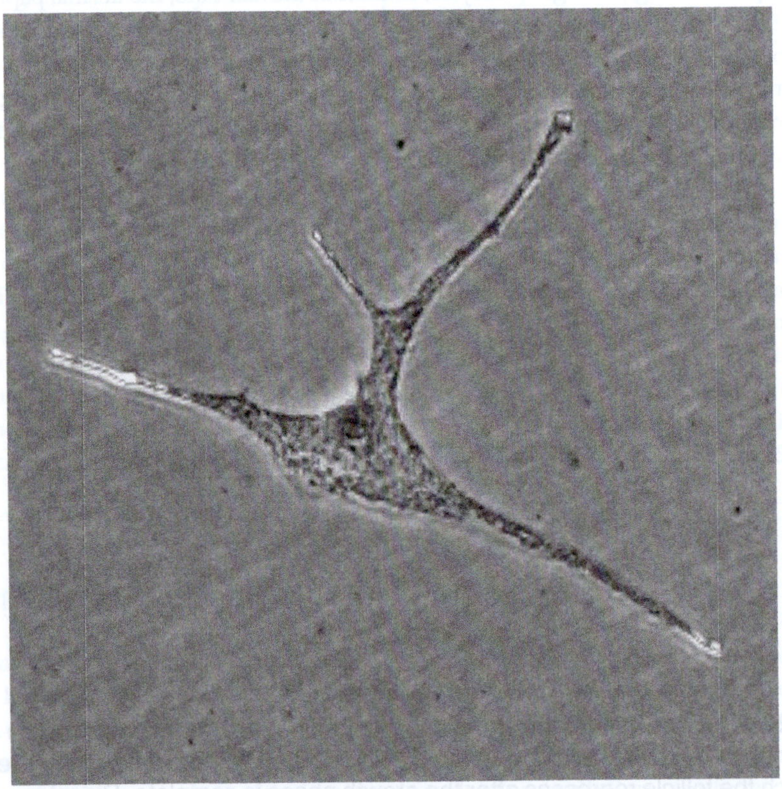

Figure 1.13 A eumelanocyte in cell culture.

pleated to form the basic structure upon which pigment is deposited in later stage melanosomes (Fig. 4.8). As the melanosome matures beyond stage II to stage III, catalytic enzymes are delivered to it and initiate deposition of melanin upon and within the matrix of the melanosome until it finally matures as the fully pigmented, chemically inert stage IV melanosome. In the living cell, the catalytic enzymes are active only briefly in the controlled environment of the melanosome, when they are needed for pigment deposition. During other stages of the cycle, their locations and functions are controlled tightly by other molecules in the pigment cell, as discussed in Chapter 4. Stage IV melanosomes are transported to the periphery of the melanocyte and transferred to keratinocytes as discussed in Chapter 4 (Scott 2007; Byers 2006).

1.7 Melanin

Melanogenesis occurs within the melanosome. Failure of melanogenesis results in albinism. The substrate molecule required for initiation of melanogenesis, tyrosine, is found in all cells; it is not a limiting factor. The enzyme tyrosinase is normally the rate-limiting factor, and is sufficient to catalyze melanogenesis *in vitro*. The biochemical pathway is relatively simple, as shown in Figure 1.14. To prevent premature activity of tyrosinase *in vivo*, the cell maintains tight control over its processing and routing to the melanosome, as described in Chapter 4. Genetic defects in tyrosinase itself, encoded at the *Albino* or *Color* (*Tyr*, *c*) locus, constitute one class of albinism. Other causes of albinism most often relate to the loci involved with processing and routing. As mentioned, mammals are able to produce two chemically distinct types of melanin pigment: (1) black or brown eumelanin and (2) pheomelanin in shades ranging from yellow to red (Chapter 4; Brilliant 2006; Hearing 2006; Oetting & Setaluri 2006; Solano & García-Borrón 2006; Figure 1.15).

Pheomelanosomes differ from eumelanosomes in structure, as well as color and the chemical nature of the pigment (Ito & Wakamatsu 2006; Sarna & Swartz 2006). These differences result at least in part from the switching off of several of the gene loci that are responsible for the production of eumelanosomes, as discussed in Chapter 5, pigment-type switching (and see Barsh 2006), and probably also in the availability of cysteine substrate.

1.8 Hair growth

As melanosomes mature, they are carried to the outer perimeter of the pigment cell and normally are transferred into neighboring keratinocytes of the growing hair or feathers in mammals (described in Chapter 4) or birds, and within the skin of humans. In the hair follicle, melanogenesis occurs during the growth of a new hair. During anagen, the first stage of hair follicle development, which lasts about 17 days in mice, the hair follicle extends deeper into and below the dermis, while melanocytes in the bulb area of the follicle (Figure 1.12) transfer melanosomes into the growing hair. Catagen, which lasts about a day, is a transitional second stage of the cycle, after the hair is fully grown, when a basal club is formed that anchors the hair in the upper part of the hair follicle. Telogen is the resting stage between cycles of hair growth. The length of telogen varies between a month and several months, depending upon the age of the mouse and other factors such as genetics, nutrition, or damage to the hair.

During the anagen or growth stage of the hair follicle, keratinocytes flow upward in the hair shaft from the lower hair bulb where they are generated, while melanocytes in the upper hair bulb synthesize melanosomes and deposit them from their long dendritic processes into the living

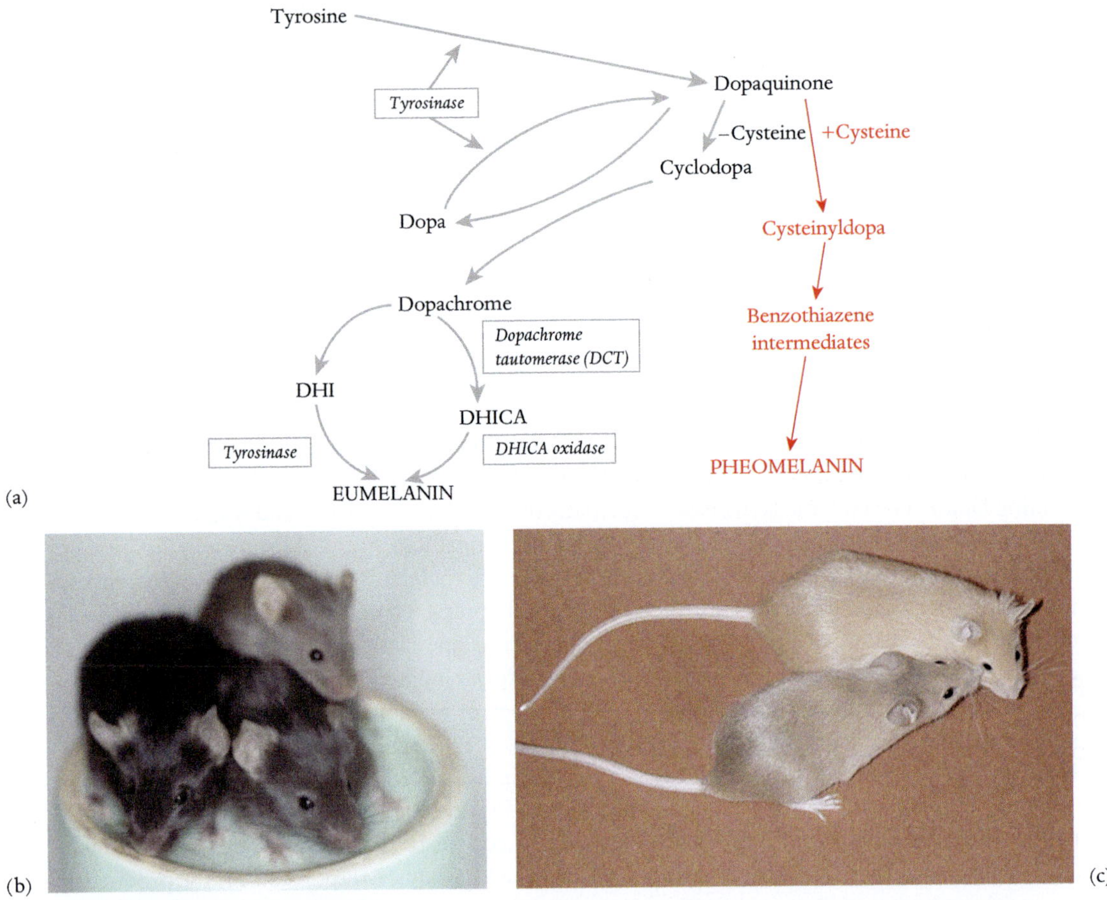

Figure 1.14 (a) The melanogenic pathway functions in the pigment granule (pathway simplified from Ito & Wakamatsu 2008; reproduced with permission of Wiley-Blackwell). Enzymes are shown in boxes. The rate-limiting enzyme is tyrosinase; mutation at the *Tyr* (*Albino*) locus reduces the amount of melanin that the pigment cell can make. In nonagouti mice that are otherwise wild type at pigment loci, mutation of the gene that encodes dopachrome tautomerase (DCT) results in slaty mice (below left, the one on the lower right is Slaty, that is, mutant at *Dct*). Mutation of the gene with 5,6-dihydroxyindole-2-carboxylic acid (DHICA) oxidase activity (*Tyrp1*) results in brown mice (b, rear) (see also Chapter 4). The mouse on the left in this picture is wild type. Pheomelanogenesis is primarily regulated by a cell-surface receptor (MC1R) and its ligand, melanocyte-stimulating hormone (MSH), that control whether or not the cell will make yellow pigment. Pheomelanogenesis is epistatic to eumelanogenesis. The two pheomelanic mice (c) are (front) *Recessive yellow*, in which the MC1R is mutant, and (back) *Lethal yellow*, in which the *Agouti* locus that encodes MSH is mutant (see also Chapter 5).

keratinocytes. The hairs become pigmented in an exquisitely reproducible fashion that reflects the genotype of the mouse, type of hair, and its location on the body. The resulting hair consists of variations on the theme of medullary segments, within which the melanosomes are arranged in a ladder-like formation, surrounded by a cortex in which the melanosomes are more or less evenly

Figure 1.15 Pheomelanic cat (known by some as 'red' and others as 'orange' and even others as 'marmalade'). Notice that the darker markings on this yellow cat are pheomelanic, whereas the darker markings on the agouti cat in Figure 1.7 are eumelanic.

distributed (Montagna & Ellis 1958; Searle 1968; Quevedo & Holstein 2006). By the time the hair emerges from the shaft of the hair follicle, the keratinocytes are pigmented, cornified, nonliving appendages to the skin.

1.9 Hair growth cycles

In mice, new hairs replace the old in more or less predictable cycles of growth. The first hairs begin to grow around the time of birth. After the first coat has grown, hair growth cycles can be induced artificially by plucking or shaving. Normally, the second cycle of hair growth begins at about 7–9 weeks of age. Thereafter, cycles of hair growth occur every few months. It may take about 2 weeks to fully replace the coat of hair. The process begins at the head of the mouse and proceeds caudad as a wave of new growth. During this time a continually changing pattern of pigmentation is usually more or less obvious, depending upon how much the newer hairs differ in appearance from the old.

Natural molt patterns may be caused by age-related genetic changes in the new hairs, or simple wear and tear of the old. In some mutant phenotypes or strain backgrounds, it is normal for pigmentation to change with age, sometimes resulting in rather dramatic molt patterns, especially at first molt (for example, Lamoreux & Galbraith 1986, and Fig. 5.4). Even if there are no developmental differences, the black or brown hairs can become rusty-looking with age, presumably because of the bleaching effect of saliva through grooming, and urine in the cage, so that a mouse in the middle of a hair cycle may have a pattern of shiny black or brown new hairs that contrast with the rusty-looking hairs, farther back on the body, that are yet to be shed. Similar patterns can be created artificially by injection (during the hair growth cycle) of substances that alter pigment phenotype (Chapter 5). The artificially created pattern will be maintained until the subsequent natural molt, when the mouse returns to its genetically determined phenotype. Patterns created by hair growth cycles are secondary to the hair cycle itself. Therefore, strictly speaking, they are not primarily the result of action of 'pigmentary genes,' but are nevertheless important to interpretation of pigmentary genetics for a couple of reasons. First, it is important to recognize the difference between phenotypes that are under direct genetic control, and those such as the hair cycle patterns that may not be directly relevant. Second, the hair cycle pattern of mice illustrates the need to be aware of differences between the physiology of humans and of mice that must be taken into consideration when interpreting data. The molt cycle of mice differs significantly from the condition in humans.

1.10 Embryonic development of the pigment cell lineage

Development has no beginning, and we hope it will have no end, technology notwithstanding. For an individual organism, development begins with the zygote, product of the fusion of a haploid sperm with an egg that carries a haploid set of chromosomes plus the extensive egg cytoplasm that is rich in nutrients and proteins. These proteins, encoded in the genome of the mother, make possible the maturation of offspring by providing enzymes and cellular biochemical structures that support development, from organizing the zygote to mitotic cell divisions and the early steps of cleavage, with the formation of the primary germ layers.

The pigment cell lineage of melanoblasts that mature into melanocytes arises after these earliest events, during the transient emergence of the neural crest as described in Chapter 3. The odyssey of neural-crest-derived pigment cells begins when neural-crest cells delaminate from the dorsolateral portion of the neural tube, in the mouse just before or as the neural tube closes. Melanoblasts (immature pigment cells) migrate, while replicating rapidly, along complex pathways. In mice, as the hair follicles develop concurrently with pigment-cell migration, the melanoblasts home to the follicles, leaving the interfollicular areas nearly devoid of pigment cells. Some melanoblasts then enter the bulge area of the follicle and others differentiate in the bulb region to pigment the growing hair as described above. Melanoblasts also migrate through the head and to the nose, eyes, ears, and even the heart following equally specific migratory pathways. The retinal pigment epithelium (see Chapter 3) is exceptional because it is not neural-crest-derived, but is induced directly in the neural tube.

1.11 Pigment cells in culture

Pigment cells of course develop normally in the tissue environment, not in cell culture, and there are differences between the cell culture requirements and biochemical properties of mouse and human melanocytes. Additionally, the mature pigment system of mice differs somewhat from the human pigment system *in vivo*, as do some of its functions and associated functional genetics (Sundberg 1994; Green 1966; Boissy et al. 2006; Montoliu et al. 2009). Therefore, it is wise to fully define the genetics of biological materials, human or murine, and to use caution when attributing results across species or genomic lines. That said, the unparalleled availability of murine pigmentary tissues and melanocytes that are inbred and genetically controlled provides a powerful tool for dissecting the basic mechanisms of melanogenesis at all levels, from the cell to the organism (Nordlund et al. 2006). Comparative mammalian genetics, especially at the molecular and phenotypic levels, has become highly sophisticated; information from the human and other species is informing mouse studies, and vice versa.

1.12 Conclusion

The remarkably choreographed accomplishments that result in development and differentiation of pigment phenotypes are made possible by an intricate web of tissue interactions and cellular signaling pathways that direct the functions of the pigment cells at every level of organization from the gene to the whole organism. Through chemical messages received from the environment and recognized by its own information-processing pathways, the pigment cell regulates the genes that direct the necessary cellular processes. Because pigment cells are easily observed, and their absence is not lethal, the pigment phenotypes provide an unparalleled model for the study of cellular communications and interacting signaling pathways as they function to bring about observable phenotypes summarized above. With the genomes of several species now sequenced, and others rapidly joining the databases, we are in a position to evaluate the pigmentary system using reductionist, phenotypic, and comparative methods in parallel.

Figure 1.16 Two dogs that are mutant at different loci influencing the choice between eumelanic and pheomelanic pigmentation (discussed in Chapter 5). The dog to the rear (but leading the pack) is a chimeric mix of eumelanic and pheomelanic pigmentation resulting from mutation at the *K* (*Canine β-defensin 103*) locus. In this case the eumelanin/pheomelanin switch is regulated via the keratinocytes. See also Figure 5.13 for a better view of the pattern. The dog in the foreground is mutant at the *Agouti* (*A*) locus that regulates the eumelanin/pheomelanin switch over the body and on each hair via the genotype of the mesoderm. Other loci that influence the switch are listed in the Appendix of this chapter, and in Chapter 5. Photograph courtesy of TheOneCreation.

Appendix: color loci of the mouse

Introduction

This appendix lists all of the loci that have been reported to contribute to pigmentation to date. For further information and updates we refer you to:

- the Mouse Genome Informatics (MGI) database maintained by The Jackson Laboratory (JAX), www.informatics.jax.org/;
- the International Federation of Pigment Cell Societies (IFPCS) Color Genes initiative, www.espcr.org/micemut/, maintained by the European Society for Pigment Cell Research (ESPCR) (Montoliu et al. 2009).

Specific (mostly classical) mutant loci are discussed throughout the book in the context of their contribution to phenotype. These sections can be identified in the text because they have headings that look like the one shown here (e.g. see Chapter 4), and they are listed below in alphabetical order by locus name.

***Silver* (*Si*, *Pmel17*, *Gp100*)**
MGI lists one spontaneous phenotypic allele
Availability: JAX, CMMR, MMRRC

In such headings, 'Availability' refers to the facilities where various resources (mice, frozen embryos, etc.) are available. We use abbreviations for the facilities, and these are listed below. Besides the main ones (JAX, HAR, ORNL, MMRRC), quite a few other centers that have proliferated internationally with the availability of modern genetic technology (see Chapter 6). These are listed in full by the International Mouse Strain Resource (IMSR), sponsored by the MGI: www.findmice.org/fetch?page=imsrStrainRepositories.

CARD	Center for Animal Resources and Development
CFG	Consortium for Functional Glycomics
CMMR	Canadian Mouse Mutant Repository
EMMA	European Mouse Mutant Archive
HAR	Mammalian Genetics Unit, Harwell
JAX	The Jackson Laboratory
KOMP Repository	Knockout Mouse Repository
MMRRC	Mutant Mouse Regional Resource Centers
ORNL	Oak Ridge National Laboratories
RBRC	RIKEN BioResource Center
TAC	Taconic
TIGM	Texas A&M Institute for Genomic Medicine

Following is an index to the specific mutant loci, listed by phenotype-based name, that are discussed in the text of this book.

Index of mutant loci discussed in text

Silver (*Si, Pmel17, Gp87*)	Chapter 4, page 140
Slaty (*Slt, Dct, Dopachrome tautomerase*, formerly *Trp2, Tyrosinase-related protein 2*)	Chapter 4, page 152
Splotch (*Sp, Pax3, Paired box gene 3*)	Chapter 3, page 91
Steel (*Sl, Kitl, Kit ligand, SLF, Steel factor; MGF, Mast cell growth factor; SCF, Stem cell factor*)	Chapter 3, page 100
Subtle gray (*Sut, Slc7a11*)	Chapter 5, page 204
Tyrp1 (*Brown, B, Tyrp1, Tyrosinase-related protein 1*)	Chapter 4, page 150
Underwhite (*Uw, Matp, Membrane-associated transporter protein, Slc45a2, Solute carrier family 45, member 2*)	Chapter 4, page 158

Cloned and uncloned loci

Table A1.1 shows a summary of the cloned mouse color genes and Table A1.2 gives a summary of the uncloned mouse color genes.

Table A1.1 Summary of the cloned mouse color genes

Symbol (old symbol)	Name (old name)	Mouse chromosome	Mutant phenotype	Molecular/biological functions	Human symbol	Human chromosome	Human syndrome
(A) Melanocyte development – involving integument							
Acd	adrenocortical dysplasia	8	Hyperpigmented skin, adrenal hyperplasia, other organ disorders	Telomere capping; may affect pigmentation through excess ACTH	*ACD*	16q22.1	N
Adam17	a disintegrin and metalloprotease domain 17	12	Irregular pigmentation in hairs	Protease, processing various bioactive proteins	*ADAM17*	2p25	N
Adamts20 (*bt*)	a disintegrin and metalloprotease domain (reprolysin type) with thrombospondin type 1 motif, 20 (belted)	15	Lumbar white belt	Metalloprotease, melanoblast migration?	*ADAMTS20*	12q12	N
Apc	adenomatous polyposis coli, allele *tm2Rak*	18	Prenatal dorsal dark stripe and head patch	Wnt pathway mediator; transcription factor	*APC*	5q22.2	Adenomatous polyposis coli
Brca1	breast cancer 1, allele *tm2Arge*	11	Abnormal skin pigmentation	DNA repair; tumor suppressor	*BRCA1*	17q21	Breast cancer
Dock7 (*m, mnlt*)	dedicator of cytokinesis 7 (misty, moonlight)	4	Distal white spotting, hypopigmented fur, but melanocytes *in vitro* hyperpigmented	widely expressed Rho-family guanine nucleotide exchange factor	*DOCK7*	1p31.3	N
Ece1	endothelin converting enzyme 1, allele *tm1Reh*	4	No melanocytes in uvea nor dorsal skin at birth (perinatal lethal)	Endothelin synthesis	*ECE1*	1p36.12	N
Edn3 (*ls*)	endothelin 3 (lethal spotting)	2	White spotting, megacolon and other neural crest defects	Melanoblast/neuroblast growth and differentiation factor	*EDN3*	20q13	Hirschsprung disease, Waardenburg–Shah syndrome
Ednrb (*s*)	endothelin receptor type B (piebald spotting)	14	White spotting, megacolon and other neural crest defects	EDN3 receptor	*EDNRB*	13q22	Hirschsprung disease, Waardenburg–Shah syndrome

Table A1.1 (*Cont'd*)

Symbol (old symbol)	Name (old name)	Mouse chromosome	Mutant phenotype	Molecular/biological functions	Human symbol	Human chromosome	Human syndrome
Egfr (*Dsk5*)	epidermal growth factor receptor (dark skin 5)	11	Dark skin	Growth factor receptor	*EGFR*	7p12.3	N
En1	engrailed 1	1	Hyperpigmentation of digits (polydactyly etc)	Transcription factor	*EN1*	2q14.2	N
Fgfr2	fibroblast growth factor receptor 2	7	Lighter skin (many other defects)	Growth factor receptor	*FGFR2*	10q26	Crouzon syndrome, Pfeiffer syndrome
Foxn1 (*tw*)	forkhead box N1, allele *tw* (traveling wave)	11	Hairless. Waves of dark/light travel slowly over skin (poss. normal hair cycle + very short hairs)	Transcription factor	*FOXN1*	17q11.2	N
Frem2	Fras1 related extracellular matrix protein 2, allele my-F11	3	Microphthalmia/anophthalmia, patches of discolored or white fur	Extracellular protein; possibly epithelial–mesenchymal interactions at basement membrane	*FREM2*	13q13.3	Fraser syndrome
Fzd4	frizzled homolog 4 (*Drosophila*), allele *tm1Nat*	7	Many abnormalities including light or silvered coat	WNT receptor, putatively for WNT5A and/or NDP (see *Ndp*)	*FZD4*	11q14.2	Exudative vitreoretinopathy 1
Gata3	GATA binding protein 3, allele *tm3Gsv*	2	Extra stem-like cells in hair follicles; abnormal hair, irregular pigment deposition	Transcription factor	*GATA3*	10p14	Hypoparathyroidism, sensorineural deafness, and renal disease syndrome, Barakat syndrome
Gli3	GLI-Kruppel family member GLI3	13	White belly patch or lumbar belt; nervous system defects (homozygous postnatal lethal)	Signaling in hedgehog pathway; modifies SOX10 expression	*GLI3*	7p14.1	Pallister–Hall syndrome and others
Gnaq (*Dsk1*, *Dsk10*)	guanine nucleotide-binding protein subunit Gaq (dark skin 1, dark skin 10)	19	Dark skin (hyperproliferation of melanocytes)	Signal transduction, possibly from an EDNR(s) to PLCβ	*GNAQ*	9q21	Possible platelet defect

Gna11 (*Dsk7*)	guanine nucleotide-binding protein subunit Ga11 (dark skin 7)	10	Dark skin (hyperproliferation of melanocytes)	Signal transduction, possibly from an EDNR(s)	*GNA11*	19p13	N
Gpc3	glypican 3, allele *tm1Arge*	X	Dominant distal and belly spotting	GPI-linked extracellular membrane protein; putative SHH-binding (competitive with PTCH)	*GPC3*	Xq26.2	Simpson-Golabi-Behmel syndrome type 1
Gpr161 (*vl*)	G protein-coupled receptor 161 (vacuolated lens)	1	Vacuolated lens, occasional belly spot, spina bifida, others	Signal transduction	*GPR161*	1q24.2	N
Hells	helicase, lymphoid specific	19	Early ageing includes graying by 15d old. p16 overexpression	DNA methylation, gene silencing	*HELLS*	10q23.33	N
Itgb1	integrin β1, allele *tm1Ref*	8	Transient patchy hypopigmentation, crest migration defect	Fibronectin receptor–cell attachment, migration	*ITGB1*	10p11.22	N
Kit (*W*)	Kit oncogene (white spotting)	5	White spotting, anemia and germ-cell deficiency	Receptor for KIT ligand/stem cell factor; required for melanoblast survival	*KIT*	4q11–q12	Piebald syndrome
Kitl (*Sl*)	Kit ligand (steel)	10	White spotting, anemia and germ-cell deficiency	Growth and differentiation factor	*KITLG*	12q22	N
Krt1 (*Dsk12*)	keratin 1 (dark skin 12)	15	Dark skin	Cytoskeleton	*KRT1*	12q13	Epidermolytic hyperkeratosis
Krt2 (*Krt2-17, Dsk2*)	keratin 2 (Keratin 2-17, dark skin 2)	15	Dark skin	Cytoskeleton	*KRT2A*	12q11–q13	Ichthyosis bullosa of Siemens
Krt4	keratin 4	15	'Bright' diluted coat color	Cytoskeleton	*KRT4*	12q13.13	White Sponge Nevus of Cannon
Krt17	keratin 17, allele *tm1Cou*	11	Abnormal hairs with clustered melanin granules	Cytoskeleton	*KRT17*	17q1.2	Pachyonychia congenita type 2 Steatocystoma multiplex
Krt75	keratin 75, allele *tm1Der*	15	Hair defects with variable pigment clumping	Cytoskeleton	*KRT75*	12q13.13	N

Table A1.1 (*Cont'd*)

Symbol (old symbol)	Name (old name)	Mouse chromosome	Mutant phenotype	Molecular/biological functions	Human symbol	Human chromosome	Human syndrome
Lef1	lymphoid enhancer binding factor 1, allele *tm1Rug*	3	Underdeveloped hair follicles lacking melanin	Transcription factor; Wnt/β-catenin mediator	*LEF1*	4q25	Sebaceous adenomas
Lmx1a (*dr*)	LIM homeobox transcription factor 1α (dreher)	1	Partial or complete white belt and/or belly spot	Transcription factor	*LMX1A*	1q22-23	N
Mbtps1	membrane-bound transcription factor peptidase, site 1, allele wrt	8	Diluted hair with white base (melanocyte death?)	Peptidase involved in regulation of membrane lipid composition	*MBTPS1*	16q23.3–q24.1	N
Mcoln3 (*Va*)	mucolipin 3 (varitint-waddler)	3	Patches of normal, diluted and white hair (and behavioral defects)	Cation channel	*MCOLN3*	1p22.3	N
Mitf (*mi*)	microphthalmia-associated transcription factor (microphthalmia)	6	White spotting and small or absent eyes	Transcription factor: master regulator of melanocyte lineage	*MITF*	3p12-14	Waardenburg syndrome type 2
Mpzl3 (*rc*)	myelin protein zero-like 3 (allele rough coat)	9	Hair follicle loss, black pigment changes to light brown	Putative adhesion protein, expressed in keratinocytes	*MPZL3*	11q23.3	N
Myc	myelocytomatosis oncogene (when KO targeted by Wnt1 promoter-Cre)	15	Pigmentary spotting, not head	Transcription factor, regulator of cell proliferation	*MYC*	8q24.21	
Notch1	Notch gene homolog 1 (*Drosophila*)	2	Scattered gray hairs, when KO targeted to melanocytes	Receptor for ligands in Delta and Jagged families	*NOTCH1*	9q34.3	N
Notch2	Notch gene homolog 2 (*Drosophila*)	3	Scattered gray hairs, when KO targeted to melanocytes; all gray with Notch1 KO, eventually white	Receptor for ligands in Delta and Jagged families	*NOTCH2*	1p12	Alagille Syndrome 1

Ntrk1 (*TrkA*)	neurotrophic tyrosine kinase, receptor, type 1	3	Mottled coat (also neural defects, skin lesions)	Coreceptor for nerve growth factor	*NTRK1*	1q23.1	Insensitivity to pain, congenital, with anhidrosis
Pax3 (*Sp*)	paired box gene 3 (splotch)	1	White belly splotch, neural crest defects	Transcription factor	*PAX3*	2q35	Waardenburg syndrome type 1, Waardenburg syndrome type 3
Pax6 (*Sey*)	paired box gene 6 (small eye)	2	Eye abnormalities can include reduced RPE, also distal/ventral white spotting	Transcription factor	*PAX6*	11p13	Aniridia Other eye disorders
Pcbd1	pterin 4α carbinolamine dehydratase/ dimerization cofactor of hepatocyte nuclear factor 1α (TCF1)	10	Mild hypopigmentation, belly spot, mild microphthalmia	Both phenylalanine metabolism and binding partner of TCF1 (HNF1), hence WNT pathway interaction	*PCBD1*	10q22.1	Hyperphenylalaninemia with primapterinuria
Rb1	retinoblastoma 1 (targeted deletion)	14	Melanocyte hyperproliferation in culture	Growth-inhibitor, suppresses E2F transactivation activity	*RB1*	13q14.2	Retinoblastoma
Rbpj (*RBP-JK*)	recombination signal binding protein for immunoglobulin κ J region (Tyr targeted KO)	5	Hair whitening; other melanocytes not affected	Transcription factor, mediator of Notch signaling and cell fate	*RBPJ* (*RBPSUH*)	4p15.2	N
Sema3c	sema domain, immunoglobulin domain (Ig), short basic domain, secreted, (semaphorin) 3C	5	Some skin hypopigmentation, ectopic pigment in internal organs	Secreted signaling factor, can mediate axon repulsion	*SEMA3C*	7q21.11	N
Sfxn1 (*f*)	Sideroflexin 1 (flexed tail)	13	Belly spot (and flexed tail, anemia etc)	Tricarboxylate carrier	*SFXN1*	5q35.3	N
Snai2	Snail homolog 2/Slug	16	Spotting, head blaze, pale hair and skin, neural crest and other organ defects	Transcription factor	*SNAI2*	8q11	Waardenburg syndrome type 2
Sox10 (*Dom*)	SRY-box-containing gene 10 (dominant megacolon)	15	White spotting, megacolon and other neural crest defects	Transcription factor	*SOX10*	22q13.1	Waardenburg syndrome types 2E, 4

Table A1.1 (*Cont'd*)

Symbol (old symbol)	Name (old name)	Mouse chromosome	Mutant phenotype	Molecular/biological functions	Human symbol	Human chromosome	Human syndrome
Sufu	suppressor of fused homolog (*Drosophila*)	19	CNS, dark hair, basal cell lesions on skin. (hedgehog pathway suppressor)	Cytoplasmic signaling intermediate	*SUFU*	10q24.32	Medulloblastoma
Tbx10 (*Dc*)	T-box 10 (Dancer)	19	Head spot (variable); ear, palate and neural defects	Transcription factor (ectopic expression in Dc)	*TBX10*	11q13.2	N
Tbx15 (*de*)	T-box 15 (droopy ear)	3	Ear shape; skeletal, altered dorsoventral color pattern with A^t, a^e	Transcription factor	*TBX15*	1p13	N
Tcfap2a	transcription factor AP-2α	13	Wnt1-targeted KO gives neural crest defects including pigmentary	Transcription factor	*TFAP2A*	6p24	N
Traf6	Tnf receptor-associated factor 6	2	Many effects including pale skin, few/delayed hair follicles; postnatal lethal	Signaling from IL1A to NF-κB	*TRAF6*	11p12	Ectodermal dysplasia, anhidrotic
Wnt1	Wingless-related MMTV integration site 1	15	Defects of neural crest including melanoblasts in mice lacking both Wnt1 and Wnt3a	Growth factor/morphogen	*WNT1*	12q13	N
Wnt3a	Wingless-related MMTV integration site 3A	11	Defects of neural crest including melanoblasts in mice lacking both Wnt1 and Wnt3a	Growth factor/morphogen	*WNT3A*	1q42	N
Zbtb17	zinc finger and BTB domain containing 17	4	Darkened coat (mixed strain background); dark skin, dark dermis around hairs. Abnormal follicles	Transcription factor	*ZBTB17*	1p36.13	N
Zfp53	zinc finger protein 53	17	Abnormal skin pigmentation	Resembles transcription factor	?	?	N
Zic2 (*Ku*)	Zinc finger protein of the cerebellum 2 (Kumba)	14	Belly spot, curly tail, hindbrain	Transcription factor	*ZIC2*	13q32	Holoprosencephaly 5

(B) Development: predominantly eye and/or ear							
Bmpr1a	bone morphogenetic protein receptor, type 1A, allele *tm1Bh*	14	Abnormal prenatal RPE with discontinuity in pigmentation	Receptor	*BMPR1A*	10q22.3	Juvenile polyposis syndrome
Bmpr1b	bone morphogenetic protein receptor, type 1B, allele *tm1Kml*	3	Abnormal prenatal RPE with discontinuity in pigmentation	Receptor	*BMPR1B*	4q23–q24	Brachydactyly, types A2, C; chondrodysplasia
Fkbp8	FK506 binding protein 8, allele *tm1Tili*	8	Microphthalmia/anophthalmia	Endogenous calcineurin inhibitor; can inhibit apoptosis	*FKBP8*	19p13.11	N
Gas1	growth arrest specific 1, allele *tm1Fan*	13	RPE transdifferentiates to neural retina	Can enhance hedgehog signaling, inhibit growth	*GAS1*	9q21.33	Holoprosencephaly
Gnpat	glyceronephosphate O-acyltransferase, allele *tm1Just*	8	Abnormal RPE morphology, microphthalmia	Enzyme	*GNPAT*	1q42.2	Rhizomelic chondrodysplasia punctata, Type 2
Grlf1	glucocorticoid receptor DNA binding factor 1 (p190 RhoGAP)	7	RPE hyperplasia, microphthalmia	Transcriptional repressor	*GRLF1*	19q13.32	N
Jmjd6	Jumonji domain containing 6, allele *tm1Gbf*	11	Lack of one/both eyes, ectopic RPE in nose	Demethylates histones; transcriptional regulator	*JMJD6*	17q25.2	N
Mab21l2	Mab-21-like 2 (*Caenorhabditis elegans*), allele *tm1Nao*	3	lack of RPE by time of embryonic lethality	Cell fate determination, TGFβ signaling	*MAB21L2*	4q31.3	N
Med1	mediator complex subunit 1	11	Low retinal pigmentation (before embryonic lethality)	Binds methylated DNA; DNA repair	*MBD4*	17q12	N
Ndp	Norrie disease homolog (allele *tm1Wbrg*)	X	Many defects including hyperpigmentation of RPE and overgrowth of strial melanocytes	TGFβ-like extracellular factor; FZD4 and LRP5 also associated with human Norrie disease	*NDP*	Xp11.4	Norrie disease
Nf1	neurofibromatosis 1	11	Small, unpigmented eyes, microphthalmia (Ras pathway)	Ras GTPase-activating protein (neurofibromin)	*NF1*	17q11.2	Neurofibromatosis 1

Table A1.1 (*Cont'd*)

Symbol (old symbol)	Name (old name)	Mouse chromosome	Mutant phenotype	Molecular/biological functions	Human symbol	Human chromosome	Human syndrome
Nr2e1 (*frc*)	nuclear receptor subfamily 2, group E, member 1 (allele fierce)	10	Brain and eye defects; asymmetrical and mottled RPE	Transcriptional repressor, recruits HDAC to DNA, stem cell maintenance	*NR2E1*	6q21	N
Otx2	orthodenticle homolog 2 (*Drosophila*)	14	Many effects including RPE hyperplasia	Hox-like transcription factor, can induce RPE identity in neural retina	*OTX2*	14q23.1	Dysgnathia complex
Pax2	paired box gene 2	19	Many effects including RPE cells extending into optic nerve	Transcription factor	*PAX2*	10q24.31	N
Pdgfb	platelet derived growth factor, B polypeptide	15	Cardiovascular and eye defects include abnormal RPE, microphthalmia	Growth factor	*PDGFB*	22q13.1	Meningioma
Pdgfc	platelet-derived growth factor, C polypeptide	3	Depigmented spots in the retina	Growth factor	*PDGFC*	4q32.1	N
Phactr4 (*humdy*)	phosphatase and actin regulator 4, allele humpty dumpty	4	Neuroblast overgrowth; outgrowths in RPE	Regulator of protein phosphatase 1 and its dephosphorylation of RB1	*PHACTR4*	1p35.3	N
Pitx3 (*ak*)	paired-like homeodomain transcription factor 3	19	Eye abnormalities including hyperpigmentation around embryonic pupil	Transcription factor; CNS neuronal differentiation	*PITX3*	10q24.32	Congenital cataract
Pygo1	pygopus 1	9	Eye and other defects including folded RPE	Cofactor for β-catenin- LEF-mediated transcription	*PYGO1*	15q21.3	N
Rs1	retinoschisis (X-linked, juvenile) 1 (human), allele *tmgc1*	X	Small patches of depigmentation in RPE	Retinal protein; homologies to cell-adhesion proteins	*RS1*	Xp22.13	Retinoschisis 1, X-linked, juvenile
S1pr2 (*Edg5*)	sphingosine-1-phosphate receptor 2	9	Ear defects include thickening and hyperpigmentation of stria vascularis	Receptor	*S1PR2* (*EDG5*)	19p13.2	N

Sema4a	sema domain, immunoglobulin domain (Ig), transmembrane domain (TM) and short cytoplasmic domain, (semaphorin) 4A	3	Abnormal RPE, postnatal depigmentation of eye	Transmembrane juxtacrine signaling protein	*SEMA4A*	1q22	Retinitis pigmentosa
Timp3	tissue inhibitor of metalloproteinase 3	10	Abnormal RPE morphology	Protease inhibitor and can block VEGF binding to receptor	*TIMP3*	22q12.3	Fundus dystrophy, pseudoinflammatory, of Sorsby
Tub	tubby candidate gene	7	Obese; eye and ear abnormalities; degeneration and loss of RPE	Anti-apoptotic; downstream mediator of $G\alpha_q$ signaling	*TUB*	11p15.4	N
Unc119	unc-119 homolog (*Caenorhabditis elegans*)	11	Retinal degeneration; mottling of RPE	Proposed receptor-associated activator of SRC-family kinases	*UNC119*	17q11.2	N
Vsx2	visual system homeobox 2	12	Microphthalmia, reduced eye pigmentation	Pax-like transcription factor	*VSX2*	14q	Corneal dystrophy, keratoconus
(C) Components of melanosomes and their precursors							
Dct (*slt*)	dopachrome tautomerase (slaty, TRP2)	14	Dilution of eumelanin color	Melanosomal enzyme	*DCT*	13q31–q32	N
Gpnmb	glycoprotein (transmembrane) NMB	6	Glaucoma, iris pigment epithelium disorders, especially with $Tyrp1^{b/b}$	Apparent melanosomal component	*GPNMB*	7p15	N
Si (*si*)	silver (gp100, gp87, Pmel-17 etc)	10	Silvering with postnatal melanocyte loss in eumelanic animals (varying with strain background)	Melanosome matrix; trapping of melanin intermediates?	*SILV*	12q13–q14	N
Slc24a5	solute carrier family 24, member a5 (NCKX5)	2	Skin, eye color	Calcium transporter, ?melanosomal	*SLC24A5*	15q21.1	Skin, hair, eye color

Table A1.1 *(Cont'd)*

Symbol (old symbol)	Name (old name)	Mouse chromosome	Mutant phenotype	Molecular/biological functions	Human symbol	Human chromosome	Human syndrome
Slc45a2 (*uw*, *Matp*)	solute carrier family 45, member a2 (underwhite, membrane-associated transporter protein)	15	Severe dilution of coat and eye pigment	Solute transporter	*SLC45A2*	5p	Oculocutaneous albinism type 4
Tyr (*c*)	tyrosinase (color, albino)	7	No pigment in null mice	Melanosomal enzyme	*TYR*	11q21	Oculocutaneous albinism type 1
Tyrp1 (*b*)	tyrosinase-related protein 1 (brown, TRP1)	4	Brown eumelanin. Allele *isa* can contribute to glaucoma	Melanosomal enzyme	*TYRP1*	9p23	Oculocutaneous albinism type 3; glaucoma-related pigment dispersion syndrome
(D) Melanosome construction/protein routing (HPS-related)							
Ap3b1 (*pe*)	adaptor-related protein complex AP-3, β1 subunit (pearl)	13	Pigmentary dilution	Organellar protein routing	*AP3B1* [HPS2]	5q14.2	Hermansky–Pudlak syndrome, type 2
Ap3d (*mh*)	adaptor-related protein complex AP-3, delta subunit (mocha)	10	Pigmentary dilution	Organelle biogenesis; AP-3 component	*AP3D1*	19p13.3	N
Bloc1s3 (*rp*)	biogenesis of lysosome-related organelles complex 1, subunit 3 (reduced pigmentation)	7	Pigmentary dilution	Organelle biogenesis; BLOC1 component	*BLOC1S3*	19q13.32	Hermansky–Pudlak syndrome, type 8
Cno (*cno*)	cappuccino	5	Pigmentary dilution	Organelle biogenesis; BLOC1 component	*CNO*	4p16-p15	N

Dtnbp1 (*sdy*)	dystrobrevin binding protein 1 (sandy, dysbindin)	13	Sandy colored coat, platelet defect	Organelle biogenesis; BLOC1 component	*DTNBP1*	6p22.3	Hermansky–Pudlak syndrome, type 7
Fig4 (*plt1*)	FIG4 homolog (*Saccharomyces cerevisiae*) (pale tremor); phosphatidylinositol-(3,5)-bisphosphate 5-phosphatase	10	Pale color with tremor	Endosome-lysosome axis; clumped melanosomes (+immune effects, etc.)	*FIG4*	6q21	Charcot-Marie-Tooth disease
Gpr143 (*Oa1*)	G-protein-coupled receptor (GPR143); mouse homolog of human ocular albinism 1	X	Reduced eye pigmentation, giant melanosomes	Melanosome biogenesis and size; signal transduction	*OA1*	Xp22.3	Ocular albinism 1 (Nettleship-Falls)
Hps1 (*ep*)	Hermansky–Pudlak syndrome 1 homolog (pale ear)	19	Pale skin, slight coat dilution, giant melanosomes; platelet defect	Organelle biogenesis and size; BLOC3 component	*HPS1*	10q24	Hermansky–Pudlak syndrome, type 1
Hps3 (*coa*)	Hermansky–Pudlak syndrome 3 homolog (cocoa)	3	Hypopigmentation, platelet defect	Organelle biogenesis; BLOC2 component	*HPS3*	3q24	Hermansky–Pudlak syndrome, type 3
Hps4 (*le*)	Hermansky–Pudlak syndrome 4 homolog (light ear)	5	Pale skin, slight coat dilution, giant melanosomes; platelet defect	Organelle biogenesis and size; BLOC3 component	*HPS4*	22q11–q12	Hermansky–Pudlak syndrome, type 4
Hps5 (*ru2*)	Hermansky–Pudlak syndrome 5 homolog (ruby-eye 2)	7	Pigmentary dilution and red eyes; platelet defect	Organelle biogenesis; BLOC2 component	*HPS5*	11p14	Hermansky–Pudlak syndrome, type 5
Hps6 (*ru*)	Hermansky–Pudlak syndrome 6 homolog (ruby-eye)	19	Pigmentary dilution and red eyes; platelet defect	Organelle biogenesis; BLOC2 component	*HPS6*	10q24.31	Hermansky–Pudlak syndrome, type 6
Lyst (*bg*)	lysosomal trafficking regulator (beige)	13	Pale coat, immunodeficiency	Organelle biogenesis and size	*LYST*	1q42	Chediak-Higashi syndrome

Table A1.1 *(Cont'd)*

Symbol (old symbol)	Name (old name)	Mouse chromosome	Mutant phenotype	Molecular/biological functions	Human symbol	Human chromosome	Human syndrome
Oca2 (*p*)	oculocutaneous albinism 2 (pink-eyed dilution)	7	Severe loss of eumelanin in hair, skin and eyes	?Glutathione transport in ER; melanosomal protein processing and routing	*OCA2*	15q11–q12	Oculocutaneous albinism type 2
Pldn (*pa*)	pallidin (pallid)	2	Pale coat, platelet defect	Organelle biogenesis; BLOC1 component	*PLDN*	15q15.1	N
Rab38 (*cht*)	RAB38, member RAS oncogene family (chocolate)	7	Brown eumelanin	Routing of Tyrp1 protein to melanosome	*RAB38*	11q14	N
Rabggta (*gm*)	Rab geranylgeranyl transferase, α subunit (gunmetal)	14	Coat dilution	Organelle biogenesis	*RABGGTA*	14q11.2	Choroideremia
Trappc6a	trafficking protein particle complex 6A	7	Pale patches in the coat and RPE	Protein trafficking	*TRAPPC6A*	19q13.32	N
Txndc5 (*mu*)	thioredoxin domain containing 5 (muted)	13	Pale pigment; platelet defect	Organelle biogenesis; BLOC1 component	*TXNDC5*	6p24–p25	N
Vps33a (*bf*)	vacuolar protein sorting 33a (buff)	5	Coat dilution	Organelle biogenesis	*VPS33A*	12q24.31	N
(E) Melanosome transport							
Mlph (*ln*)	melanophilin (leaden)	1	Silvery coat dilution, melanosomes cluster around nucleus	Melanosome transport	*MLPH*	2q37	N
Mreg (*dsu*, *Wdt2*)	melanoregulin (dilute suppressor, whn-dependent transcript 2)	1	Suppresses dilute phenotype (*Myo5a*$^{d/d}$)	Melanosome transport (interacts with Myo5a)	*MREG*	2q35	N
Myo5a (*d*)	myosin Va (dilute)	9	Silvery coat dilution, melanosomes cluster around nucleus	Melanosome transport	*MYO5A*	15q21	Griscelli syndrome
Myo7a (*sh-1*)	myosin VIIa (shaker-1)	7	Deafness, balance, head shaking	Melanosome transport (pigmented retina; ear?)	*MYO7A*	11q13.5	Usher syndrome, type 1B
Rab27a (*ash*)	RAB27A, member RAS oncogene family (ashen)	9	Silvery coat dilution, melanosomes cluster around nucleus	Melanosome transport	*RAB27A*	15q21	Griscelli syndrome

(F) Eumelanin and pheomelanin							
a	agouti (or nonagouti)	2	Different alleles alter eumelanin/pheomelanin balance, either way	Eumelanin/pheomelanin switch	*ASIP*	20q11.2	N
Atrn (*mg*)	attractin (mahogany)	2	Pheomelanin darkened	Eumelanin/pheomelanin switch (among others)	*ATRN*	20p13	N
Drd2	dopamine receptor 2, allele *tm1mok*	9	Agouti color darkened; POMC level raised	Receptor	*DRD2*	1q23.1–23.2	N
Eda (*Ta*)	ectodysplasin-A (tabby)	X	Darkening of agouti pigment; striping in +/– females; deficient sweat gland and hair morphogenesis	Membrane-bound, ?TNF-related ligand	*ED1*	Xq12–q13	Ectodermal dysplasia, anhidrotic/hypohidrotic ectodermal dysplasia
Edaradd (*cr*)	ectodysplasin A receptor-associated death domain (crinkled)	13	Delayed hair growth, agouti coat darker dorsally, yellower laterally	Receptor	*EDARADD*	1q43	N
Ggt1	γ glutamyltranspeptidase 1	10	Reduced pheomelanin	Glutathione metabolism	*GGT loci* (several)	22q11	Glutathionuria
L1cam	L1 cell adhesion molecule, allele *tm1Sor*	X	Black fur patches on agouti	Cell adhesion	*L1CAM*	Xq28	X-linked hydrocephalus, MASA/Crash syndrome
Mc1r (*e*)	melanocortin 1 receptor (extension)	8	Different alleles alter eumelanin/pheomelanin balance, either way	Receptor	*MC1R*	16q24.3	Red hair
Mgrn1 (*md*)	mahogunin, ring finger 1 (mahoganoid)	16	Melanin color, CNS effects	E3 ubiquitin ligase	*MGRN1*	16p13.3	N
Ostm1 (*Gl*)	osteopetrosis associated transmembrane protein 1 (Grey-lethal)	10	Loss of pheomelanin; osteopetrosis	Pheomelanin and osteoclast function	*OSTM1*	6q21	Severe recessive osteopetrosis
Pomc	pro-opiomelanocortin-α	12	Minimal or no effect on phenotype in black mice	MSH precursor	*POMC*	2p23.3	Obesity and red hair

Table A1.1 (*Cont'd*)

Symbol (old symbol)	Name (old name)	Mouse chromosome	Mutant phenotype	Molecular/biological functions	Human symbol	Human chromosome	Human syndrome
Slc7a11 (*sut*)	solute carrier family 7 (cationic amino acid transporter, y+ system), member 11 (subtle gray)	3	Required for normal pheomelanin levels	Cystine transporter	*SLC7A11*	4	N
Smarca5	SWI/SNF related, matrix associated, actin dependent regulator of chromatin, subfamily a, member 5, allele MommeD4	8	Dominant mottled coat with A^{y}/–	Transcriptional regulator	*SMARCA5*	4q31.1–q31.2	N
Smchd1 (*MommeD1*)	SMC hinge domain containing 1	17	Affects the percentage of female yellow pups on A^{vy} background	Modifies imprinting, X-inactivation	*SMCHD1*	18p11.32	N
Sox2 (*ysb*)	SRY-box-containing gene 2 (yellow submarine)	3	Yellow hair, neural, deafness	Transcription factor; Sox2 regulates Notch1 in eye	*SOX2*	3q26.33	Microphthalmia, anophthalmia, neural and pituitary defects
Sox18 (*rg*, *Dcc1*)	SRY-box-containing gene 18 (ragged, dark coat color 1)	2	Dark coat in agouti mice, sparse hair	Transcription factor	*SOX18*	20q13.33	N
(G) Systemic effects							
Atox1	Antioxidant protein 1 homolog 1 (yeast)	11	Hypopigmentation	Copper transport	*ATOX1*	5q32	N
Atp7a (*Mo*)	ATPase, Cu^{2+}-transporting, α polypeptide (mottled)	X	Pale fur in hemizygous males, striped in heterozygous females	Copper transport	*ATP7A*	Xq13.2–q13.3	Menke's disease
Atp7b (*tx*)	ATPase, Cu^{2+}-transporting, β polypeptide (toxic milk)	8		Copper transport	*ATP7B*	13q14.3–q21.1	Wilson disease
Bcl2	B-cell leukemia/lymphoma 2	1	Early graying, loss of melanocyte stem cells	Inhibitor of apoptosis	*BCL2*	18q21.3	B-cell lymphoma

Casp3	caspase 3, allele *tm1Flv*	8	Abnormal RPE	Effector of apoptosis	*CASP3*	4q35.1	N
Dst (*dt*, *ah*)	dystonin, allele *dt-J*; (dystonia musculorum; athetoid)	1	Pale skin		*DST*	6p12.1	N
Elovl3	elongation of very long chain fatty acids (FEN1/Elo2, SUR4/Elo3, yeast)-like 3, allele *tm1Jaco*	19	Abnormal hairs with scattered hyperpigmentation	Enzyme: fatty acid biosynthesis	*ELOVL3*	10q24.32	N
Elovl4	elongation of very long chain fatty acids (FEN1/Elo2, SUR4/Elo3, yeast)-like 4	9	Abnormal retinae including RPE	Enzyme: fatty acid biosynthesis	*ELOVL4*	6q14.1	Stargardt Disease 3
Ercc2	excision repair cross-complementing rodent repair deficiency, complementation group 2	7	UV-sensitivity, graying hair, reduced lifespan	DNA excision repair	*ERCC2*	19q13	Xeroderma pigmentosum, group D trichothiodystrophy, Cockayne syndrome
Fas	Fas (TNF receptor superfamily member 6)	19	Fewer strial melanocytes on certain background	Receptor mediating apoptosis	*FAS*	10q23.31	N
Heph (*sla*)	hephaestin; sex-linked anemia	X	abnormal pigment location in RPE	Regulates iron levels, including in RPE	*HEPH*	Xq12	N
Hs2st1	heparan sulfate 2-O-sulfotransferase 1	3	Abnormal RPE differentiation	Enzyme in heparan sulfate biosynthesis	*HS2ST1*	1p22.3	N
Oat	ornithine aminotransferase	7	Abnormal RPE cell morphology	Enzyme	*OAT*	10q26.13	Gyrate atrophy
Pah	phenylalanine hydroxylase	10	Effects include hypopigmentation, worsening with age	Tyrosine synthesis	*PAH*	12q3.2	Phenylketonuria
Pdpk1	3-phosphoinositide dependent protein kinase-1, allele *tm1Bcol*	17	Abnormal eye pigmentation	Phosphorylates and activates AKT kinase	*PDPK1*	16p13.3	N
Polg	polymerase (DNA directed), γ	7	General premature ageing including coat graying	DNA polymerase γ	*POLG*	15q26.1	Alpers syndrome

Table A1.1 *(Cont'd)*

Symbol (old symbol)	Name (old name)	Mouse chromosome	Mutant phenotype	Molecular/biological functions	Human symbol	Human chromosome	Human syndrome
Polh	polymerase (DNA directed), η (RAD 30 related)	17	Pigment (melanocyte?) accumulation in ear skin following UV irradiation	DNA polymerase η	*POLH*	6p21.1	Xeroderma pigmentosum, variant type
Pts	6-pyruvoyl-tetrahydropterin synthase	9	Coat dilution (low biopterin, high Phe)	Tetrahydrobiopterin synthesis	*PTS*	11q23.1	Phenylketonuria III
Rbp1	retinol binding protein 1, cellular	9	Abnormal RPE morphology	Intracellular transport of retinol	*RBP1*	3q23	N
Rpl24 (*Bst*)	ribosomal protein L24 (Belly spot and tail)	16	Eye, coat, skeletal	Protein synthesis	*RPL24*	3q12.3	N
Rps19 (*Dsk3*)	ribosomal protein S19 (Dark skin 3)	7	Defect gives p53 stabilization and SCF [KITL] synthesis in keratinocytes, hence dark skin with extra melanocytes	Protein synthesis	*RPS19*	19q13.2	Diamond-Blackfan anemia (DBA)
Rps20 (*Dsk4*)	ribosomal protein S20 (Dark skin 4)	4	As Rps19	Protein synthesis	*RPS20*	8q12.1	N
Rxra	retinoid X receptor α	2	Premature hair graying then hair loss	Retinoid receptor Regulation of diverse pathways	*RXRA*	9q34.2	N
Slc31a1	solute carrier family 31, member 1	4	Copper deficiency, hypopigmentation	Copper uptake into cells	*SLC31A1*	9q32	N
Vldlr	very low density lipoprotein receptor	19	Thickening and disruption of RPE	Lipid uptake into cells	*VLDLR*	9p24.2	Cerebellar hypoplasia and mental retardation

ACTH, adrenocorticotrophic hormone, corticotropin; AKT, oncogene isolated from transforming murine retrovirus AKT8 found in a thymoma cell line (=RAC=protein kinase B, PKB); CNS, central nervous system; EDNR, endothelin receptor; ER, endoplasmic reticulum; GPI, glycosylphosphatidylinositol; HDAC, histone deacetylase; KO, knockout; LEF, lymphoid-enhancer-binding factor; N, none known; NF-κB, nuclear factor κB; PLC, phospholipase C; POMC, pro-opiomelanocortin; RPE, retinal pigment epithelium; SCF, stem cell factor, also known as Kit ligand; TGFβ, transforming growth factor β; TNF, tumor necrosis factor; UV, ultraviolet; VEGF, vascular endothelial growth factor. We are greatly indebted to the Mouse Genome Informatics (MGI) and Online Mendelian Inheritance in Man (OMIM) websites for much of the information presented here (www.informatics.jax.org/, www.ncbi.nlm.nih.gov/sites/entrez?db=omim). A table like this placed in a book will inevitably become outdated quite soon; however, we hope that updated information about the mouse color genes will continue to be available at the IFPCS Color Genes resource (www.espcr.org/micemut). The site also has links to human and zebrafish orthologs, and to other useful resources.

Table A1.2 Summary of the uncloned mouse color genes

Symbol	Name	Chromosome	Effect or possible function
(A) Development?			
Alm	anterior lenticonus with microphthalmia	?	Defects of eye, coat, others
ao	apampischo	?	Hair loss then regrowth of darker, sparser hair
baln2	balance 2	?	Defects of eye, coat, neurological
baw	black and white	18	White ventrum, scattered white hairs dorsally
Bswt	belly spot with white toes	1	Belly spot, white hind toes
bt2	belted 2	?	White belt
crsp	cryptorchidism with white spotting	5	Coat and skin pigment, male reproductive system
cw	curly whiskers	9	CBA mice go darker. (Some lymphoma)
dds	dorsal dark stripe	15	Dorsal dark stripe
dkd	darkened dorsal	2	Dorsal dark stripe
Dph1	DPH1 homolog	11	Delayed embryonic eye pigmentation
Dwh	dispersed white hair	2	White hairs and patches throughout coat
Ednrbm1	endothelin receptor type B modifier 1 (QTL)	10	Modifies extent of spotting with $Ednrb^s$
Eed	embryonic ectoderm development	7	Diluted coat (dwarfism etc)
Exma	exencephaly and severe microphthalmia or anophthalmia	X	Patchy coat pigmentation, microphthalmia
fc	flecking	2	Belly spot, head spot
Fk	fleck	?	white on belly, tail, feet
gand	gandalf	?	Diluted coat, +ataxia etc, delayed lethal
Gn	gentoo	10	Belly spot, head spot
Gsfali029	gsf abnormal limbs mutant 029	?	Belly spot, polydactyly
Gsfali19	gsf abnormal limbs mutant 019	?	Belly spot, polydactyly
Gsfali20	gsf abnormal limbs mutant 020	?	Belly spot, polydactyly
Gsfdcc2	gsf dark coat colour 2	?	Belly spot, darker coat
Gsfkta19	gsf kinked tail 19	?	Belly spot, kinked tail
Gsfsco6	gsf spotted coat 6	?	Small head blaze
Gsfsco7	gsf spotted coat 7	?	Distal depigmentation, belly spot, white vibrissae

Table A1.2 *(Cont'd)*

Symbol	Name	Chromosome	Effect or possible function
Gsfund3	gsf undefined 3	?	White ring around tail
Gsfwbs011	gsf white belly spot 011	?	Belly spot
Gsfwbs1	gsf white belly spot 1	?	Belly spot, kinked tail
Gsfwbs3	gsf white belly spot 3	?	Large belly spot
Gsfwbs5	gsf white belly spot 5	?	Belly spot
Gsfwbs9	gsf white belly spot 9	?	Belly spot
Gsfwnw	gsf white nose and whiskers	?	White nose and vibrissae
Gsfwt	gsf white tail	?	White tail
Hpt	hair patches	4	Hair patchy and skin has patches of pigment; also cardiovascular defects, etc.
hs	head spot	?	Head spot
Idc	iris dysplasia with cataract	?	Eye abnormalities, microphthalmia, belly spot
Mtu	Montu	12	Fewer neural crest cells; belly spot, curly tail
Mwfh	modifier of white forelock hypopigmentation	10	Modifies phenotype of $Sox10^{Dom}$
Pbdl	piebald-like	?	Spotting, progressive dilution, megacolon
Ph	patch deletion region (patch)	5	White spotting; responsible gene in deletion uncertain; not *Pdgfra*; but KIT expression is altered
pwk	patchwork	10	Patchwork of pigmented and unpigmented hairs; autocrine growth of melanocytes?
rg	rotating	?	Ear development, neural, sometimes belly spot
Rgsc58	RIKEN Genomic Sciences Center (GSC), 58	?	White hairs, spots or band in dorsal lumbar region, varying with genetic background
Rgsc117	RIKEN Genomic Sciences Center (GSC), 117	?	Belly spot, domed skull, dominant (not all mice)
Rgsc257	RIKEN Genomic Sciences Center (GSC), 257	?	Scattered white hairs dorsally
Rgsc269	RIKEN Genomic Sciences Center (GSC), 269	?	White digits and tail tip, some white patches on belly
Rgsc288	RIKEN Genomic Sciences Center (GSC), 288	?	Patches of paler fur
Rgsc394	RIKEN Genomic Sciences Center (GSC), 394	?	Variable white digits and tail tip
Rgsc398	RIKEN Genomic Sciences Center (GSC), 398	?	Variable white digits and tail tip
Rgsc444	RIKEN Genomic Sciences Center (GSC), 444	?	Variable number of white spots on tail

Rgsc510	RIKEN Genomic Sciences Center (GSC), 510	?	Variable white digits and tail tip
Rgsc662	RIKEN Genomic Sciences Center (GSC), 662	?	Abnormal digit pigmentation and white tail tip
Rgsc713	RIKEN Genomic Sciences Center (GSC), 713	?	White distal feet and tail tip, sometimes belly spot
Rgsc755	RIKEN Genomic Sciences Center (GSC), 755	?	White toe tips and tail tip
Rgsc767	RIKEN Genomic Sciences Center (GSC), 767	?	White toe tips and tail tip
Rgsc990	RIKEN Genomic Sciences Center (GSC), 990	?	White toe tips and tail tip
Rgsc1246	RIKEN Genomic Sciences Center (GSC), 1246	?	Coat dilution especially ventrally; some white spotting.
Rgsc1461	RIKEN Genomic Sciences Center (GSC), 1461	?	White belly spot, skeletal changes
Rgsc1513	RIKEN Genomic Sciences Center (GSC), 1513	?	Dilution dorsally, gray fur ventrally (even in *A*/–), some belly spotting
Rgsc1520	RIKEN Genomic Sciences Center (GSC), 1520	?	White digits, tail tip, belly spot
Rgsc1545	RIKEN Genomic Sciences Center (GSC), 1545	?	Belly spot
Rgsc1554	RIKEN Genomic Sciences Center (GSC), 1554	?	Belly spot
Rgsc1658	RIKEN Genomic Sciences Center (GSC), 1658	?	Scattered white hairs, some color dilution
Rgsc1742	RIKEN Genomic Sciences Center (GSC), 1742	?	Belly spot
Rgsc1843	RIKEN Genomic Sciences Center (GSC), 1843	?	Variable scattered white spots in females
Rgsc1855	RIKEN Genomic Sciences Center (GSC), 1855	?	Belly spot
rn	roan	14	Micro-spotting, whole coat
rs	recessive spotting	5	Micro-spotting (reduced melanocyte numbers); interacts with Kit
rslk	recessive spotting-like	5	Gray coat, head and/or belly spot, white spots
Shmu	shamu	9	White feet, belly spot, head spot
Ska7	skeletal/axial 7	?	Large belly spot and skeletal defects
Skc42	skin/coat color 42	?	Sharply delineated white belly
Skc43	skin/coat color 43	?	White belly patch and tail
skc44	skin/coat color 44	?	White belt, sometimes spotting
Sls	semidominant lethal spotting	2	Semidominant spotting; may be allelic to Edn3
smk	smoky	?	Gray coat on a/a, with reproductive system defects
stn	stunted	19	Belly spot, altered facial skeleton
Stol	stripy oily	X	Stripy in heterozygote, dark in homo- and hemizygote; oily hair; microphthalmia
Strx2	striated, X-linked 2	X	Striped fur, thick skin in +/–, lethal in –/Y
Strx3	striated, X-linked 3	X	Striped fur, thick skin in +/–, lethal in –/Y
Strx4	striated, X-linked 4	X	Striped fur, scaly skin in +/–, lethal in –/Y

Table A1.2 (*Cont'd*)

Symbol	Name	Chromosome	Effect or possible function
Tcm	total cataract with microphthalmia	4	Microphthalmia, abnormal iris, lens, retina
tga	transposition of the great arteries	4	Ectopic pigmentation in heart and thoracic cavity
tmgc17	Tennessee Mouse Genome Consortium 17	X	Belly spot, syndactyly
Tmgc19	Tennessee Mouse Genome Consortium 19	?	Belly spot, other spotting, postnatal dominant lethal
Tmgc21	Tennessee Mouse Genome Consortium 21	?	Belly spot, other spotting, white feet
tp	taupe	7	Diluted color, female reproductive system
Ts	tail-short	11	Belly spot, white distal forelimbs; deficiencies of skeleton, blood, growth, etc.
vs	variable spotting	9	Spotting on belly, head, tail, feet
Vss	variable spot and size	2	Variable belly spot, small size
Wbct	white belly, claws and tail	1	Variable spotting of belly, feet, tail
Whto	white toes	7	Color, digit development
wn	white nose	15	White nose, ventral streak
Wtgr	wavy tiger	X	Coat striped and wavy; reproductive defects
Xls	*X*-linked stripe	X	Coat striping, 1 white spot on left flank
Xs	extra-toes spotting	7	Color, digit development
Xsl	extra-toes spotting-like	7	Extra toes, belly spot
(B) Melanocyte function only?			
brwd	brownoid	?	Melanin color (brown)
Cal7	caracul-like 7	15	Pale skin and eyes
dj	dilution Japan	?	Pink skin, gray coat (a/a)
dp	dilution-Peru	15	Pale coat
plto	platino	?	Off-white coat, black eyes; not Tyr
powder	powder	?	Pale coat
rgsc1820	RIKEN Genomic Sciences Center (GSC), 1820	?	No pigment, albino
Rgsc1904	RIKEN Genomic Sciences Center (GSC), 1904	?	Coat color dilution
ru2l	ruby-eye 2-like	7	Like ru2; gray coat, pale skin, red eyes

sea	sepia	1	Coat color dilution
skc14	skin/coat color 14	?	Pale coat
skc19	skin/coat color 19	?	Pale coat
skc21	skin/coat color 21	?	Pale coat when young; white guard hairs
skc22	skin/coat color 22	?	Pale coat when young; white guard hairs, red eyes
skc23	skin/coat color 23	?	Pale coat when young; dark red eyes
skc28	skin/coat color 28	?	Silver coat
skc29	skin/coat color 29	?	Silver coat
skc30	skin/coat color 30	?	Coat silvering, melanocyte death
skc31	skin/coat color 31	?	Coat silvering, melanocyte death
Skc36	skin/coat color 36	?	Mottled coat in female
titanm	titanium	?	Pale coat
tmgc14	Tennessee Mouse Genome Consortium 14	?	Diluted coat
Tmgc18	Tennessee Mouse Genome Consortium 18	?	Diluted coat, dominant
uwl	underwhite-like	15	Ruby-eyed cream; possible uw (Slc45a2) allele
(C) Eumelanin and pheomelanin?			
Cdif	color difference (QTL)	15	Modifies yellow coat color
da	dark	7	Pheomelanin deficient
dal	dark-like	7	Like dark
din	dense incisors	16	Loss of pheomelanin/white belly on agouti background; loss yellow on ears
Dmyaq1	darker modification of yellow agouti QTL 1	1	Darkening of yellow and agouti coats
Dmyaq2	darker modification of yellow agouti QTL 2	1	Darkening of yellow and agouti coats
Dmyaq3	darker modification of yellow agouti QTL 3	15	Darkening of yellow and agouti coats
dwg	dwarf grey	10	Multiple effects, including deficient pheomelanin, osteoclasts
gdn	golden	?	Eumelanin deficient
gr	grizzled	10	Pheomelanin deficient; tail
gri	grey intense	11	Pheomelanin deficient
gt	gray tremor (probably extinct)	15	Pheomelanin deficient; spotting, neurological
Och	ochre	4	Deficient eumelanin, balance, other

Table A1.2 (*Cont'd*)

Symbol	Name	Chromosome	Effect or possible function
Pheo	pheomelanin (QTL)	15	Modifies color of A^{y}/– mice; more eumelanic allele dominant
Reph8	reduced pheomelanin 8	?	Agouti mice darker
skc17	skin/coat color 17	?	Dark back
U	umbrous	?	Dorsal pheomelanin darkened
Up	umbrous-patterned	?	Dorsal pheomelanin darkened (patchy)
Ym	yellow mottled	X	Yellow mottling, hemizygous lethal
Yv	yellow value (QTL)	15	Modifier of eumelanin/pheomelanin ratio
(D) Organelle biogenesis or transport?			
dill	Dilute-like	?	Like dilute
rgsc80	RIKEN Genomic Sciences Center, 80	?	Diluted coat, interacts with *Myo5a*^{d}
skc6	skin/coat color 6	?	Pale coat, lysosomal storage defect, long hair
skc9	skin/coat color 9	?	Pale coat, lysosomal storage defect
skc10	skin/coat color 10	?	Pale coat, lysosomal storage defect
skc12	skin/coat color 12	?	Pale coat, lysosomal storage defect
skc15	skin/coat color 15	?	Pale skin
(E) Dark skin			
Dfp	dark foot pads	?	Dark skin
Dfp2	dark foot pads 2	4	Dark skin
Dsk6	dark skin 6	3	Dark skin
Dsk8	dark skin 8	3	Dark skin
Dsk9	dark skin 9	11	Dark skin
Rgsc45	RIKEN Genomic Sciences Center, 45	?	Gray-pigmented footpads with thickened epidermis
Rgsc63	RIKEN Genomic Sciences Center, 63	?	Slightly dark hind footpads
Rgsc150	RIKEN Genomic Sciences Center, 150	?	Dark footpads
Rgsc183	RIKEN Genomic Sciences Center, 183	?	Dark footpads

Rgsc194	RIKEN Genomic Sciences Center, 194	?	Slightly dark footpads
Rgsc207	RIKEN Genomic Sciences Center, 207	?	Pigmented dermatoglyphs of footpads
Rgsc372	RIKEN Genomic Sciences Center, 372	?	Dark footpads
Rgsc515	RIKEN Genomic Sciences Center, 515	?	Dark footpads
Rgsc526	RIKEN Genomic Sciences Center, 526	?	Dark footpads
Rgsc715	RIKEN Genomic Sciences Center, 715	?	Slightly dark footpads
Skc39	skin/coat color 39	?	Dark skin
Skc41	skin/coat color 41 (blackfoot)	?	Dark skin
soo	sooty foot	2	Dark skin

(F) Unknown

Dlp1	dominant lightened pigment 1	?	Lighter coat, sometimes slight belly spot
Dlp2	dominant lightened pigment 2	?	Lighter coat
Dlp3	dominant lightened pigment 3	?	Much lighter coat
fe	faded	6	Progressive coat fading
fnld	faint lined	X	Hemizygous lethal; fine dorsal striping
Fw	fawn	?	Lightens *Rn* mutant mice
ge	greige	1	Paler coat and skin in dilute, brown mice
Gsfbcc2	gsf bright coat colour 2	?	Complex coat color variation
lgr	London grey	?	Gray coat, systemic effects
Lgt (*Li*)	light	?	Light coat (dominant), light skin (recessive)
Li	lined	X	Hemizygous lethal; fine striping; deletion that includes *Rsk2*
Mch	modifier of chinchilla	?	$Tyr^{c\text{-}ch}$ mice look browner
Mchm1	modifier of chinchilla-mottled 1	?	Lightens $Tyr^{c\text{-}m}$ mice
Mchm2	modifier of chinchilla-mottled 2	?	Lightens $Tyr^{c\text{-}m}$ mice
Mfs	mutant fur is striped	?	'Striped fur'
nmf192	neuroscience mutagenesis facility, 192	?	Spotted or mottled retinae
nur15	neurological 15	?	Diluted coat (and neural effects, early lethal)
nur16	neurological 16	?	Diluted coat (and neural effects, early lethal)
nur17	neurological 17	?	Diluted coat (and neural effects)
Rd4	retinal degeneration 4	4	Eye defects include pigmented spots in the fundus

Table A1.2 *(Cont'd)*

Symbol	Name	Chromosome	Effect or possible function
Rd9	retinal degeneration 9	X	Retina mottled and degenerates
rdp	reduced pigment (not same as rp/Bloc1s3)	?	Marked pigment dilution, red eyes
Rgsc71	RIKEN Genomic Sciences Center, 71	?	Dilution dorsally (dominant), gray belly, sometimes belly spot
Rgsc212	RIKEN Genomic Sciences Center, 212	?	Black rostrally and brown caudally of a mid-trunk demarcation line
Rgsc547	RIKEN Genomic Sciences Center, 547	?	Patch of brown fur between the eyes, by 8 weeks old
Rgsc796	RIKEN Genomic Sciences Center, 796	?	Slight coat dilution (dominant) and behavioral changes
skc18	skin/coat color 18	?	Abnormal coat color, small size
Skc45	skin/coat color 45	?	'Shading of coat in animals expected to be white'
Sta	autosomal striping	X	Striping in both sexes
Sta2	striping, autosomal 2	?	Striped coat
Sta3	striping, autosomal 3	?	Striped coat
Strg	striped greasy	X	Hair texture and color
tmgc22	Tennessee Mouse Genome Consortium 22	7	Abnormal RPE, choroid
tmgc23	Tennessee Mouse Genome Consortium 23	7	RPE hyperpigmented around optic disk
tmgc25	Tennessee Mouse Genome Consortium 25	7	Abnormal RPE, hypopigmented fundus
tmgc29	Tennessee Mouse Genome Consortium 29	7	Like tmgc25
wuf	white under fur (extinct?)	?	Underfur white
Xmo2	X-linked mottled 2	X	Mottled coat; hemizygous lethal
Xmo3	X-linked mottled 3	X	Mottled coat; hemizygous lethal
Xmo4	X-linked mottled 4	X	Mottled coat; hemizygous postnatal lethal, pale coat
Xmo5	X-linked mottled 5	X	Mottled coat; hemizygous lethal
Xmo6	X-linked mottled 6	X	Mottled coat; hemizygous lethal

QTL, quantitative trait locus; RPE, retinal pigment epithelium. Categories and functions are generally provisional or speculative.

Figure 1.17 Studs Terkel, Mahonia, and Postdoc (left to right). Terk may have a mutation at the *Silver* locus (see Chapter 4). Mahonia is pheomelanic, often known as chestnut or sorrel (probable genotype $Mc1r^{e}/Mc1r^{e}$). Postdoc is the horse equivalent of wild type, known as bay or brown, depending on the amount of eumelanic pigment. Note that the mane, tail, and legs are eumelanic, and the pheomelanic pigmentation on the body varies from bright red/yellow to dark and nearly eumelanic, in response to modifying genes.

Part II
The Pigmentary Loci

Figure 2.1 *Merle* Australian Shepherd dog illustrates a mutation involving the *Silver* gene locus (Chapter 4). This dog is heterozygous. The mutation is caused by a transposable element and affects skin and eyes.

2

Introduction to Mutant Pigmentary Genes

In Part II we discuss the basic processes that are impacted by mutations in the murine pigmentary system, focusing on the classical pigmentary mutations and the genic loci that are involved with regulation of these processes. These generally fall into four major categories: (1) white spotting and progressive graying (cellular development and survival); (2) melanosomal defects and albinism (cellular differentiation); (3) melanosomal transport in the pigment cell and transfer to keratinocytes; and (4) pigment-type switching. These four are briefly outlined below and discussed in the following chapters. See Tables A1.1 and A1.2 in the Chapter 1 Appendix for lists of the mutant loci that affect pigmentation.

2.1 Defects of normal melanocyte development: white spotting and graying with age

White spotting and progressive graying are terms used here to describe any situation in which the skin or hair lacks viable melanocytes (pigment cells). In white spotting, the defect is evident at birth; in progressive graying (or vitiligo in humans), the organism is born pigmented and the melanocytes later disappear progressively. In either case, affected animals may be completely unpigmented, although usually they are not. Albino animals are not included in this category because the melanocytes of albino animals are viable, whereas white-spotted animals lack melanocytes in the affected areas. Albino mammals lack pigment because of defective melanogenesis (synthesis of melanin pigment, which is discussed below). We will not use the term piebald as a synonym for white spotting, because *Piebald* is the name of a specific murine white-spotting locus, the *Ednrb* locus (Fig. 2.2), which encodes endothelin receptor B. The *Piebald* locus is one of many loci that are required to support melanocyte survival. Mutation at any of these loci might cause white spotting.

Normal development of melanocytes (pigment cells) progresses in a continuum of increasingly restricted cell fates, beginning approximately at mid-gestation in mice, and includes the following processes:

The Colors of Mice: A Model Genetic Network, 1st edition. By M. Lynn Lamoreux, Véronique Delmas, Lionel Larue, and Dorothy C. Bennett. Published 2010 by Blackwell Publishing Ltd.

Figure 2.2 This *Piebald* mouse (JU/CtLm-$Ednrb^s/Ednrb^s$) illlustrates white spotting.

- specification of pluripotent neural-tube cells to the neural-crest lineage and then to the pigment-cell lineage;
- epithelial–mesenchymal transition of neural-crest cells to melanoblasts;
- proliferation of melanoblasts and migration across the face of the somites;
- migration of melanoblasts through the mesenchyme and across the basement layer of the epidermis;
- homing of melanoblasts to final sites of differentiation in the hair follicles, ears, and heart;
- establishment (in the case of the hair follicle) of a population of pluripotent stem cells and differentiation of melanocytes in the hair bulb to form the first generation of melanocytes;
- maintenance or replacement of differentiated melanocytes throughout life.

The genic loci that are known to be involved in the above processes are listed in Tables A1.1 and A1.2 in the Chapter 1 Appendix. The classical white spotting loci are discussed in Chapter 3 in the context of their contribution to our understanding of the development of the pigmentary system.

2.2 Defects in normal melanosome development: albinism

Once established in the skin and hair follicles of mice, at about the time of birth, murine pigment cells begin to differentiate to produce pigmented organelles (melanosomes) that the cell can control, move about, and deliver to growing hairs or transfer to other types of cells, most often keratinocytes. The melanosome is a melanocyte-specific, membrane-bound organelle of the lysosomal family of organelles. It is composed of pigmentary proteins that function primarily in a structural role, and other proteins whose most important contribution to pigmentation is enzymatic. The processes required in the construction of a normal melanosome include:

- the origin as a membrane-bound multi-vesicular body that separates from the endoplasmic reticulum;

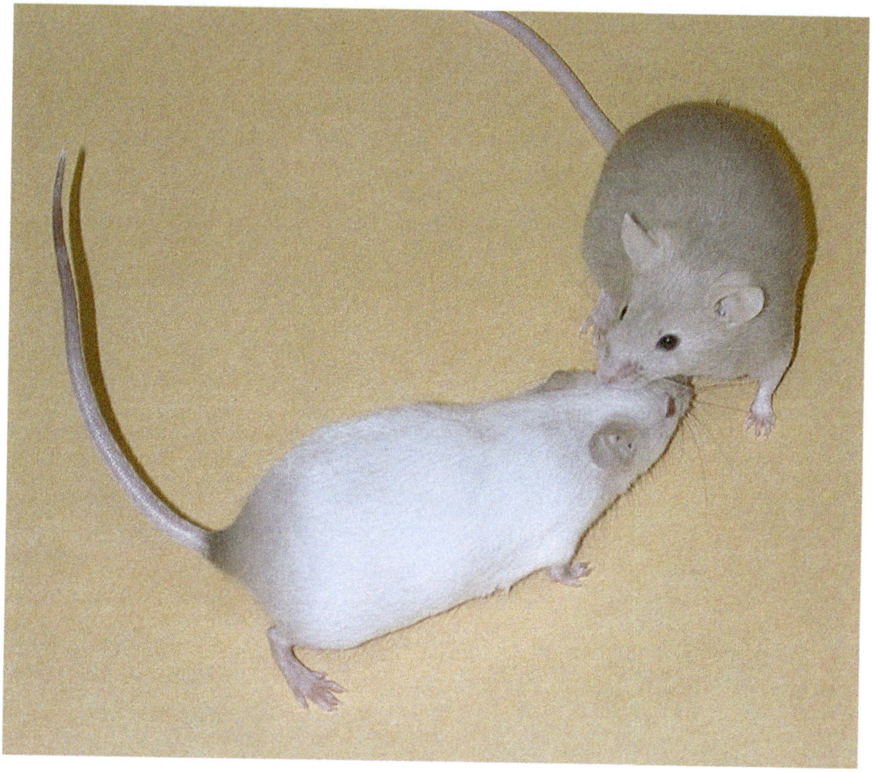

Figure 2.3 Mice mutant at the *Tyrosinase* (*Albino*) locus (Chapter 4). C57BL/6J-$Tyr^{c\text{-}h}$/$Tyr^{c\text{-}h}$ (*Himalayan*) (left). C57BL/6J-$Tyr^{c\text{-}e}$/$Tyr^{c\text{-}e}$ (*Extreme dilution*) (right).

- the production of the proteins that will be important to creating the melanosome:
  - structural proteins such as silver,
  - enzymes such as tyrosinase (Fig. 2.3), tyrosinase-related protein 1, and dopachrome tautomerase,
  - proteins involved in processing melanosomal proteins, such as underwhite,
  - proteins involved in routing the melanosomal proteins to the developing melanosome, such as the BLOC proteins;
- the assembly of the melanosome;
- ultimately, the relatively simpler process of melanogenesis, depositing the melanin pigment upon and within the maturing melanosome.

The genic loci that are known to be involved in the above processes are listed in Tables A1.1 and A1.2 in the Chapter 1 Appendix. The classical loci that are involved in generating the melanosome are discussed in Chapter 4 in the context of their contribution to our understanding of pigment-cell differentiation.

2.3 Transport of melanosomes to other cells: the 'dilute' phenotype

When assembly of the melanosomes is complete in the murine hair follicle, or as assembly progresses, the melanosomes are transported within the cytoplasm to the tips of the dendritic

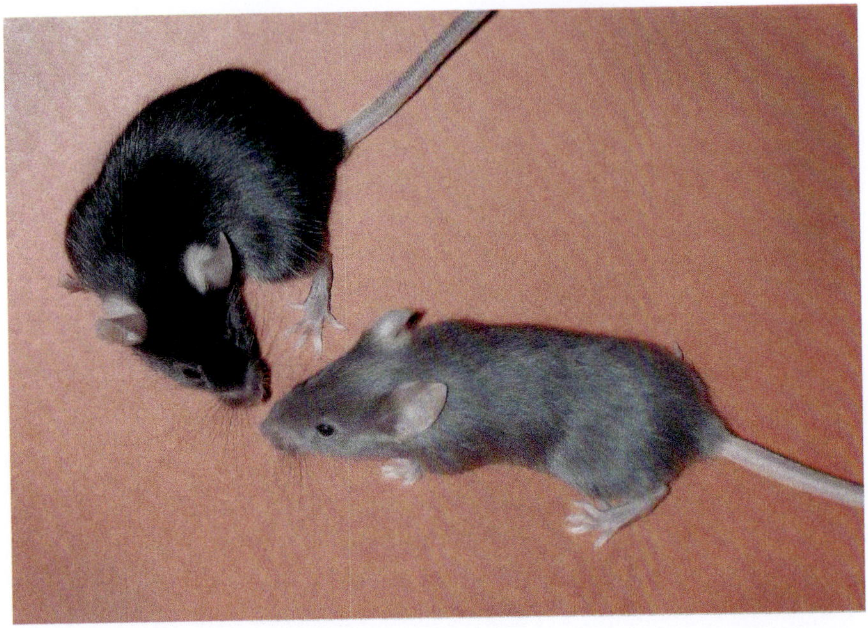

Figure 2.4 Dilute mouse and nondilute mouse illustrate the phenotype that results from failure of normal melanosome transport and insertion into the growing hair (Chapter 4).

processes of the pigment cell, from which they are deposited into the keratinocytes of the epidermis or the growing hair, as described in Chapter 4. Failure of this process in the melanocyte results in mice of a characteristic phenotype that is commonly referred to as the 'dilute' phenotype (after which the *Dilute* locus is named; Fig. 2.4). The basis for the phenotype is abnormal transport of melanosomes from the melanocyte into the keratinocytes so that they arrive in clumps and blobs, or entire melanocytes, as well as in the normal finely granular uniformity. The result is a characteristic grayish or beigey appearance (depending upon the color of the pigment) that results from abnormal scattering of the light reflected from the hair coat of 'dilute' animals. Several other loci of the mouse mimic or influence this 'dilute' phenotype, and their study has clarified the processes involved in organelle transport. The genic loci that are known to be involved in these processes are listed in Tables A1.1 and A1.2 in the Chapter 1 Appendix. The classical loci involved are discussed in Chapter 4 in the context of their contribution to our understanding of melanosome transport.

2.4 Pigment-type switching: from eumelanogenesis to pheomelanogenesis

The melanocyte is capable of producing two major types of melanin pigment within the melanosome: eumelanin is dark black or brown; pheomelanin is variously referred to as red, yellow, or orange, depending upon species, and in mice is referred to as yellow. Apparently, the cell does not make both types of pigment at the same time, but it can rapidly switch from one to the other in mid-melanosome. Pigment-type switching, from eumelanin to pheomelanin and the reverse, is a complicated process in which several proteins must be upregulated to produce eumelanin or downregulated to produce pheomelanin as described in Chapter 5. The wild-type phenotype of the mouse is agouti. Agouti colored animals have hairs that are banded with patterns of eumelanic

Figure 2.5 This agouti ground squirrel is presumably *A/A* but this is not known. Agouti involves pigment-type switching from eumelanin to pheomelanin, as discussed in Chapter 5.

and pheomelanic pigmentation (Fig. 2.5). This patterning is tightly controlled within individual hair follicles (Galbraith 1964) as yellow or nonyellow portions. The majority of dorsal hairs of the *Agouti* mouse are normally tipped with eumelanin, banded below the tip with pheomelanin, and then have eumelanin to the base of the hair. Most often the intensity of pigmentation decreases gradually near the base of the hair. The *Agouti* locus also controls spatial patterns, as in a *Black-and-tan* mouse or dog, for example.

Complex control over this type of pigment patterning is exerted over time and space by what has often been referred to as the 'eumelanin/pheomelanin switch mechanism.' In the mouse, primary control over this switch is exerted by the *Agouti* (also known as *nonagouti*, *A*) locus that encodes a ligand (ASP, agouti signaling protein) to the cell-surface receptor melanocortin 1 receptor (MC1R), encoded at the *Mc1r* locus, also known as *Recessive yellow* or *Extension*. In mice, these loci together determine which of the two major types of melanin pigment will be manufactured by the pigment cell. Several other loci contribute to the so-called switch mechanism, including *Mahogany* and *Mahoganoid*. An additional yellow/nonyellow locus has been described in dogs and cattle. The identified loci that function in the switch mechanism are listed in Tables A1.1 and A1.2 in the Chapter 1 Appendix. Genic interactions among classical pigmentary loci, as they contribute to the yellow/nonyellow switch mechanism, are discussed in Chapter 5.

Figure 2.6 Pheomelanic calves are shown in the front row. The two cows at the back are brindle in phenotype, although it is difficult to see because of the way the photograph is taken. They are eumelanic and pheomelanic in stripes as is shown more clearly in Figure 5.11. The three calves in front are pheomelanic, the various shades of color caused by other segregating genes, one of which is the *Silver* mutation from the Charolais breed that is discussed in Chapter 4. Photograph courtesy of Britten Farms.

3

White Spotting and Progressive Graying

Figure 3.1 White spotted mouse genotype *Mitf*$^{mi\text{-}rw}$/*Mitf*$^{mi\text{-}rw}$.

The goal of this chapter is to describe the developmental journey of murine pigment cells as they generate normal pigmentation. The chapter is organized into five sections reflecting the biological history of pigmentation, each at a different level of organization, together describing the pigmentary phenotypes.

- Section 3.1 gives definitions of white spotting and vitiligo, plus some general background.
- Section 3.2 outlines the timeline of development of pigmentation, with an overview of the relevant developmental anatomy and physiology of the embryo.

The Colors of Mice: A Model Genetic Network, 1st edition. By M. Lynn Lamoreux, Véronique Delmas, Lionel Larue, and Dorothy C. Bennett. Published 2010 by Blackwell Publishing Ltd.

- Section 3.3 gives an overview of molecular signaling events relevant to the development of the pigment-cell lineage.
- Section 3.4 describes the pigment phenotypes and the classical white-spotting genes, with a time-line of development of the murine pigmentary phenotype. Gene functions during embryogenesis are also discussed. This section uses the above information to describe the processes set in motion by pigmentary genes, and describe their effects upon the pigmentary phenotype.
- Section 3.5 looks at specific cases in the head, heart, ears, and eyes.

Sections 3.2 and 3.3 describe processes that regulate and/or are regulated by pigmentary genes, and as such they define the intermediate steps between gene functions and their phenotypic manifestations. Readers who are fluent in development and/or cell biology of pigment cells will probably prefer to skip directly from section 3.1 to section 3.4.

3.1 Definitions and general background

White spotting, vitiligo, and progressive graying are phenotypes that result from pigment-cell death or failure to develop. Thus, in studying these phenotypes we learn about the developmental processes required for pigment-cell development. Developmental failure may involve any process required for renewal or survival of the pigment-cell lineage, from its origin in the neural crest, to differentiation of the pigment cell at its final destination. It may be studied at any level of organization, from the gross phenotype to its molecular genetic cause.

In this chapter we provide a general overview of the processes that are necessary for normal pigment-cell development. We begin by describing the phenotypes, follow with a look at pigment-cell genesis at the tissue/cellular level of developmental biology, and then provide an overview of intra- and intercellular mechanisms of communication that may function in pigment-cell survival. Finally, we attempt to explain the relationships among these interacting factors and their expression in the organismal phenotype, as they are regulated by the functions of the major white-spotting gene loci.

3.1.1 Background

To recap on the discussion in Chapter 1, melanocytes of mammals are specialized cells that produce melanin-based pigment and are responsible for coloration of the eye, skin, and hair of mammals. The majority of melanocytes are found in the epidermis of the skin, the hair follicles, the inner ear, retinal pigment epithelium (or RPE), and choroidal layer of the eye. Vertebrate melanocytes are derived from two embryonic origins. Those in the RPE arise from the optic cups, being induced from the neural tube by the overlying ectoderm, whereas those in the integument, inner ear, and choroid are derived from neural-crest cells that arise in the closing neural tube and migrate extensively through the embryonic tissues to reach their final locations.

Melanoblasts are precursors of melanocytes that are specified but not yet pigmented. They develop from the neural-crest cells of the developing mouse embryo. The neural-crest cells are a pluripotent population of ectoderm-derived cells that originate around the site and time of dorsal neural-tube closure, migrate along defined pathways within the embryonic tissues, and differentiate into numerous cell types, including melanocytes of the pigment-cell lineage and the neurons and glia of the peripheral nervous system. The processes of cell lineage restriction and the paths of cell migration are regulated by a network of environmental influences, intercellular interactions, and intrinsic factors. Defects in the migration, survival, or differentiation of pigment cells results in white spotting and/or vitiligo/progressive graying.

Figure 3.2 *Piebald/piebald lethal*, white-spotted mice ($Ednrb^{s}/Ednrb^{s\text{-}l}$). One (left) is *Recessive yellow* ($Mc1r^{e}/Mc1r^{e}$) (Chapter 5). The other is not ($Mc1r^{E}/Mc1r^{e}$).

3.1.2 White spotting defined

White spotting is defined as the congenital absence of viable melanocytes from some or all of the areas where they are normally present. In the mouse, pigment cells are normally found in the hair follicles, skin, eyes, inner ear, and Harderian glands, and in the heart. In general, white spotting is the result of the failure of pigment cells to survive at some stage, from their origin in the neural crest until they accomplish their final differentiated fate. White spotting in mice and humans is sometimes referred to as hypopigmentation, or piebaldism. We will avoid the former term in this context, because it is not specific to the absence of viable melanocytes. White spotting differs from albinism (Chapter 4), which can also be considered hypopigmentation. In albinism, the absence or reduction of pigment results from inability of the viable melanocytes to make normal pigment. Similarly, we will avoid the term piebaldism because piebald is the classical name of only one (*S*, *Ednrb*; see Fig. 3.2) of the numerous gene loci that can determine white spotting.

In general, congenital, inherited white spotting results from defect(s) in one or more of multiple gene loci whose function is required for the establishment and maintenance of the melanocyte lineage from its origin in the neural crest until birth. Therefore, and because absence of pigment is not lethal to the organism, white spotting is very useful in studying these processes, both within the pigment cells themselves and in their interactions with other cell types.

3.1.3 Vitiligo defined

Whereas white spotting is congenital, vitiligo is a progressive condition that arises after birth, and is defined by some as the progressive loss of viable melanocytes with age and by others as the acquired, patchy loss of pigment from interfollicular skin, with the hair follicles sometimes also involved. Progression is often limited; in humans, regression sometimes occurs. Vitiligo is a medical term; therefore it is based in human phenotypes (see Chapter 30 of Nordlund et al. 2006).

In the case of vitiligo, the human phenotype differs from that of the mouse in ways that make direct comparisons difficult and suggest that different processes may sometimes be involved. In human skin, at birth, most of the visible melanocytes are located on the basement membrane of the epidermis. From here they transfer melanosomes into the surrounding keratinocytes of skin. Mice do not demonstrate patchy loss of melanocytes from the interfollicular epidermis because in mice, after about 1 week of age, most of the differentiated melanocytes are located in the bulb of the hair follicles, where they transfer melanosomes into the keratinocytes of the growing hairs (see Chapter 1). Therefore, the progressive loss of melanocytes in mice is normally seen as progressive graying of the hairs that is expressed at each successive molt.

Furthermore, in humans, the skin – rather than the hair – is the primary organ of surface protection; replacement of pigment in the basal layer of the epidermis is a continual process, and hairs are replaced as they are lost. In mice, the hair coat, rather than skin itself, provides the outer barrier between animal and environment, and replacement is not continual. Mice molt their hairs on a more or less regular schedule and in wave patterns over their bodies. During each molt, the pigment cells that reside at the root of the hair are replaced by progeny of stem cells that reside in the niche or bulge area of the follicle. If the new hairs contain fewer pigment cells, or are of a different color or shade, then the anterior–posterior wave of hair replacement creates shifting coat color patterns. These are not homologous to human pigment patterns (Figs 3.3 and 3.4).

Another problem in identifying vitiligo in the mouse is that human vitiligo is multifactorial; that is, most cases of human vitiligo involve the participation of more than one gene locus, thus more than one primary genetic process, in addition to environmental influences. One of the advantages of inbred mice is that we can control variability at unknown loci; a disadvantage may arise when we need variability to identify multifactorial phenotypes. Perhaps mice might exhibit a human-like

Figure 3.3 Two genetically co-isogenic mice heterozygous for two alleles at the *Mitf* locus (C57BL/6J-*Mitf*$^{mi\text{-}vit}$/*Mitf*$^{mi\text{-}or}$). These are characterized by both white spotting (congenital absence of pigment cells from the white areas) and progressive loss of pigmentation in the pigmented areas during successive molt cycles. The mouse in front has his initial hair coat, and is less than 3 months old, while the mouse behind has molted three times and is more than a year old.

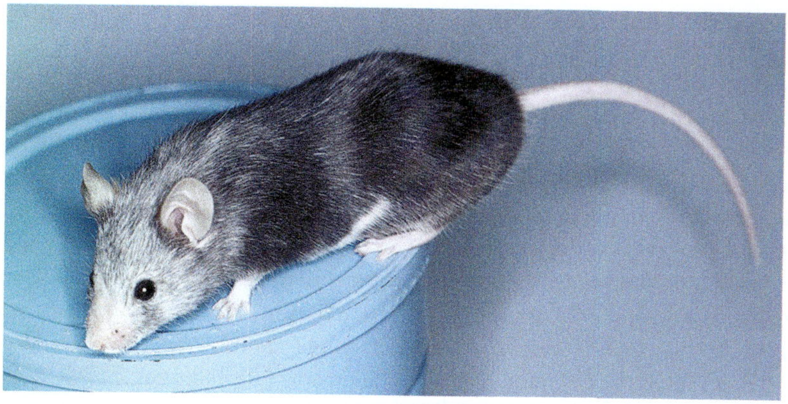

Figure 3.4 Another 'vitiligo' mouse of the genotype C57BL/6J-$Mitf^{mi\text{-}vit}$/$Mitf^{mi\text{-}rw}$. This mouse is in the midst of her first molt cycle. She already has her new coat on the head and is in process of losing the older one from the back part of her body.

'vitiligo' phenotype if we had the appropriate combination of mutations. We simply don't know. Thus, the human model (Spritz 2007), the pig model (Tissot et al. 1993), which has skin more similar to human skin, and the well-studied chicken model (Wang & Erf 2004) have become the best models for human vitiligo. Nonetheless, the mouse, largely because of the availability of inbred strains and modern techniques of genetic manipulation (see Chapter 6), remains the most broadly useful model for understanding basic principles of pigment-cell survival.

Many people reserve the term vitiligo for phenotypes similar to those in humans that involve the postnatal expansion of defined white spots. Others, especially in earlier literature, use the term vitiligo more generally to include non-localized depigmentation of follicular melanocytes; that is, progressive graying. Progressive graying is caused by loss of viable melanocytes from the hair follicles, and is characteristic of several mutants of mice, including *Silver* (*Si*), some *Microphthalmia*-locus combinations (see for example Figs 3.3 and 3.4), and some genotypes at the *Brown* (*Tyrp1*) and *Slaty* (*Dct*) loci. Thus, in the literature, we have two somewhat different applications of the same term, vitiligo. Therefore, we will use the term progressive graying, and apply it to questions of why and how viable pigment cells are found primarily in the hair follicles of mice and why and how they are, in some genotypes, lost from the hair follicles during successive molt cycles. These defects often relate to the maintenance of melanocyte stem cells in the bulge region of the hair follicle (see section 3.4.7).

3.1.4 Research applications

White-spotting phenotypes in the whole organism may result from defects in:

- specification of pluripotent neural-tube cells to the pigment-cell lineage;
- epithelial–mesenchymal transition (EMT) of neural tube to neural-crest cells;
- proliferation of the neural-crest cells / melanoblasts and their migration through the mesenchyme and across the basement membrane of the epidermis;
- cell survival;
- homing, in mice, predominantly to the hair follicle;
- establishment of a population of stem cells in the bulge area of the hair follicle;
- differentiation of melanocytes.

After birth, progressive graying helps us to understand maintenance of stem cells and repopulation of the hair follicle with functional melanocytes during each successive molt.

All of our questions about white spotting and vitiligo relate to the viability of pigment cells and the cells from which they originate in the neural crest. By definition, the classical white-spotting genes and the graying genes represent gene loci (and functions) that are specifically necessary for survival and development of the pigment-cell lineage. However, white-spotting genes and progressive-graying genes are often required also for development of tissues other than pigment cells. That is, white-spotting and progressive-graying genes are often pleiotropic; pleiotropy is defined here as the phenotypic impact upon several tissues or organs as a result of mutation at one locus, and of course it results from expression of the locus in more than one type of cell, tissue, or organ, or interaction with other gene products so expressed. Therefore, white spotting and progressive graying are useful for studying these other tissue types. The pleiotropic effects of the *Kit* (*W*, *Dominant white spotting*) locus, for example, have been especially useful in evaluating gene function in red blood cells, mast cells, and germ cells. Another example is *Mitf* (*Mi*, *Microphthalmia*, *Microphthalmia-associated transcription factor*), encoding a transcription factor with functions that are necessary at different times for normal development of eyes, ears, mast cells, bone marrow stem cells, osteoclasts, and germ cells. These two loci alone impact eight major tissue types. Many or most of the other white-spotting gene loci similarly influence the development of more than one tissue or organ type. Therefore, various white-spotting loci have been used as viable models to inform many or most of the other developing systems of the body (Searle 1968; Silvers 1979; Nordlund et al. 2006).

White spotting and progressive graying also illustrate the most active points of integration and communication, via the interactions among the biochemical regulatory pathways, between the pigment system and other systems of the body. During their remarkable developmental journey from the neural crest to the skin and hair follicles, the pigment cells are continuously surrounded by other sorts of tissues with which they must communicate. Genetic defects that interfere with any necessary process or interaction may result in congenital absence of pigment cells from all or part of the mouse – that is, white spotting – or their postnatal loss in progressive graying. Thus, combined with the availability of congenic inbred strains and the fact that most pigmentary mutations are not lethal, these characteristics of pigment-cell development make white spotting a uniquely useful model of specific gene functions and of genic interactions during development, and an elegant method of confirming or rejecting the applicability to living systems of findings *in vitro* (see Chapter 6).

These sorts of relationships are now being pursued beyond physiological genetics (Silvers 1979) to the biochemical and molecular levels, as will be discussed for specific loci in Chapter 4. The goal of this chapter is to present an overview of the processes that are necessary to achieve the normal distribution of viable pigment cells in the mouse, as we currently understand them, using information gained from mice with the classical mutations that affect pigment-cell survival and development, and occasionally from newer mouse mutants and other species such as chickens, fish, and domestic pets and livestock.

3.2 Pigment-cell development: developmental biology

A discussion of melanocyte viability must begin with the life history of melanocytes and the neural-crest cells from which melanoblasts and many other cell types are embryonically derived (Fig. 3.5).

The neural crest derives from the dorsal part of the neural tube in the early embryo (Box 3.1). The neural crest represents an evolutionary advance unique to vertebrates (Huang & Saint-Jeannet 2004) and gives rise to a wide variety of tissues. Structures derived, at least in part, from neural crest

Figure 3.5 Photograph of 13.5-day embryo that has been stained to show the migrating pigment cells. This is a transgenic mouse that has a *LacZ* structural gene attached to the promoter of *Dct*. Dct is a melanogenic enzyme and is expressed throughout the life of the pigment cell. LacZ is an enzyme that causes the blue pigment to form when provided with its substrate. The superficial punctate stained cells are pigment cells. The brain and some nerves also stain blue.

include: conotruncal structures of the heart, as well as diverse skeletal and connective-tissue structures of the head that have been deemed mesectodermal in origin; the peripheral and enteric nervous systems and their interfaces with the central nervous system and the endocrine system; and the melanogenic tissues of the body, including probably all melanocytes except the RPE. It is because of its complexity that abnormal development of the neural crest and its derivatives can have such widespread medical consequences, referred to as neurocristopathies (Bolande 1997), that include Waardenburg–Shah syndrome (white spotting with aganglionic megacolon), frontonasal dysplasia (multiple craniofacial defects), DiGeorge syndrome (craniofacial and heart defects), and others.

The neural tube is formed as the neural plate rolls itself into a hollow tube in the process called neurulation (Box 3.1a–c). To form this tube, the lateral edges of the neural plate push upward, beginning centrally and proceeding anteriorly and posteriorly, as bilateral neural folds. The neural

Box 3.1 Melanoblast development in the mouse, showing expression of some key genes

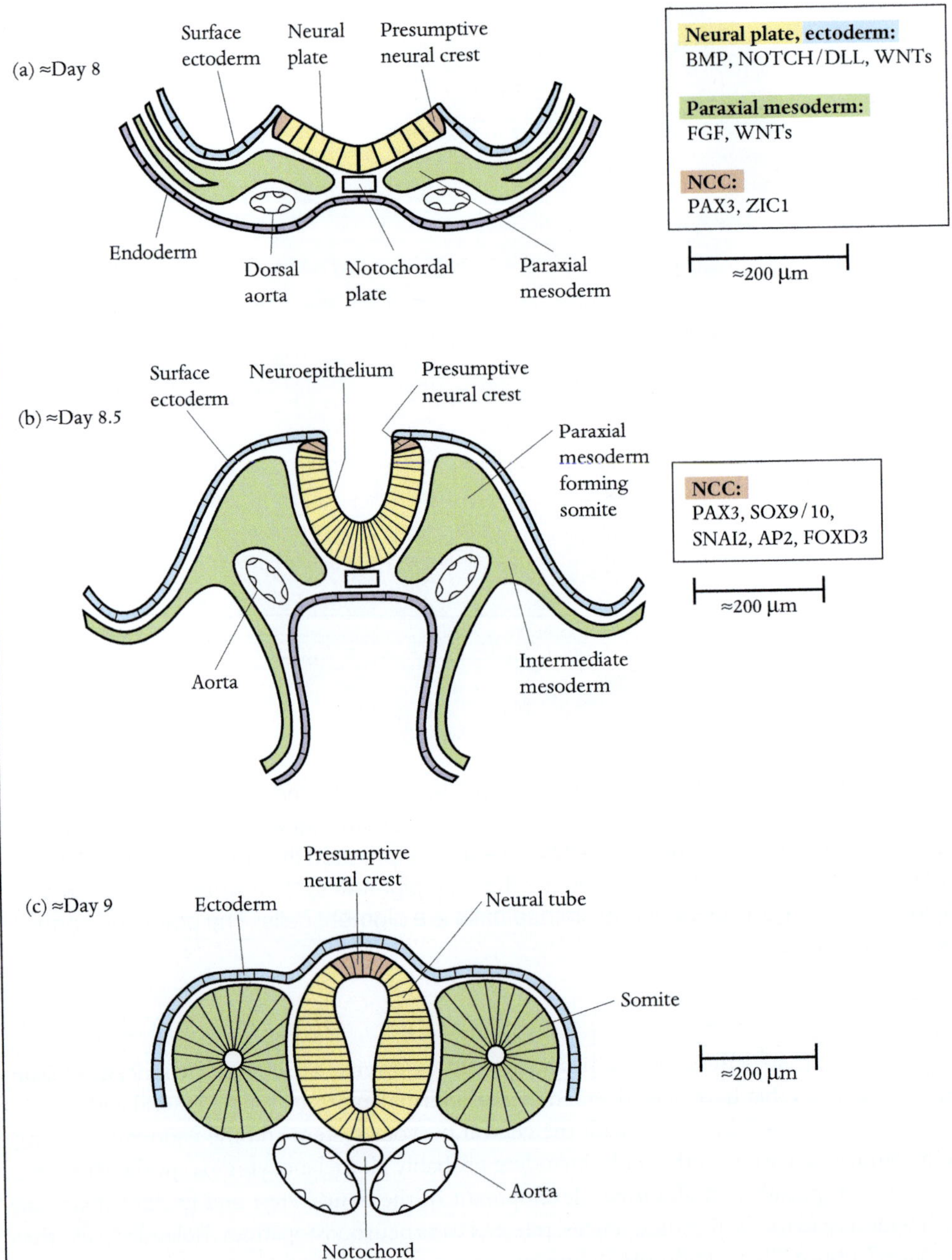

Diagrams apart from (f) show sections through the cervical end of the trunk region of a mouse embryo, at successive stages of development, with emphasis on neural crest and melanoblasts. Boxes indicate sites and times of expression of some key color genes –

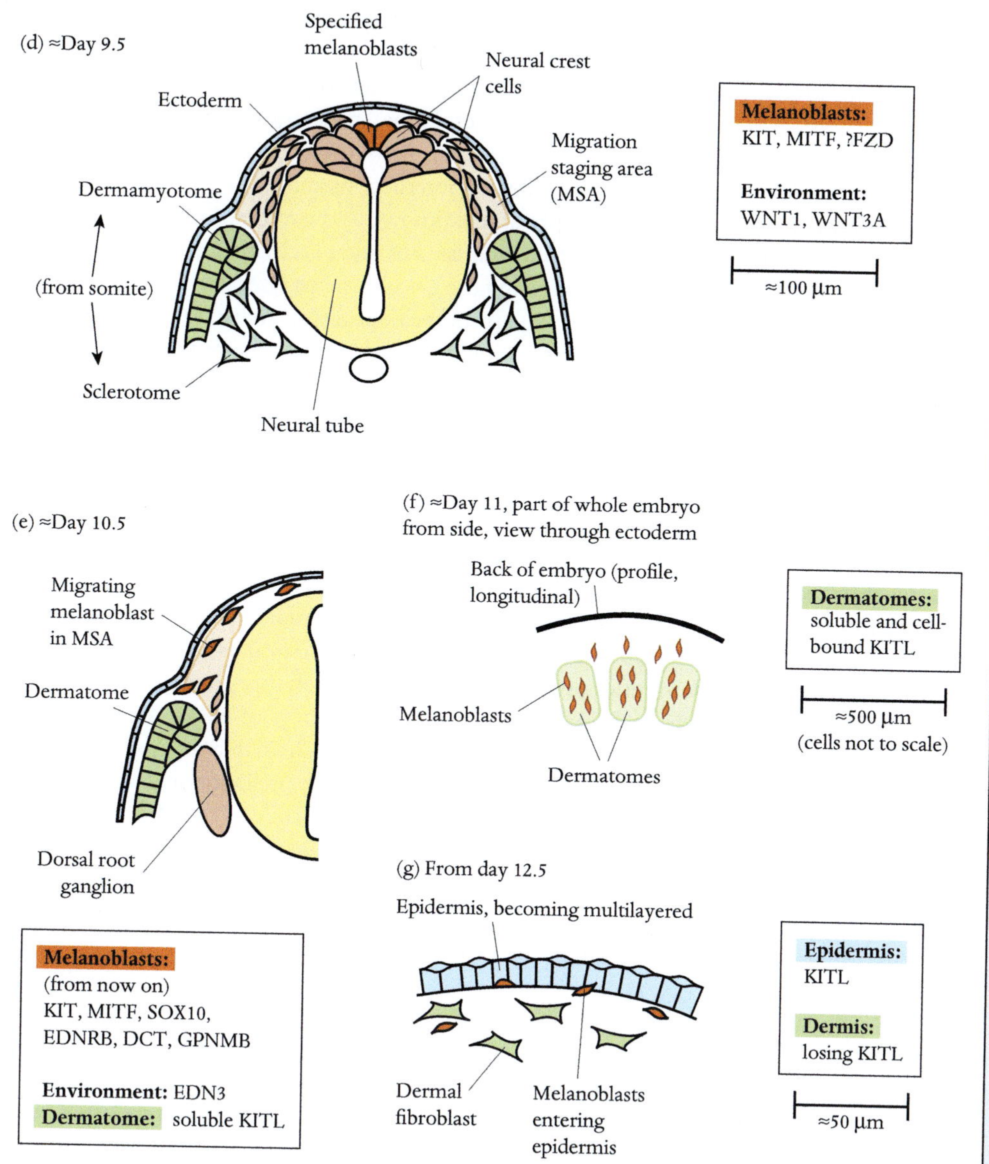

regulators and markers – as discussed in the text (Sections 3.2 and 3.4). Sources of information on anatomy and cell locations: Erickson and Weston (1983), Kaufman (1992), Wehrle-Haller and Weston (1995), and Wilson et al. (2004). Sources for gene expression: see Section 3.4. For definitions of abbreviations see the Figure 3.9 legend.

folds consist of the raised edges of the embryonic neural plate, along with the adjacent ectoderm, which bilaterally push upward, forming an elongated peak of tissue, the neural fold. The neural folds on the right and left sides of the embryo continue to move toward each other and fuse together over the midline, converting the original neural plate into a hollow tube overlaid by ectoderm (eventually epidermis). The original neural plate becomes the neural tube; the lateral (outer) sides of the neural fold become the covering of surface ectoderm (Box 3.1c). Thus, the surface ectoderm and the neural plate are brought close together, initiating inductive interactions and leaving, on either side of the neural tube, a wedge-shaped empty area (empty of cells) known as the migration staging area (MSA) (Box 3.1d). At the same time that the neural folds are rising to meet each other, somites are forming from the paraxial mesoderm (the tissue underlying the ectoderm just lateral to the notochord). In the mouse, neural-crest cells arise initially from parts of the neural plate margin and unfused neural folds in the head, and later from the dorsal parts of the fused neural tube in the head and trunk (Nichols 1981; Erickson & Weston 1983; Wilson et al. 2004). They delaminate into the MSA.

Neurulation occurs in a wave from the future neck area toward both ends of the embryo. In the rostral portions (head end) of the embryo, the neural tube is already well formed before the process begins in the caudal (tail) end.

The derivatives of neural-crest cells vary in different rostral–caudal regions of the embryo (Le Douarin & Kalcheim 1999; Trainor & Krumlauf 2001). For example, only cranial neural crest gives rise to cartilage, while sensory neurons and glia are trunk-specific neural-crest derivatives, those in the head coming from different placodes or outgrowths. The rostral to caudal regions of the neural crest are: cephalic (to somite 8), truncal (somites 8–28), and lumbosacral (somites 28 onward). A cardiac region has been described at the anterior portion of the truncal region that gives rise to neural-crest cells that migrate to the heart and take part in its development, some of which differentiate into melanocytes (Kirby et al. 1983; Mjaatvedt et al. 2005). Neural-crest cells in different parts of the body differentiate into a wide variety of tissues in addition to pigment cells (Le Douarin et al. 2004).

The pigment-cell lineage is generated over the whole length of the neural tube, but we will focus on its development in the trunk, the area between the forelimbs and the hind limbs, which is best understood. Other pigment-related manifestations of neural-crest cells, in the head, eyes, ears, and heart pathways of melanoblast migration will be briefly discussed at the end of this chapter, along with the RPE which derives directly from the neural tube.

3.2.1 Specification of neural-crest cells

Specification is the process by which embryonic cells become functionally different from each other (Davidson 1990; Kimmel et al. 1991) as they mature toward their differentiated fate. Cellular differentiation is the production of different cell types in a multicellular organism. When one cell type is differentiating from its precursor, specification is the beginning of that process, and can be defined by the first detectable change (molecular or biological) that is characteristic of the development of that cell type. It is important to emphasize that specification does not necessarily imply irreversible commitment to the change; however, it is generally believed that embryonic precursor cells undergo cell-fate restrictions before they overtly express cell-type-specific phenotypes (Henion & Weston 1997).

Specification of the neural-crest cells occurs in *Mus* (but not identically in every species) just before or just as the neural-tube fusion takes place. Data from birds and amphibians suggest that the adjacent paraxial mesoderm also participates in the induction and specification of pre-migratory neural-crest cells in the dorsolateral portion of the forming neural tube (Huang & Saint-Jeannet 2004). Specification involves the expression of dorsal neural-tube markers (Dickinson et al. 1995; Yanfeng et al. 2003; Sakai et al. 2006) consisting of molecules such as NOTCH/DELTA, WNT1 and 3a,

SLUG/SNAIL (SNAI1/SNAI2) and bone morphogenetic protein (BMP) and its inhibitors including NOGGIN (Thomas & Erickson 2008). There are species-specific differences in this process. In mice, the process begins before the mid-dorsal fusion of the left and right neural 'crests' (Kulesa et al. 2004).

From this early population of pre-migratory neural-crest cells, the neural-crest cells delaminate and migrate into the MSA and then along specific pathways into and through the mesenchymal tissues of the embryo (Box 3.1). The earliest biochemical markers of migratory melanoblasts are KIT and microphthalmia-related transcription factor (MITF). Cells expressing KIT appear at about day 9 (Wehrle-Haller & Weston 1995; Opdecamp et al. 1997; Baxter & Pavan 2003; Wilson et al. 2004) at the midbrain/hindbrain junction of the embryonic neural tube, from where they migrate subsequently into the head region. The specification of neural-crest cells continues caudally, but not necessarily uniformly or in precise sequence, over the course of about 24 h (Dickinson et al. 1995).

3.2.2 Delamination

Upon specification, the neural-crest cells transform rapidly. They change morphologically from the phenotype of an ectodermal tissue to that of individual migrating cells. Interactively, they change their ability to respond to chemical messages: cytokines, ligands, and growth factor molecules that are present in the surrounding extracellular fluid and on neighboring cells (Dupin & Le Douarin 2003; Moore & Larue 2004; Kalcheim & Burstyn-Cohen 2005). The neural-crest cells delaminate, bilaterally, from the internal wedge of the merging neural folds, or from the dorsal portion of the newly formed neural tube (Box 3.1). They emerge from this tissue into the adjacent acellular space as individual migratory neural-crest cells. The process of delamination is an example of EMT, which involves loss of cell–cell adhesions (Erickson & Reedy 1998; Thiery 2003; Harris & Erickson 2007). During the EMT, the delaminating cells lose their tight relationship with other cells within the neural tube, coincident with loss of expression of N-cadherin (Pla et al. 2001; Bolós et al. 2002) and other changes in their cellular communications, and become separated neural-crest cells with the ability to migrate. In the truncal region, the delaminating neural-crest cells consist of rapidly replicating, cells that migrate along strictly defined pathways.

In vitro clonal analysis, in which single crest cells were isolated by limiting dilution, has demonstrated that the initial neural-crest population contains pluripotent precursors, as well as precursors that generate a limited range of derivatives and nearly 50% of apparently committed precursors that give rise to a single cell type (Henion & Weston 1997; Sieber-Blum and Cohen 1980; Baroffio et al. 1988, 1991; Sieber-Blum 1989; Dupin et al. 1990; Ito and Sieber-Blum 1991; Sextier-Sainte-Claire Deville et al. 1992; Stemple and Anderson 1992).

3.2.3 Migration and differentiation

Cranial neural crest gives rise to craniofacial cartilage and connective tissue (participating in structuring the unique vertebrate head), to ganglia, and to cells of the melanocyte lineage. The latter migrate rostrally and to the ears and eyes (Baxter & Pavan 2003; Wilson et al. 2004) in a pattern that is illustrated by the pigmentation patterns of mammals that are mosaics of two viable pigment types, such as the cat in Figure 3.6. Mosaicism is discussed in Secton 3.4.

Beginning at the level of the midbrain/hindbrain of the 9-day mouse embryo, the neural crest produces cells of the pigment lineage, as well as dorsal root ganglia, sympathetic ganglia, and adrenal medulla (Luo et al. 2003). In the trunk, the neural-crest cells of all these lineages migrate first into the MSA (Erickson et al. 1992; Wehrle-Haller & Weston 1995). In transverse section, the MSA is a triangular space flanked by the neural tube, the surface ectoderm and the somite. The cells

Figure 3.6 Blue-cream ('dilute') tortoiseshell cat. This cat is white spotted around the neck where pigment cells are absent. Additionally, she is a tortoiseshell as a result of Lyonization. Lyonization is chimerism that is caused by heterozygosity for genes that are located on the X chromosome. In this case, the cat is heterozygous for the *Yellow* (*Orange*) allele at the X-linked locus that determines whether or not the pigment will be yellow. One or other of the X chromosomes is inactivated at an early developmental stage in female mammals, generating chimerism. In this case, the cat is chimeric for yellow and nonyellow clones of pigment cells (Section 3.3). At the same time, all of the pigmented areas are affected by mutation at the *Melanophilin* locus that influences melanosome transport (Chapter 4) and results in a 'dilute' phenotype. In sum, her genotype is O^{+}/O at the X-linked *Orange* (*Yellow*) locus, $Mlph^{ln}/Mlph^{ln}$ (homozygous mutant at the *Melanophilin* locus that cat fanciers refer to as *Dilute*, an autosomal locus that affects all pigment cells), and white spotted of unknown genetic origin.

proceed, migrating and replicating, along their appropriate pathways (Box 3.1). The migratory pathways of the neural-crest cells are defined by biochemical characteristics that may inhibit or attract the migrating cells. (Dorsky et al. 2000; Krull 2001; Harris & Erickson 2007). The neural-crest cells that first emerge migrate ventrolaterally through the MSA, then through the rostral sclerotome of each somite, establishing the segmental nature of the sympathetic ganglia to which they give rise. The later, ventromedial, migration of neural-crest cells proceeds through the MSA and then ventrally, internal to the somite, along the lateral surface of the neural tube (Erickson & Goins 1995; Wehrle-Haller et al. 2001). These cells then participate in the formation of dorsal root ganglia and Schwann cells. Finally, neural-crest cells of the pigment-cell lineage emerge from the neural crest into the MSA and proceed laterally, over the top edge of the forming somite and down its lateral surface, beneath the ectoderm.

Some prospective glial and neuronal cells of the trunk region migrate ventrally and then proceed caudally along the developing gut, directed by the substrate through which they migrate, the population pressure of replicating cells, and the availability of a nurturing environment. Somewhat later in the developmental process, the enteric neural-crest cells from more caudal regions of the embryo also migrate to the caudal end of the gut, and then proceed in a rostral direction. In this way, in the normal mouse, the gut is fully populated by neural-crest cells when the two streams meet somewhere along the forming colon. These neural-crest cells differentiate into glia and neurons of the enteric nervous system, and the innervation of the gut is accomplished.

Neural-crest cells from the vagal region migrate along a specific cardiac pathway to the developing heart where they can be found in the valves and septa.

By the time specified melanoblasts arrive at the MSA, they are expressing a cell-surface receptor protein known as KIT, encoded at the *Kit* (*W*, *Dominant white spotting*) locus. Melanoblasts are defined as specified but unpigmented precursors of melanocytes, and specification is defined by the sustainable expression of KIT and/or MITF. KIT is absent from other neural-crest cell-derived lineages. Neural-crest cell-derived melanoblasts of the trunk region first arise at the rostral and caudal portions of the trunk, and very soon thereafter in the mid portions (Wilson et al. 2004; Baxter et al. 2004). This sequence of appearance of specified melanoblasts is reminiscent of that observed in medaka, zebrafish, and birds (Dupin & Le Douarin 2003; Lamoreux et al. 2005) and correlates with what is otherwise the curious phenotypic observation that adult white-spotted mice (and other vertebrates; Figs 3.7 and 3.8) consistently exhibit a higher probability of pigmented hair across the shoulders and hips, and on the head, than on the mid-trunk (Schaible 1969).

Clonal descendants of these first truncal melanoblasts then migrate laterally, over the top of the adjacent developing somite and then laterally and ventrally, tight against its external surface, replicating rapidly as they migrate so as to fill all the available surface areas (which are successively becoming dermamyotome, then dermatome) in normal mice.

As the melanoblasts begin to fill the available surface area of the somites, the dermatome fragments into mesenchyme or dermis, through which the melanoblasts next migrate peripherally, to and into the basal layer of the overlying ectoderm, which is becoming epidermis. Movement into the dorsolateral ectoderm occurs mainly from embryonic days 12–14 (Mayer 1973b). As these areas are occupied, and after birth, the melanoblasts move toward and into the developing hair follicles. In mice, most of the melanoblasts then leave the interfollicular epidermis and come to reside in the hair follicles, where some remain as stem cells in their niche in the bulge region and others go on to differentiate in the bulb of the hair follicle beginning at about the time of birth. After birth, the stem cells replenish newly developing follicles through successive molt cycles.

Some investigators define development as ending at birth, followed by homeostasis; others include the entire cycle of life as a process of development. Either way, the classical theories of developmental biology center around two major concepts of cellular descent and/or development. First is the common origin of diverse cell types from the same precursor cell types; for example, development of the glial lineage and the neuronal lineage from the same precursor neural-crest cells as described above. Second are the functional interactions (crosstalk) between cells of the same or different lineage; for example, induction of one tissue type by another, as happens when the skin ectoderm induces formation of the first pre-migratory neural-crest cells within the neural tube.

Two extreme hypothetical developmental models, both of which are well supported by data, parallel the concepts, respectively, of relationship by lineage and of induction by crosstalk between cells or tissues and their environments. As applied to development of neural-crest cells, the first model implies preprogramming: 'in the sense that particular neural-crest subpopulations acquire a fate bias before leaving the dorsal neural tube. This bias confers unique, cell-autonomous migratory properties on that particular subpopulation and allows those cells to access defined paths within the embryo' (Harris & Erickson 2007). The second model proposes that multipotent neural-crest cells respond to elements of their environment by progressive lineage restriction, and suggests that their final colonization of a particular niche and their differentiation within the niche is dependent on the instructions provided by that niche. Both these models are well discussed and referenced in Harris and Erickson (2007). These hypotheses have been tested from the beginnings of

Figure 3.7 A white-spotted cat.

developmental genetics studies using mouse, chicken, and amphibian models, and more recently zebrafish, medaka, and other model systems. In the mouse, such studies form a strong foundation of the 1979 'bible' of mouse pigmentary genetics written by W.K. Silvers, *The Coat Colors of Mice*. In the 30 years since this book's publication developmental biology has gained access to additional elegant biochemical and molecular technologies, genetic manipulation techniques (Chapter 6), and the genomes of several vertebrate species. Thus the question of nature/nurture is now being studied actively at the level of cell biology and molecular interactions. As our understanding of the processes expands, it becomes increasingly difficult to argue that either preprogramming or the cell's interaction with its environment is the primary method of genetic control over developmental processes, including the development of pigment cells, although of course one of them may be primary to one specific step in development.

In part, such either/or questions are artificial, and arise from the limitations of the scientific method, which requires the elimination of variables that appear to be irrelevant to the specific question being asked. It is difficult to ask two questions at the same time, and it is not possible to obtain scientifically valid data unless variables are controlled tightly and wisely. In other words, the scientific method itself biases results toward linear hypotheses. It has become clear, however, that developmental processes are not linear. They are the result of multiple preprogrammed and environmental cues, often imposed concurrently upon the target cell or tissue, and sometimes

Figure 3.8 A white-spotted dog.

reciprocal. Thus, the power of the scientific method also imposes limitations to evaluation of the data, a fact that sometimes confounds our understanding of development as a fluid, evolving, interacting, self-regulating whole, in principle rather like an ecosystem. The answer to how this is accomplished lies in the flow of information within the system. Development of the embryo requires the constant flow of information through a network that consists primarily of biochemical signaling pathways and other biochemical interactions that transmit information within cells and between and among cells and tissues. The next section provides a summary overview of the relationship between pigment-cell development and cellular communications.

3.3 Cellular signaling pathways for melanocytes

By definition, the classical white-spotting genes and the graying genes represent gene loci (and functions) that are specifically necessary to survival and development of the pigment-cell lineage. At the same time, these genes do not function in a vacuum; they cohabit and interact with other genes that are not classical white-spotting genes but are necessary to and supportive of the functions of the white-spotting and graying genes in ensuring the survival of the pigment-cell lineage. Thus white-spotting and graying phenotypes, in addition to defining the loci necessary for survival of pigment cells, also provide information about the interface of communication between the pigment-cell lineage itself and the environment upon which it depends for survival. In Section 3.4 we will discuss the white-spotting genes themselves, and their roles in development of the mouse phenotype.

In this section we will describe the web of cellular information transfer that is known to be required for the pigmentary genes to function, and specifically gene loci for which there is evidence that they influence pigmentary phenotype. This topic lies at the intersection where molecular and cellular biology (which are not our primary topics) meet the study of murine pigmentary phenotypes. Because of the rapid development of our understanding of cell biology, this section of the book will become quickly outdated. Nevertheless, it is important here to recognize the complexity of the background network of biochemical information pathways that is necessary for the pigmentary genes to function.

At the cell biology level, we see a few differences in neural-crest cell development among the classes of vertebrates. Embryonic development of birds, fishes, mammals, etc. sometimes utilize the same inter- and intracellular communications pathways to create quite different phenotypic results. These differences are not relevant to this study of mouse pigmentation, except to point out that white spotting in the inbred mouse may also be a useful model to study the ways in which the same biochemical pathways may be used differently – perhaps with timing differences or networking differences – to bring about differences in species and classes of vertebrates; that is, evolution.

3.3.1 Basic mechanisms of cellular signaling

The lipid cell membrane imposes a potential barrier to communication between the liquid interior of the cell and the extracellular fluids, as well as an opportunity for the cell to regulate traffic across the membrane. Thus the cell membrane contains a battery of receptors and facilitators that control cell signaling across the membrane. As a broad generalization, two types of signal are transmitted across the cell membrane:

1. substances such as nitrous oxide and lipid-soluble molecules, including some hormones, that are able to cross the membranes unaided;
2. transduced signals from water-soluble molecules, including peptide hormones such as melanocyte-stimulating hormone (MSH), as well as cytokines, chemokines, and growth factors that cannot cross the membrane on their own.

In cellular biology, transduction refers to any process by which a cell converts one kind of signal or stimulus into another. This is most often accomplished by an ordered series of biochemical reactions that begins with a signaling molecule (the ligand) outside the cell that is recognized by and binds to a receptor floating in the lipid membrane of the cell. When the ligand binds to the receptor, the biochemical 'message' is activated inside the cell as a series of chemical reactions in the cytoplasm that is often also communicated to the nucleus. The general goal of this information flow is often or usually to turn genes on or off in the nucleus. In development, the specific object of this process is often to turn on (or off) a transcription factor that will then direct the functions of other genes as they must adapt or react to the local tissue environments as they follow the general program of development of the organism.

In the developing embryo, the focal point of this interface, where the environment is able to or not to influence the 'programming' of the cell by transducing a stimulus into the cell, is most often the cell-surface receptor. Cell-surface receptors are specific proteins that are transcribed from genes within the nucleus of the responsive cell, usually modified in the cytoplasm, and then delivered to the lipid surface membrane of the cell, where they remain as long as needed to do their function. Their function is to recognize their specific chemical message molecule, or ligand. A ligand is an effector molecule that can bind reversibly to a specific site on a target protein receptor.

Cell-surface receptor molecules or complexes that are embedded in the cell membrane consist of three basic structural subunits that relate to different functions: the receptor site is exposed to the extracellular fluids outside the cell and available to bind its specific ligand; the mid portion of the receptor floats in the lipid membrane of the cell, keeping the receptor properly located/oriented; and the effector site is exposed to the cytoplasm inside the cell membrane. The effector site is associated with a signaling pathway in the cell. Receptor sites are specific to their ligands; effector sites are specific to the signaling pathways they activate. Thus, a specific receptor–ligand complex transduces information into a cell by identifying a specific molecule on the outside of the cell and activating it. This in turn activates the effector site inside the cell which usually results in a cascade of biochemical interactions inside the cell. If a cell lacks the specific receptor protein for detection of a particular stimulus (ligand), then the cell can be said to be 'blind' to the stimulus. It cannot respond. So, it could be said that the cell 'decides' to which stimuli it is willing to respond by generating the appropriate receptor. It does this by activating the gene that encodes the receptor molecule.

Often, the biochemical message generated by a ligand–receptor complex is transduced through a series of biochemical reactions from the receptor through the cytoplasm and then into the nucleus. Along the way, the message can be influenced by conditions (the biochemical milieu or crosstalk from other signaling cascades) inside the cell. As an example of how such a pathway can cause a cell to change or evolve, receptor–ligand binding could result in activation of a gene that encodes a new receptor that is then made available at the cell surface, thus changing the programming of the cell by making the cell sensitive to new (additional) messages.

Cells normally support many kinds of receptor at their surfaces and multiple copies of each type; therefore, they can interact with more than one type of ligand, and they can react to a balance among the signals they receive (Pawson & Nash 2000).

The result of all these delicate controls will be to either maintain the programming status of the target cell or to alter it by changing the status of target genes: turning them on or off. Thus the programming of a cell and its response to its environment are inseparably intertwined in a fluid web of biochemical interactions that often result in some modification of the cell in response to conditions of its environment. The outcome is exquisitely controlled and might be anything in a range between death of the cell or its differentiation. Below we briefly describe some of the cellular signaling pathways that are implicated in pigment-cell development and functional regulation (for a broader introduction to the basic ideas and terminology of cellular signaling, see Alberts et al. 2008). The pathways are most often classified and referred to by the type of receptor, as the basic decision-maker of the communication cascade. For reference, Figure 3.9 summarizes the main signaling pathways that are implied to be important in mouse pigment-cell development, by the effects on the pigmentary phenotype of mouse germline mutations (or transgenes) in pathway components.

3.3.2 Pathways from seven-pass transmembrane receptors

The seven-pass transmembrane receptors are so named because they are transmembrane proteins that wind seven times back and forth through the plasma membrane. The following pathways important for melanocytes involve this class of receptor: endothelin 3 (EDN3)/endothelin receptor B (EDNRB), MSH/melanocortin 1 receptor (MC1R), WNT/frizzled (FZD), and L-dopa/ocular albinism 1 protein (OA1). These are G-protein-coupled receptors: that is, binding of the ligand activates an associated intracellular GTPase-binding protein (or G-protein), which dissociates to

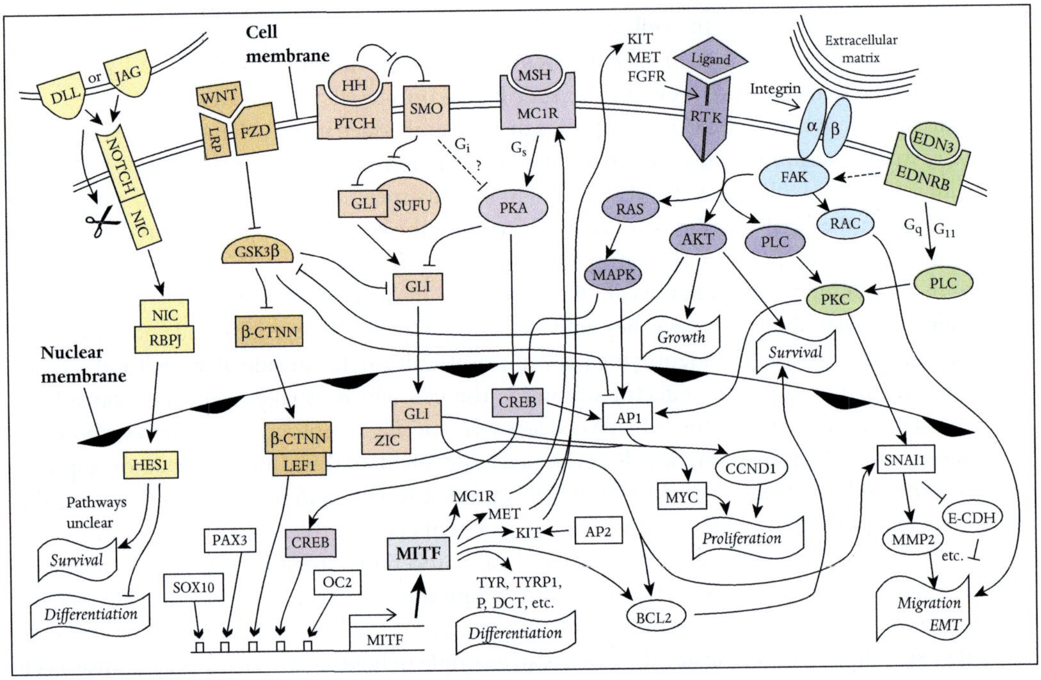

Figure 3.9 Signaling pathways important in normal melanocyte development. Germline mutations in each of these pathways can produce developmental pigmentary defects (Table A1.1, Chapter 1 Appendix). They are simplified here to emphasize key components, interactions, convergence, and biological outcomes (shown as banners). Colors are used to distinguish the different pathways. Components are often shown as generic families (e.g. WNT); see text and Table A1.1 for specific family members important in pigment cells. More detail follows on specific pathways: see text and further pathway diagrams. Compiled from sources cited in the text and from Alberts et al. (2008), Bennett (2008) (general), Aoki and Moro (2002) (MITF, KIT), Bar-Eli (2001) (AP2), Cadigan and Liu (2006) (WNT pathway), Levy et al. (2006) (MITF promoter), Ogden et al. (2008) (SMO, G_i), Ruiz i Altaba et al. (2007) (HH pathway), and Svärd et al. (2006) (GLI, PKA). Arrows, upregulation; T-bars, inhibition; rectangles, transcription factors. Abbreviations (note: expansions are omitted for proteins best known by their abbreviations, as in AKT): AP1, AP2, activating proteins 1, 2; β-CTNN, β-catenin; CCND1, cyclin D1; CREB, cyclic AMP-response-element-binding protein family; DCT, dopachrome tautomerase; DLL, delta-like; E-CDH, E-cadherin; EDN3, endothelin 3; EDNRB, endothelin receptor B; EMT, epithelial–mesenchymal transition; FAK, focal adhesion kinase; FGFR, fibroblast growth factor receptor; FZD, frizzled; G_{11}, G_i, G_s, G_q, G-protein families, of which i=inhibitory, s=stimulatory; GSK3B, glycogen synthase kinase 3β; HES1, hairy/enhancer of split, homolog 1; HH, hedgehog; JAG, jagged; LEF1, lymphoid-enhancer-binding factor 1; LRP, low-density-lipoprotein-receptor-related protein; MAPK, mitogen-activated protein kinase; MC1R, melanocortin 1 receptor; MITF, microphthalmia-related transcription factor; MMP2, matrix metalloproteinase 2; MSH, melanocyte-stimulating hormone; NIC, Notch intracellular component; OC2, onecut2; P, pink-eyed dilution protein (OCA2); PAX3, paired-box 3; PKA, protein kinase A; PKC, protein kinase C; PLC, phospholipase C; PTCH, patched; RAC, a RAS-related G-protein; RBPJ, recombination signal-binding protein for immunoglobulin κJ region; RTK, receptor tyrosine kinase family; SMO, smoothened; SNAI1, snail 1; SOX10, SRY-box 10; SUFU, suppressor of fused; TYR, tyrosinase; TYRP1, tyrosinase-related protein 1; ZIC, zinc-finger protein of cerebellum.

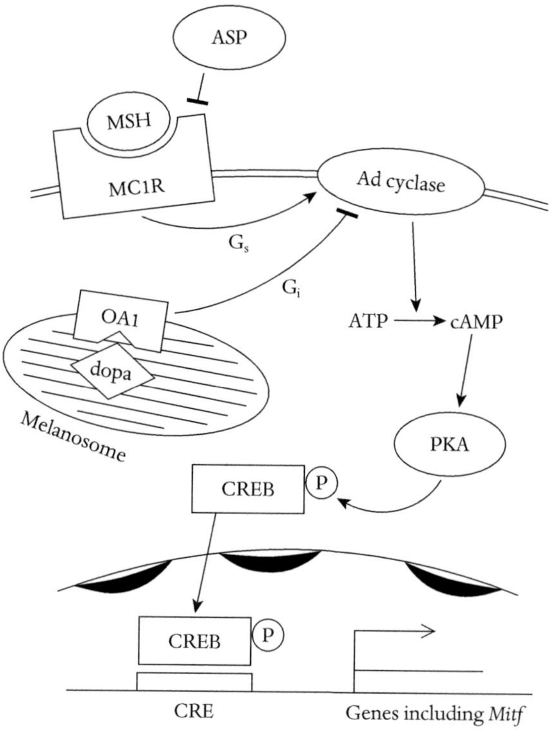

Figure 3.10 Basic components of cyclic AMP (cAMP) signaling in the melanocyte lineage. Symbols and abbreviations are as in Figure 3.9. See text for more details and sources. Ad cyclase, adenylate cyclase; ASP, agouti signal protein; CRE, cAMP-response element; CREB, CRE-binding protein; OA1, ocular albinism 1 protein (a receptor for L-dopa); P in circle, phosphate group.

release a soluble G_α subunit which then activates other signaling, according to the type of α subunit. OA1 is one such receptor of melanocytes, but, unusually, is located in the melanosomal membrane and is discussed elsewhere in the context of melanogenesis (Chapters 5 and 6).

The MSH/ASP/MC1R pathway

MC1R is a G-protein-coupled receptor encoded at *Mc1r* (*Recessive yellow, E*) (see Fig. 3.10). In mice it has at least two natural and antagonistic peptide ligands which impact type and color of melanin, as well as melanoblast survival and differentiation, namely the hormone MSH and the paracrine factor agouti signal protein (ASP), encoded at the *Agouti* (*A, Nonagouti*) locus. However, the main discussion of these will be in the chapter on pigment-type switching (Chapter 5).

The EDN3/EDNRB pathway

Endothelin 3 or EDN3 is a peptide hormone encoded at the classical white-spotting locus *Edn3* (*Lethal spotting, Ls*). Its principal receptor, endothelin receptor B or EDNRB, is encoded at *Ednrb* (*Piebald, S*), also a classical white-spotting locus. In addition to their importance in melanoblasts, EDNRB and EDN3 also participate in the earlier development of neural-crest cells and during both the ventrad and dorsolateral migrations, and so are also necessary for normal development of some truncal neuronal and glial lineages (Nataf et al. 1996). EDNRB is coupled to G-proteins that, in hypermorphic (overactive) mutants, when mutant, result in dark skin (originally referred to as Dsk mutants) (Van Raamsdonk et al. 2004). Dsk1 and Dsk10 are mutations in Gnaq, and Dsk7 in Gna11, loci encoding two $G_{\alpha q}$ (G_q) proteins. The known function of this type of G-protein is to activate phospholipase Cβ, which in turn activates protein kinase C (PKC) (Fig. 3.9) (Alberts et al. 2008), a kinase known to be important in melanoblast and melanocyte proliferation. The dark skin was not seen when these mutants were also null for *Ednrb*, showing that these G-proteins are only or predominantly coupled to EDNRB in the melanocyte lineage (Van Raamsdonk et al. 2004). Mice normally have very little pigment in the skin outside of hair follicles, so this dark skin is phenotypically somewhat the reverse of white spotting, a phenotype that represents absence of pigment cells.

Frizzled receptors and WNT signaling

By analogy, G-proteins are also likely to be elements of WNT/frizzled-1 signaling to the β-catenin/lymphoid-enhancer-binding factor (LEF)/T-cell-specific transcription factor (TCF) pathway (Liu et al. 2001; Malbon et al. 2001; Yasumoto et al. 2002) (see Fig. 3.11). The WNT family of secreted

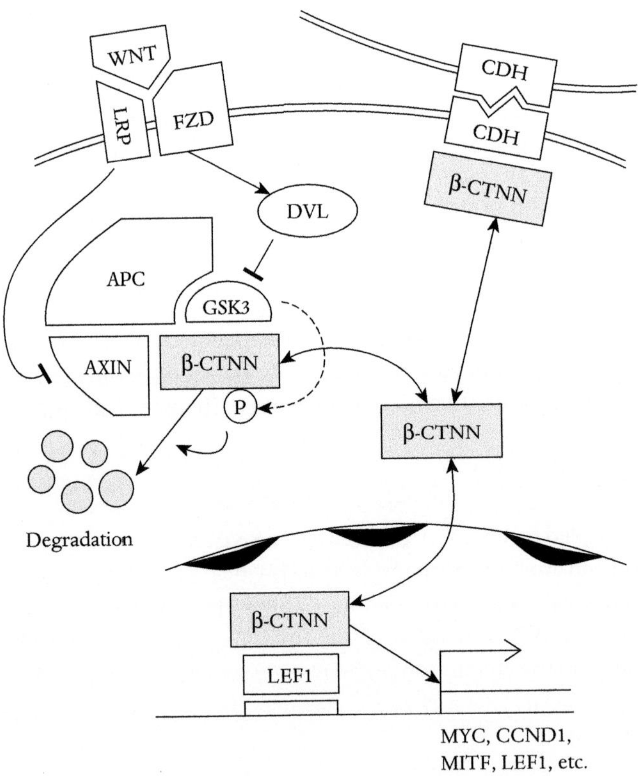

Figure 3.11 Main known elements of WNT/β-catenin signaling in the melanocyte lineage. Symbols and abbreviations as in Figure 3.9. Double-headed arrows indicate alternative sites for β-catenin. See text for more details and sources. APC, adenomatous polyposis coli protein; CDH, cadherin; DVL, disheveled.

glycoprotein morphogens bind receptors of the frizzled (FZD) family of seven-pass transmembrane receptors, and are required during the earliest stages of neural-crest formation and at other stages of melanocyte development. WNT signaling is sometimes classed into canonical and atypical pathways.

The WNT ligands and their receptors are found in many tissues; they are not specific to the pigment-cell lineage. Nevertheless, homozygous germline deletion of both *Wnt1* and *Wnt3a* (an embryonic lethal combination) leads to loss of neural-crest derivatives including melanoblasts, showing that while these two WNTs appear to be mutually redundant, one of them is required for melanocyte development (Ikeya et al. 1997). Moreover, mice that are deficient for other components of the WNT signaling pathway also show abnormal pigmentation. Of the several WNT signaling pathways, we will first introduce the canonical pathway.

WNT proteins bind to receptors in the frizzled (FZD) family (named after a *Drosophila* mutant), and may activate a number of different pathways, depending in part on which FZD and which cofactor associates with it. In 'canonical' WNT signaling (see Fig. 3.11), a frizzled receptor is associated with the low-density-lipoprotein-receptor-related protein 1 (LRP1) co-receptor. When the ligand is not present, FZD/LRP1 activates disheveled which, through glycogen synthase kinase 3β (GSK3β), tags β-catenin for destruction through the proteasome (Cadigan & Liu 2006). In the presence of WNT signaling, β-catenin is not degraded, accumulates, and so can participate in functions such as transcriptional activation and cell–cell adhesion (Fig. 3.11). β-Catenin is remarkable for its large number of functional interactions with other proteins (Fig. 3.11).

In its transcriptional function, WNT signaling activates members of the LEF/TCF family of transcriptional regulators and so participates in the upregulation of target gene products, including the melanocytic isoform of MITF (Hecht & Kemler 2000; De Strooper & Aennert 2001; Huelsken & Birchmeier 2001) (Fig. 3.11). This seems the most direct relationship between WNT signaling and establishment of the pigment-cell lineage, as MITF is a central organizing factor in melanoblasts and melanocytes.

Deficient pigmentation is seen in mice with mutations at the *Fzd4* or *Lef1* loci. Combined null mutations in *Lef1* and *Tcf7* result in a defect in the formation of paraxial mesoderm that is virtually identical to that seen in *Wnt3a*-deficient mice (Galceran et al. 1999), implying probable WNT3A signaling through LEF1 and TCF7. Adenomatous polyposis coli protein (APC) is an antagonist of the pathway; *Apc* mutants have abnormal precocious dorsal and head pigmentation appearing before birth (Kuraguchi et al. 2006). On the other hand, mice with the WNT pathway overactivated in the melanocyte lineage, by expression of constitutively activated β-catenin from the tyrosinase (TYR) promoter, paradoxically have reduced numbers of melanoblasts (Delmas et al. 2007). These findings support a model in which canonical WNT/FZD signaling is critical to development of the neural crest and the pigment-cell lineage, but there is an optimal level beyond which melanoblast development may also be impaired.

The cell–cell adhesion function of β-catenin is mediated by its interactions with the large cadherin (meaning calcium-dependent cell adhesion) family of molecules. The cytoplasmic domain of cadherins is linked to the cytoskeleton through the catenins (α, β, and γ) (Butz & Larue 1995; Pla et al. 2001; Vleminck and Kemler 1999; Gumbiner 2000). Apparently WNT1 and WNT3A signaling in the neural crest, originating in cells adjacent to the prospective neural-crest cells, prior to specification of the pigment-cell lineage, participates in the release of cell–cell adhesions during delamination of the neural-crest cells, and also acts to expand the population of neural-crest cells (Ikeya et al. 1997; Dunn et al. 2000, 2005; Logan & Nusse 2004). Nakagawa and Takeichi (1995, 1998) showed in chick embryos that regulation of cadherin expression is required for delamination of neural-crest cells, and specifically those of the pigment-cell lineage (see also Hirai et al. 1989; Larue et al. 1996). Using the Cre-loxP system to ablate β-catenin specifically in neural-crest cells, Hari et al.

(2002) demonstrated the importance of β-catenin to all of the neural-crest cells, which failed to develop beyond the stage of delamination.

WNT signaling is not identical in all vertebrate embryos (Dickinson et al. 1995), but in all cases is necessary for the normal establishment of neural-crest cells and the pigment-cell lineage. WNT and BMP signaling seem to have antagonistic functions in specification of the trunk neural crest (Jin et al. 2001) just at the time the neural-crest cell lineages are defined. Both in cell cultures and in avian and zebrafish embryos, activation of the WNT signaling pathway in neural-crest cells promoted the formation of melanoblasts (Dorsky et al. 1998; Jin et al. 2001). WNT signaling directly activates Mitf transcription in zebrafish and in cultured mouse melanocytes, leading to pigment-cell differentiation (Dorsky et al. 2000; Takeda et al. 2000). Dorsky et al. (1998) demonstrated, by injection of β-catenin into cranial neural-crest cells of developing zebrafish, that WNT signaling can specify multipotent neural-crest cells to the melanoblast, glial, and neuronal lineages before they leave the neural tube. Lewis et al. (2003), using zebrafish transgenic for a WNT inhibitor, defined two periods of WNT signaling that are critical to the pigment-cell lineage in zebrafish: first in the initiation of neural-crest cells within the neural tube, and second in the specification of the pigment-cell lineage.

In vitro studies of mouse neural-crest cells demonstrated that overexpression of canonical WNT signaling genes [*Ctnnb1* (*β-catenin*), *Wnt3a*, or *Wnt1*] can increase, in an endothelin-dependent manner, numbers of neural-crest cells expressing pigment-cell markers. WNT1 and WNT3A promote expansion of melanocytes through distinct modes of action, suggesting interaction with more than one FZD receptor. Both can increase the number of neural-crest cells prior to expression of pigment-cell markers, but only WNT3A induces a SRY-box 10 (SOX10)-dependent maintenance of MITF expression, in cells already expressing it, and a dependence on EDN3 and KIT ligand (KITL) (all properties of melanoblasts and melanocytes) (Dunn et al. 2005). *Sox10*, *Mitf*, *Edn3*, and *Kitl* are classical white-spotting genes. This study suggested that early melanoblasts are capable of differentiating into other cell types, but MITF expression – perhaps at some critical level – can function to stabilize pigment-cell determination.

Other studies have shown that normal melanoblasts continue to experience changes in their expressed cadherins and the associated catenins, including β-catenin, as they migrate through the dermis and epidermis and eventually home to the hair follicles. In early mouse embryos, cadherin 20 is expressed rather widely, including in cells likely to develop into the pigment-cell lineage (Moore et al. 2004). Jouneau et al. (2000) demonstrated that murine melanocytes express E- and P-cadherin when in contact with keratinocytes, while N-cadherin is the principal cadherin expressed by melanoblastic cell lines *in vitro*. In the migratory pathway of melanoblasts, E-cadherin expression begins in the melanoblasts of the dermis, prior to their migration into the epidermis, when P-cadherin expression begins. Keratinocytes in the epidermis and hair follicles express both E- and P-cadherin.

Overall, the WNT/β-catenin signaling pathway, regulating both cell adhesion and transcription, influences many processes in development, including establishment of the anteroposterior axis of murine embryos, establishment of pre-migratory neural-crest cells, delamination of neural-crest cells, correct development of the hair follicle, and maintenance of pigmentary stem cells within the follicle (Cadigan & Nusse 1997; Huelsken et al. 2000, 2001; Larue et al. 2003; Schepsky et al. 2006).

The WNT/β-catenin pathway is also part of a signaling network with other pathways such as transforming growth factor α (TGF-α) and Notch (Hecht and Kemler 2000; De Melker et al. 2004). Notch (see below) apparently serves as a negative regulator of the WNT pathway (De Strooper & Annaert 2001). Apparently WNT1 transcription is stimulated by BMP signaling from the developing somites (Burstyn-Cohen et al. 2004).

The hedgehog pathway

Patched (PTCH; not to be confused with the white-spotted mouse *Patch*, *Ph*/+) and Smoothened (SMO) proteins complex to form a cell-surface receptor for the hedgehog ligands. PTCH proteins span the cell membrane 11–12 times (Nagao et al. 2005), whereas SMO is a seven-pass G-protein-coupled receptor, though differing from those above in its downstream signaling through G_i. Neither the hedgehog intercellular signaling peptides nor their receptors are encoded by white-spotting genes, but this may be because non-conditional mutations are embryonic lethal; both families are widely expressed in the developing embryo, and maintained in the adult. Hedgehog signaling is implicated in neural-crest specification (see below).

The hedgehog signaling pathway (Fig. 3.9) is named for its ligands (in turn named after the *Drosophila* homolog) (Ingham & McMahon 2001). The mammalian genome encodes at least two receptors of the patched family, PTCH1 and PTCH2, and three hedgehog ligands, sonic hedgehog (SHH), indian hedgehog (IHH), and desert hedgehog (DHH). The ligands bind equally to any patched receptor, but the ligands and receptors are expressed differently in different tissues.

In the absence of hedgehog ligands, the patched receptor(s) act to inhibit Smoothened. When extracellular hedgehog is present, it binds to and inhibits patched, so that Smoothened function is derepressed. At the end of the cascade of reactions, the gliotactin (GLI) family of zinc-finger transcription factors are activated: GLI1 and GLI2 are activators of transcription of their target genes, whereas GLI3 is a repressor.

Like their famous *Drosophila* cognate, the hedgehog morphogens in vertebrates are instrumental in many aspects of patterning the early embryo. Hedgehog signaling is important in neural-crest formation in the head region, and specification of the glial/neuronal neural-crest cells (Fu et al. 2004). The balance between WNT and hedgehog signaling pathways is important to the correct development of eyes and ears; development in the head regions requires a balance between fibroblast growth factor (FGF) and hedgehog signaling. Hedgehog also functions in prenatal development of hair follicles and postnatal regulation of the molt cycle. Activation of the hedgehog pathway is required for transition of the hair follicle from the resting to the growth phase (Paladini et al. 2005). Upregulation of SHH activity in postnatal skin induces resting hair follicles to enter anagen, the phase of hair growth (Sato et al. 1999).

There is evidence for crosstalk between hedgehog and WNT signaling at the level of GSK3β. Further, there may be crosstalk with phosphoinositide 3-kinase (PI3K) signaling through AKT and with PKCδ signaling through MEK1 in the melanocyte lineage (Fig. 3.9). Both of the latter are components of signaling pathways that are activated by a different category of receptor, the single-pass receptors.

3.3.3 Pathways from single-pass transmembrane receptors

These receptors reside in the cell membrane with the effector domain inside and the ligand-binding domain on the outside, linked by only one transmembrane section. This category of receptors is large and diverse and its members may not all be related in evolution. We will discuss several of the major single-pass receptors and their signaling pathways known to be important to the pigment-cell lineage. We begin with one of the odd ones.

Notch

The Notch pathway functions in the migration and maintenance of neural-crest cells and other stem cells of multiple lineages. It is one of the factors thought to contribute in the neural tube to initiation of neural-crest formation and delamination (Hari et al. 2002), and functions also at later stages in pigment-cell survival, and in maintenance of stem cells in the niche of the hair follicle. None of the Notch signaling elements are found among the classical white-spotting genes, nor are they specific to pigment cells; however, targeted ablation of Notch pathway elements in pigment cells results in pigment-cell loss and hair graying (see below).

Notch receptors are single-pass transmembrane proteins that initiate a highly conserved cell signaling system present in most multicellular organisms. Vertebrates possess four different Notch receptors, NOTCH1–4 (Kraman & McCright 2005). The Notch ligands (Delta/Jagged) are also single-pass transmembrane proteins, found on the plasma membranes of nearby cells. Thus Notch signaling is juxtacrine rather than paracrine or autocrine (Artavanis-Tsakonas et al. 1999), in that it requires cell–cell contact. Known Notch ligands in mammals are the three Delta-like (DLL) and two Jagged proteins. Notch may also respond to other ligands. The two cells need not always be adjacent; one of the cells may send out processes that contact the other at a distance. In the case of induction of neural-crest cells, Notch ligands are presumably on the surface ectodermal cells as they fold over the neurectoderm that is expressing Notch during neurulation.

Each NOTCH protein is a heterodimer composed of a large extracellular portion that associates in a calcium-dependent, non-covalent fashion with a smaller intracellular section. When the receptor is activated by a ligand, the Notch intracellular domain is cleaved and translocates into the nucleus where it complexes with a CSL-family transcriptional repressor such as RBP-Jb and converts it into an activator that promotes transcription of target genes (Moriyama et al. 2006; Das et al. 2007). One target gene product is Hairy/enhancer of split (HES1), a basic helix-loop-helix (bHLH) transcription factor (see below for bHLH factors). HES1 promotes maintenance of stem cells of several types, including neuronal stem cells and melanoblasts (Kageyama et al. 2000; Aubin-Houzelstein et al. 2008) by inhibiting apoptosis and differentiation (Hitoshi et al. 2002; Duncan et al. 2005; Schouwey et al. 2007). Notch signaling also functions in neurogenesis, gliogenesis (Morrison et al. 2000), and other developmental processes.

Targeted ablation of *Notch1*, *Notch2*, or *Rbpj* in the pigment-cell lineage in mice, using the TYR promoter, results in an overall reduction in the number of pigment cells, and also a progressive loss of pigment cells from the hair follicles of adult mice; that is, progressive graying (Moriyama et al. 2006; Schouwey et al. 2007). Targeted overexpression of HES1 rescues the pigmentation defect in Notch-ablated mice (Moriyama et al. 2006), so, in summary, all these pathway elements seem important to establish and maintain melanocyte stem cells.

Crosstalk has been identified between the Notch pathway and most other major intracellular signaling pathways. Notch signaling inhibits the WNT/β-catenin pathway, as mentioned above, probably at several steps (De Strooper & Annaert 2001; De Melker et al. 2004); this interaction has been identified in pigment-cell development and in hematopoietic stem cell maintenance (Duncan et al. 2005; Das et al. 2007). These findings support the idea of an optimal level of WNT signaling for melanoblast development; perhaps Notch signaling helps to prevent the level exceeding this optimum. Notch and TGF-β (discussed below) interact in control of HES1, through TGF-β target SMAD3 (Blokzijl et al. 2003). Notch signaling interacts with that of BMP4 during myogenesis (Dahlqvist et al. 2003), and that of KIT in development of neural stem cells in the eye (Das et al. 2004). In the neural crest, Notch is an actor in a pathway that involves FGF2 signaling (Ota & Ito 2006).

Thus the Notch pathway, combined with targeted gene manipulation, illustrates the ways that pigmentation models help to dissect the nodes of the complex web of intersecting signals that regulate cellular function and embryonic development (Hurlbut et al. 2007).

TGF-β receptor superfamily: serine/threonine phosphorylation

This superfamily includes two types of single-pass transmembrane protein that are protein serine / threonine kinases. Each of the families and subfamilies of receptor binds a different kind of ligand, and the receptors are discussed here with their ligand families.

The TGF-β family of ligands This family (TGF-β1, 2, and 3) are cytokines belonging to the TGF-β superfamily which also includes the BMP family, inhibins, activin, anti-Müllerian hormone, decapentaplegic and Vg-1. TGF-β1, 2, and 3 participate in many cellular functions, including the control of cell growth, cell proliferation, cell differentiation and apoptosis. Each binds to receptor TGF-β receptor 2 (TGFβR2). This then binds TGFβR1 and both become activated as kinases (Alberts et al. 2008). Within the cell, the activated receptor complex signals through the SMAD pathway. An activated SMAD binds a coSMAD (another SMAD-family member). The complex of SMAD and coSMAD enters the nucleus where it participates in a transcription factor complex that regulates transcription of target genes. Because there are three members of the TGF-β family and several different SMAD proteins, the message that is transmitted is very precise.

In initial studies using chick neural tubes *in vitro*, Delannet and Duband (1992) reported that TGF-β1 and 2 stimulated emigration of cultured neural-crest cells, possibly by increasing the adhesion of the cells to their substrate at the expense of intercellular interactions. TGF-β is also involved in G_1 arrest in the cell cycle and the transition from G_1 to S is necessary for delamination.

The BMP family of ligands This family interacts with BMP receptors (BMPRs) of which there are at least two. Signals from BMPR1A and BMPR1B receptors are transduced within the cell by a specific SMAD signaling pathway (different from the above), via SMAD1, SMAD5, and SMAD9. BMP is involved at the beginning of neural-crest development through the action of its competitive antagonists chordin and noggin. In early development, noggin is present along the neural crest as a gradient, thus creating a gradient of BMP signaling that establishes the rostral / caudal gradient of neural-crest development and initiates the rostral–caudal progression of neural-crest cell formation within the neural tube. Details of BMP function in neural-crest cell development differ somewhat among the animal model species: fish, birds, and mice.

Later in development, BMP2 is essential for the genesis of migratory neural-crest cells (Aybar & Mayor 2002). In murine BMP2-null mutant embryos, neural-crest cells are created, but they fail to delaminate from the neural tube. The failure to migrate is not the result of cell death, but is accompanied by abnormal closure of the neural tube (Correira et al. 2007). In normal quail and chick embryos and neural-crest cell cultures, BMP signaling seems to interact with WNT signaling in the initial specification of the glial / neuronal versus melanogenic lineages, with BMP4 favoring the glial / neuronal and WNT3a favoring the pigment-cell lineage (Jin et al. 2001).

Yet later, BMP is required for normal migration of enteric neural-crest cells (Goldstein et al. 2005) and acts as a positive regulator of neurogenin in cell culture. Neurogenin-2 (NEUROG2) is a HLH transcription factor essential for the specification of sensory neurons, much as MITF specifies the pigment-cell lineage, and is a marker of the neuronal / glial lineage of trunk neural-crest cells. FGF2 treatment has been shown to suppress NEUROG2 expression and promote formation of glial cells through activation of Notch. Thus BMP and FGF2 act as positive and negative regulators of NEUROG2 expression, respectively, within the glial / neuronal lineage (Ota & Ito 2006).

Murine central nervous system stem cells can be induced to differentiate into neural-crest cell fates *in vitro* and in chick and quail embryos, but only if they have been exposed to BMP and FGF2

(Busch et al. 2006). Clearly, BMP collaborates with a number of other factors to direct different developmental stages during neural-tube and neural-crest development.

Receptor tyrosine kinases (RTKs): tyrosine phosphorylation

Protein tyrosine kinases (PTKs) catalyze the phosphorylation of tyrosine residues of target proteins. This very large protein family includes two main classes: receptor PTKs and cytoplasmic or non-receptor PTKs. Receptor PTKs are single-pass cell-surface receptors. Their catalytic activity is activated when the ligand binds to two receptor molecules, which usually then each phosphorylate tyrosines in the other to form an active receptor dimer, initiating signaling cascades of phosphorylation (Fig. 3.12).

Some non-receptor tyrosine kinases can act directly on transcription factors in the nucleus. Others, and receptor PTKs, act through a complex but much-studied web of signaling pathways, especially the mitogen-activated protein kinase (MAPK) and PI3K pathways, both initiated through the RAS, RHO, and RAL families of small G-proteins. The MAPK cascade (Fig. 3.12) is one of the most ubiquitous signal-transduction systems of eukaryotes, and regulates numerous physiological processes including cell proliferation, differentiation, migration and death (Wilkinson & Millar 2000; Easty & Bennett 2000; Bottcher & Niehrs 2005). It involves recruitment of a RAS protein and then a RAF protein to the cell membrane, then sequential and amplifying phosphorylations of MAPK kinase (MAPKK), MAPK, and downstream targets depending on context. Figure 3.12 shows PTK signaling elements of known importance in pigment cells. The PI3K pathway works by activating the serine/threonine kinase AKT, primarily to promote cell survival and growth in cell size. Intracellular PTKs can link the same signaling pathways to other kinds of receptor.

RTK receptor/ligand combinations active in the melanocyte lineage Several RTK receptors and their ligands, although not identified as classical white-spotting genes, are nonetheless known to impact pigmentation by manipulation in culture or via transgenic mouse technology. We will include some of these pairs in the present discussion, for completeness (Table 3.1). However, KITL and its receptor KIT (*White, W, Dominant white spotting*) do qualify as classical white-spotting genes. The role of most of these ligands in development of melanocytes may be secondary to their importance in the neural crest as a whole and the various neural-crest derivatives. Striking parallels are seen, for example, between the functions of RTKs in the development of the autonomic nervous system and of the pigment-cell system. For more detail see Wehrle-Haller & Weston (1997) and Le Douarain & Kalcheim (1999).

Table 3.1 RTK receptors and their ligands

Ligand symbol	Ligand name	RTK receptor symbol
KITL/SCF	KIT ligand/stem cell factor	KIT
HGF/SF	Hepatocyte growth factor/scatter factor/hepatopoetin A	MET
GDNF	Glial-cell-derived neurotrophic factor	RET
PDGF(A,B)	Platelet-derived growth factor	PDGFRA, PDGFRB
EGF	Epidermal growth factor	EGFR
FGF family	Fibroblast growth factors	FGFR family

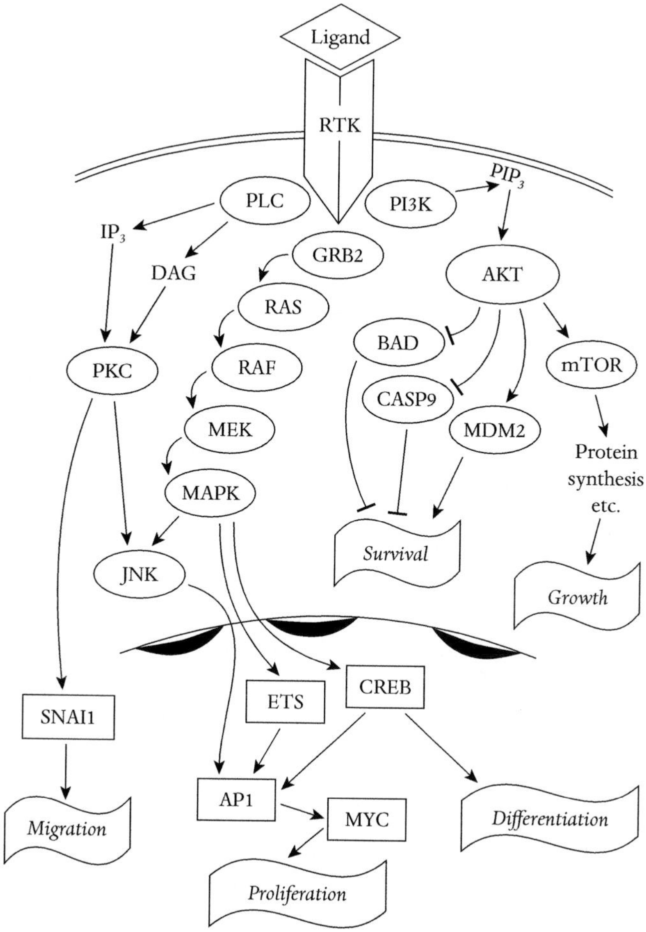

Figure 3.12 Basic elements of RTK signaling. Melanoblasts and melanocytes have numerous types of RTK receptors, and it is thought that the relative input into the three different sub-pathways (PLC, RAS, PI3K) varies between receptors. Symbols and abbreviations as in Figure 3.9. See text for more details and sources. BAD, BCL2 antagonist, cell death; CASP9, caspase 9; DAG, diacylglycerol; ETS, erythroblastosis oncogene homolog; GRB2, growth factor receptor-bound protein 2; IP_3, inositol trisphosphate; JNK, JUN N-terminal kinase; MAPK, mitogen-activated protein kinase; MDM2, mouse double minute 2; mTOR (or FRAP1), mammalian target of rapamycin; PIP_3, phosphatidylinositol 3,4,5-trisphosphate.

KIT The RTK receptor KIT is encoded by a classical white-spotting gene (*Kit, W, Dominant white spotting*) and is one of the first-expressed marker proteins, perhaps the first, that identifies melanoblasts developing from the neural crest (Price et al. 1998). KITL is encoded at another classical white-spotting locus (*Kitl, Steel, Sl*). KITL, also called stem cell factor (SCF) or mast-cell growth factor (MGF), has both soluble and cell-bound (transmembrane) isoforms, which together are required for migration, proliferation, and survival of the melanoblasts along the dorsolateral pathway in the embryo. These will be discussed in more detail in section 3.4.

Hepatocyte growth factor (HGF)/scatter factor (SF) This is the ligand for mesenchymal-epithelial transition factor, or MET (formerly c-met). In embryos, HGF is secreted by mesenchymal fibroblasts and targets epithelial and endothelial cells and pigment cells at times when they express MET. In melanocyte culture, in combination with synergistic factors such as FGF2 (basic FGF) and KITL, HGF functions as a growth and differentiation factor, maintaining high TYR activity and melanin content (Matsumoto et al. 1991; Halaban et al. 1992; Hirobe et al. 2004a).

HGF and MET These are principal mediators of mesenchymal–epithelial interactions in several different systems, including the hair follicle, where the ligand is localized in the hair follicle mesenchyme (dermal papilla fibroblasts) and the receptor, MET, in the neighboring hair bulb keratinocytes, during the active stages of the hair cycle (Lindner et al. 2000). *In vivo*, MET is not required for neural-crest formation up to embryonic day 14 (Kos et al. 1999) in $Met^{-/-}$ mice. However, overexpression of HGF in the melanocytes of transgenic mice induces an abnormal accumulation of melanocytes, in great excess throughout the dermis and epidermis of the adult mice, besides promoting a wide variety of tumors of both mesenchymal and epithelial origin (Takayama et al. 1996, 1997). Melanoma formation is increased in such mice by neonatal ultraviolet irradiation (Wolnicka-Glubisz & Noonan 2006).

RET/GDNF The RET proto-oncogene encodes a PTK receptor for members of the glial-derived neurotrophic factor (GDNF) family. During normal development, RET is expressed primarily in neural-crest-derived neural structures, sensory tissues, and the developing kidney (Avantaggiato et al. 1994). RET/GDNF is important in the development of the neuronal/glial lineage of the enteric nervous system (Taraviras et al. 1999; Heuckeroth & Pachnis 2006).

This is interesting to pigmentation devotees primarily because of the way in which the RET/GDNF signaling system parallels, in the development of the glial/neuronal lineage of neural-crest cells, the KIT/KITL system in the melanoblast lineage. It is another example of the ways in which the classical pigment mutants contributed to our current understanding of the communications web of the cell. In the early stages of neural-crest and neural-crest-cell formation, both the melanocyte and glial lineages require the ongoing activity of SOX10 and PAX3 transcription factors, as well as EDNRB and its ligand EDN3 (discussed above). These are all classical white-spotting proteins. Then, as the neural-crest cells begin their migration and development, glial/neuronal precursors, under the influence of PAX3 and RET/GDNF, transcribe their specific products, including NEUROG2, while melanoblasts, under the influence of PAX3 and KIT/KITL, transcribe their specific products, including MITF.

Epidermal growth factor (EGF)/neuregulin family Neuregulins are a family of growth and differentiation factors related to EGF. The receptors for neuregulins are the ERBB (or human EGF receptor, HER) family of tyrosine kinase transmembrane receptors, which includes the EGF receptor (EGFR). In particular, the neuregulin/ERBB signaling pathways play crucial roles in regulating the proliferation and differentiation of Schwann cells, the glial cells of the peripheral nervous system (not only those that make myelin). Human melanocytes express several HER receptors (Easty & Bennett 2000); this is largely untested in mice, but a recent dark-skinned mouse mutant was identified as having a point mutation in the EGFR, increasing its activity (Table A1.1, Chapter 1 Appendix).

Platelet-derived growth factor receptors (PDGFRs) Platelet-derived growth factor (PDGF) A, B, and C are subunits of the dimeric ligands of the PTK receptors PDGFRA and B. PDGFRA and B can bind homodimers of PDGFA, B, and C, or PDGFA-PDGFB heterodimers, with varying affinity. The PDGFs are important in angiogenesis and connective tissue; their importance to murine white spotting has to do with the finding of a deletion in the *Pdgfra* locus in *Patch* (*Ph*/+) mice, and the initial attribution of the spotting of these mice to the deletion. However, although targetted germline deletion showed that *Pdgfra* was required for development of mesenchymal neural-crest derivatives in the head, pigmentation was unaffected (Soriano 1997). The spotting phenotype in *Ph*/− mice turned out to result from a chromosomal rearrangement affecting the nearby *Kit* locus. For more information, see the *Patch* section of this chapter (section 3.4.5). Mutations in ligands PDGFB and PDGFC can impair eye pigmentation (Table 3.1).

FGF receptors (FGFRs) The FGFR family of RTKs has four members, FGFR1–4. Alternative splicing of FGFR transcripts generates a diversity of FGFR isoforms with distinct FGF-binding specificities. The FGF family of ligands is large, numbering more than 20, with functions in diverse cellular processes (Bottcher & Niehrs 2005). An FGF receptor–ligand complex is formed of two FGF peptides, two heparans, and two receptor molecules (Alberts et al. 2008). As already mentioned, FGF signaling is implicated in early neural-crest development, and FGF2 or basic FGF is a growth factor for melanocytes (Halaban et al. 1987) and melanoblasts (Sviderskaya et al. 1995; Hirobe 2001). Targeted mutations in receptor gene *Fgfr2* result in pale skin among many other defects (Table A1.1).

3.3.4 Downstream signaling in melanocyte development: transcription factors

As already mentioned, transcription factor MITF appears central to determination and differentiation of the melanocyte lineage. MITF is regulated by a complex net of signaling that includes at least five other transcription factors and a plethora of helper molecules. The transcription factors implicated to date in direct regulation of the MITF promoter include paired-box 3 (PAX3), cAMP-response-element-binding protein (CREB), SOX10, LEF1, and ONECUT (Fig. 3.9) (Goding 2000; Levy et al. 2006), conferring on MITF the capacity to respond to diverse signaling pathways. The importance of the links between MITF expression and *Pax3*, *Sox10*, and *Lef1* is underscored by the genetics of mutant mice that link mutations in each of these loci to loss of melanoblasts (white spotting) and, in the case of *Pax3* and *Sox10*, to syndromes of pigmentary and other birth defects, as discussed in Chapter 4. Other transcription factors impinge at least indirectly on this regulatory circuit, since mutations in them can alter melanocyte development. Here we introduce the known players in this transcriptional network, divided by structural class.

bHLH/leucine zipper (LZ) transcription factors

bHLH/LZ transcription factors function as homo- or heterodimers. They contain a DNA-binding basic domain, and HLH and LZ domains, which mediate dimerization.

MITF MITF itself is one such factor; it has already been introduced above and will be discussed in more detail in section 3.4, as the product of a classical white-spotting gene.

CREB family CREB-family proteins are bHLH/LZ transcription factors that bind to sequences called cAMP-response elements (CRE) in gene promoters, and thereby regulate transcription. In general, activation of certain receptors will lead to cAMP production, activating protein kinase A (PKA) which phosphorylates a CREB protein or proteins (Fig. 3.9). This then enters the nucleus, binds to a CRE region, and binds a CREB-binding protein (CBP; CREBBP or EP300) as co-activator.

The complex can then activate transcription of the gene. CREB-family proteins known to be important in the melanocyte lineage include CREB1 (formerly CREB), which can upregulate MITF in response to MSH/MC1R signaling, (Buscà & Ballotti 2000), and ATF2, which can also dimerize with JUN proteins in proliferative signaling.

Helix-turn-helix factors

BRN2 BRN2 (Brain2, POU3F2, N-Oct-3) is a POU domain factor, as it contains a bipartite DNA-binding domain called the POU domain. BRN2 serves as an early marker of neural crest in most species. BRN2 may play a part in melanocyte differentiation. Cook et al. (2003) established melanoblast and melanocyte cultures from human foreskins and found that melanoblasts (containing unpigmented melanosomes) could be grown, or generated from a melanocyte culture, using FGF2, KITL/SCF, and EDN3. BRN2 expression was high in the melanoblast cultures. When the three factors were removed, BRN2 levels decreased and the cells differentiated to melanocytes with pigmented melanosomes, suggesting a role for BRN2 in maintaining the precursor state.

Signaling pathways known to influence BRN2 activity include β-catenin, BRAF (Goodall et al. 2004a, 2004b), and probably Notch/Delta. *Pou3f2* (the BRN2 gene) is a direct target of the β-catenin–LEF1 complex. In transgenic mice that express a lacZ marker in the pigment-cell lineage, BRN2 was clearly evident in the neural tube, but was not detectable in melanoblasts or neural crest. BRN2 was also present in mature hair follicles (Goodall et al. 2004a, 2004b).

Forkhead/winged helix factors

FOXD3 The winged-helix or forkhead-box (FOX) class of transcription factors has been shown to play important roles in cell specification and lineage segregation. The forkhead box is a DNA-binding sequence of 80–100 amino acids. Some Fox genes are downstream targets of the hedgehog signaling pathway, mentioned above. FOXD3 is not expressed in the pigment-cell lineage, but plays at least two important roles in its development. First, it is involved in the segregation of the neural-crest lineage from the neuroepithelium. Second, it represses melanoblast determination, thereby allowing other neural-crest derivatives to differentiate during early neural-crest patterning (Kos et al. 2001). The human FOXD3 locus was identified as a susceptibility locus for vitiligo; a rare variant sequence in the promoter appears to cause increased FOXD3 expression (Alkhateeb et al. 2005), presumably resulting in decreased numbers of melanocytes postnatally.

Zinc-finger factors

GLI Zinc-finger protein GLI is the primary transcription-factor target of the hedgehog signaling pathway. Gli3 is a modifier of SOX10 (Matera et al. 2008).

***Snail* (*Snai1*, *Snail-1*) and *Slug* (*Snai2*)** In vertebrates, two members of the Snail family of proteins are known: SNAI1 (Snail) and SNAI2 (Slug/Slugh) (Sefton et al. 1998). These are involved in the EMT of neural-crest cells and maintenance of melanoblast stem cells in the hair follicle niche, as well as the expression of KIT early in specification of the pigmentary, hemopoietic, and germ cell lineages (Pérez-Losada et al. 2002). Snail family genes encode zinc-finger transcription factors (Carver et al. 2001) that are highly unstable. In *Drosophila*, Snail acts as a transcriptional repressor during the period of mesoderm formation by preventing expression of mesectodermal and ectodermal genes in the mesoderm anlage (Smith et al. 1992).

SNAI1 is essential to survival in the mouse (Sefton et al. 1998). SNAI2 is not; it is necessary for melanoblast migration and/or survival but not for neural-crest formation (Sanchez-Martin et al. 2002), reflecting its expression in migratory but not pre-migratory neural-crest cells (Jiang et al. 1998). Expression of SNAI2 in mice differs from that in chick embryos, and even differs among strains of mice as shown by Sanchez-Martin et al. (2002), who suggested that Slug functions downstream of MITF.

Mutations in SNAI2 contribute to human 'piebaldism', specifically Waardenburg syndrome type II, in a dose-dependent fashion (Pérez-Losada et al. 2002; Tachibana et al. 2003) and to white spotting in mice. However, Snai2-deficient mice exhibit infertility and macrocytic anemia, in addition to white spotting (Sanchez-Martin et al. 2002), and their pigmentary phenotype as shown in this paper is strongly reminiscent of mice that are heterozygous for defective KITL. They do not resemble mice mutant at *Mitf*.

SNAI2 is not expressed in $Kit^{-/-}$ cultured mouse cells, and is reduced in melanocyte cultures made from mice that lack KITL/SCF (Sanchez-Martin et al. 2002). This syndrome of defects in developing germ cells, hemopoietic cells, and pigment cells is typical of mice mutant at *Kit*, and not of *Mitf* mutant mice. Furthermore, of all the white-spotting defects in mice, *Kit* defects are the most sensitive to strain-dependent differences in size and location of the unpigmented regions, while *Mitf* genotypes are the least sensitive (Lamoreux 1999). These facts reinforce the suggestion that SNAI2 is a downstream target of signaling from KIT in mice (Inoue et al. 2002; Pérez-Losada et al. 2003). However, it has also been reported that the KIT signaling pathway and the MITF signaling pathway intersect at MAPK (Hemesath et al. 1998; Hou et al. 2000) (Fig. 3.12), and Sanchez-Martin et al. (2003) have identified a proposed binding site for MITF on the Slug promoter.

In cell culture, production of SNAI2 provokes the loss of epithelial markers, as well as changes in cell shape and the expression of mesenchymal markers, and it attenuates the cell cycle (Vega et al. 2007). It has been suggested (Barrallo-Gimeno & Nieto 2005) that the primary function of the Snail genes is to induce cell movement and survival, as a part of the EMT, rather than specifically triggering the EMT.

SNAI factors, nonetheless, have been implicated in the EMT, in mouse and human (Savagner et al. 1997; Jiang et al. 1998; Sanchez-Martin et al. 2003; Tachibana et al. 2003), including the EMT necessary for neural-crest cell delamination, and are strong repressors of N-cadherin activity (Cano et al. 2000). Just as canonical WNT signaling stabilizes β-catenin by inhibiting GSK3, SNAI1 and -2 are regulated in a similar fashion by GSK3, and GSK3 is responsive to several other signaling pathways in addition to WNT signaling, including TGF-β1 (Katoh & Katoh 2006, 2007), PI3K (Zhou et al. 2004), and probably hedgehog (Zhou & Hung 2005) (Fig. 3.9). Hedgehog signaling is important in neural-crest formation in the head region, and specification of the glial/neuronal neural-crest cells.

Beta scaffold factors

p53 is not only a transcription factor but a key tumor-suppressor and the most commonly mutated gene in human cancer, although less commonly mutated in melanoma than in most cancers (Bennett 2008). Its reported effect upon pigmentation is two-fold. If the protein is deficient in pigment cells, the result is a belly spot, suggesting reduced melanoblast numbers. Overexpression of normal p53 in the keratinocyte generates darker pigmentation of the skin through increased expression of KITL by keratinocytes (McGowan et al. 2008).

Other transcription factors

Other transcription factors include SOX10 and PAX3, which are discussed in more detail in section 3.4. Lastly, there is activating protein 2α (AP2α), encoded at *Tcfap2a*. Mice deficient in AP2α show abnormalities, including neural-crest defects, and die *in utero*; if deficiency is targetted specifically to neural crest with the WNT1 promoter some mice survive postnatally with neural-crest defects including deficient pigmentation, indicating a requirement for this factor at some point in the melanocyte lineage (Brewer et al. 2004).

3.3.5 Interactions among pathways, and emergent properties

Signaling pathways are not linear and independent units within the cell (Nusse 2003; Nelson & Nusse 2004), as can readily be seen from Figure 3.9. Several pathways may all regulate the same biological outcome, for example converging to stimulate cell proliferation or survival. Moreover, the interactions between pathways may set up some unexpected higher-level (emergent) properties, such as biological switches. At least two such potential switches can be found in Figure 3.9.

The first is a potential positive-feedback circuit between MITF and MC1R. Activation of MC1R stimulates MITF transcription through CREB, while MC1R is itself a transcriptional target of MITF (Steingrímsson et al. 2004). Thus, the activation of each component can activate the other, a self-stabilizing process that could amplify MC1R signaling in the presence of MSH, further upregulate cAMP levels and promote cell differentiation, potentially contributing to the stability of the switch from melanoblast to melanocyte.

The second is a similar circuit between MITF and RTK receptors KIT and MET. These receptors also appear to be MITF targets, while a stimulatory pathway can be seen from RTK activation through MAPK and CREB to MITF. It is tempting to speculate that this could play a part in the determination of melanoblasts from neural crest: in the presence of ligands KITL and/or HGF, MITF and KIT/MET should mutually upregulate and stabilize each other's expression.

There is always the danger, with a brief presentation of the multiple interacting pathways that participate in development, that the description will appear to be one of linear programming. Therefore, it is important to repeat, as we said in the introduction, that the processes of cell lineage restriction and the paths of cell migration are regulated by a network of environmental influences, intercellular interactions, and intrinsic factors. Pathway choices of cells during development are the result of changes in the environment that generate changes in the developing cells with continued support from the changing environment.

3.4 Pigment phenotypes and the classical white-spotting genes

In section 3.2 we described the physiological development of the melanoblast lineage within the embryo; in section 3.3 we described the cell biology of some of the information exchanges required for that development and for the development of the eventual phenotype. The phenotype of an organism is the appearance of the organism, either as a whole or relative to the function of a specific gene. The pigmentary phenotype is a direct interface between the organism and its environment and is therefore critical to the survival of the individual. In addition, the phenotype is foundational to the species and to speciation because natural selection acts upon the phenotype of the organism within its environment; the conversation between natural selection and the genome is mediated by the phenotype. From the beginning of genetics, scientists have used the phenotype as entrée to that conversation between the phenotype and the genotype, and we are now doing so in increasingly sophisticated ways (see Chapter 6) to clarify genic actions and to validate hypotheses that are based on experimentation *in vitro*.

Figure 3.13 White-spotted burro, parentage unknown.

In this section our focus will be on the classical pigmentary genes and their genotype–phenotype relationships as they illuminate development of the pigment-cell lineage in mice. Photographs of other species provide context and relevance of the mouse work (e.g. Fig. 3.13). This chapter emphasizes basic process; detailed information about specific genes can be found online (see the Reference section at the end of the book). We will use white-spotting phenotypes, which reflect failure of pigment-cell development, as well as pigmented phenotypes that help to explain how pigmentation develops. We will use information gained from other vertebrate species to illuminate, or compare with, the condition in mice; bearing in mind that there is variability among species. We will continue to emphasize the trunk region as our exemplar, while recognizing that the head region develops rather differently and is perhaps more important to our comprehension.

All of the above of course assumes that the classical white-spotting genes and their helper genes are functioning normally (Figs 3.14 and 3.15). If they are not, the result is pigment-cell death or failure to normally replicate or migrate, white spotting (congenital absence of pigment cells), and/or progressive depigmentation (graying with age).

A special advantage of the pigmentary system as a research model is the extent to which changes in the genotype are visible in the phenotype of the living animal (Baxter et al. 2004). In previous sections of this chapter we have discussed (3.2) morphogenesis of the pigmentary system and (3.3) cellular signaling. Here we will discuss specific key regulatory loci, along with several hypotheses about the process of pigment-cell development and survival (Rawles 1947; Lyon 1961; Mintz 1967, 1971a, 1971b; Mayer & Green 1968; Schaible 1969; Jenkins et al. 1981), as they relate to a developmental time line:

- the time before and during specification of the pigment-cell lineage (see section 3.4.1, below);
- the initial migratory phase in the mesenchymal layer, at 9.5–10 days post coitum (dpc; which means the age of the embryo) in the head region; at 11.5 dpc in the anterior and posterior portions of the trunk region, and a bit later in the mid-trunk region (Yoshida et al. 1996) (discussed in section 3.4.2);

Figure 3.14 White-spotted tortoiseshell guinea pig. Tortoiseshell is a phenotype that can be caused by several different mechanisms. In the guinea pig, tortoiseshell is autosomal, not X-linked as in the cat, and is caused by a mutation at the *Extension* (*E*) locus that is presumably homologous to the *Recessive yellow* (*Mc1r*, formerly *E*) locus of the mouse. Interestingly, recessive yellow mice also, but only very rarely, can exhibit somatic reversion patches to the nonyellow phenotype (L. Lamoreux, personal observation). The genotypes of tortoiseshell guinea pigs are e^p/e^p or e^p/e; the latter usually exhibit more red (=yellow pheomelanin) than the former. Tortoiseshell in the rabbit is also autosomal, is known by some as 'Japanese,' and is similarly inherited as e^j/e^j or e^j/e. It has been speculated that these autosomal chimerisms involve gene duplication. An excellent little book about guinea pig (cavy) genetics is authored by Harry Claus and distributed by the American Cavy Breeders Association, 16540 Hogan Avenue, Hastings, MN 55033-9576, USA, c/o Lenore Gergen.

Figure 3.15 White-spotted mouse mutant at the *Mitf* locus; C57BL/6J-$Mitf^{mi\text{-}sp}/Mitf^{mi\text{-}ew}$. Mice of the genotype $Mitf^{mi\text{-}sp}/Mitf^{mi\text{-}sp}$ (*Mi-spotted*) appear to be essentially wild type. $Mitf^{mi\text{-}ew}/Mitf^{mi\text{-}ew}$ mice are white and microphthalmic.

- the secondary migratory period (Wilkie et al. 2002) as the cells begin to move into the epidermis from about 12.5 dpc (Yoshida et al. 1996); until perhaps 15.5 and 16.5 dpc when the melanoblasts have colonized all the normally available areas in the epidermis (Jordan & Jackson 2000), described in section 3.4.5, and, finally,
- homing and entry into the hair follicles, which begins during the second migratory phase and continues until at least a week after birth (Hirobe 1984a) (see section 3.4.7).

3.4.1 Genes required for specification of the pigment-cell lineage in the neural crest

We can operationally define a cell specified to the pigment-cell pathway as a neural-crest-derived cell expressing MITF or KIT or both, and capable of sustaining their expression. These are melanoblasts if unpigmented, otherwise melanocytes.

Specification of the melanoblast lineage is poorly understood overall, but is known to require products of pigmentary genes and signaling genes, as well as ubiquitous housekeeping genes, and interaction among all these. MITF and KIT are encoded at classical white-spotting loci.

In chick and fish, the first step in specifying melanoblasts from the neural-crest lineage occurs before the neural-crest cells leave the neural tube, when WNT signaling competes with BMP signaling. BMP2/4 specifies neuronal and glial cells (Dorsky et al. 1998; Jin et al. 2001). In mice, KIT-expressing cells that are believed to be specified melanoblasts can be identified in the dorsal neural tube (Box 3.1d) (Wilson et al. 2004). Using a mouse cell-culture model, Dunn et al. (2005) found that WNT1/WNT3a act prior to dopachrome tautomerase (DCT) function, probably in the neural tube, to expand the percentage of cells of the melanoblast lineage before delamination. (DCT is a melanosomal enzyme discussed in Chapter 4 that is well expressed in melanoblasts as well as melanocytes, and so is a convenient marker for both.) β-Catenin (CTNNB), a component of WNT signaling, is one of the transactivating proteins known to be required for initiation of expression of MITF (Schepsky et al. 2006).

Two of the classical white-spotting loci that are expressed before melanoblast specification (Box 3.1b) are required for normal development of glial/neuronal neural-crest cell lineages, as well as the pigment-cell lineage, and interact with each other and with other signaling genes to initiate specification of all three of the neural-crest cell lineages. These loci are: *Pax3* (*Splotch*, *Sp*) and *Sox10* (*Dominant megacolon*, *Dom*).

Mutations at the *Splotch* and *Dominant megacolon* loci are lethal in the homozygous condition. They also have pleiotropic effects that indicate the importance of these genes in other lineages. Heterozygous $Pax3^{Sp}$/+ or $Sox10^{Dom}$/+ mice have similar spotting phenotypes consisting of a large belly spot with well defined outlines. Shown in Figure 3.16 is a $Pax3^{Sp}$/+ mouse. *Pax3* and *Sox10* both encode transcriptional activators that act directly, as well as synergistically, with other cellular proteins, to regulate transcription of the major organizing gene of the pigment-cell lineage, *Mitf* (Bondurand et al. 2000; Chi & Epstein 2002). Also involved in these regulatory interactions is *Brn2* (not one of the classical murine pigment genes) (Cook & Sturm 2008). Physical interactions between BRN2 and PAX3, BRN2 and SOX10, and SOX10 and PAX3 have been reported (Smit et al. 2000). Additionally, the ability of SOX10 and PAX3 to mediate synergistic transcriptional activation of the MITF promoter is well established (Bondurand et al. 2000; Potterf et al. 2000).

Splotch (*Sp*, *Pax3*, *Paired box gene 3*)

JAX lists five spontaneous phenotypic alleles.

Availability: HAR, JAX, ORNL, and MMRRC (see Appendix in Chapter 1 for details about repositories).

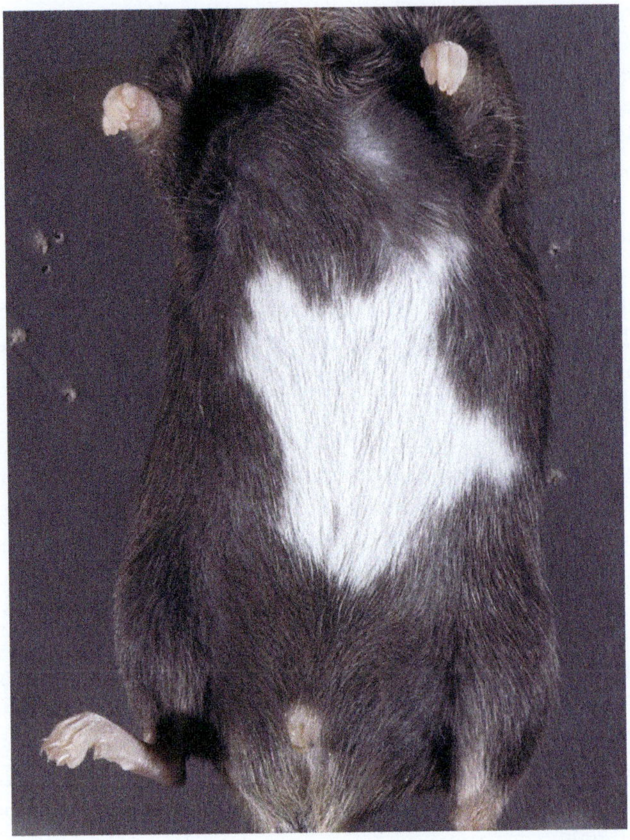

Figure 3.16 *Splotch* mouse, one example of the belly spot phenotype. The belly spot is a common phenotype, alone or in combination with other defects. When it is associated with a genetic defect that acts autonomously in the pigment cell, the belly spot suggests that numbers of melanocytes, at the end of their developmental pathway, are somewhat insufficient in quantity. This insufficiency might be the result of processes at any stage of development. We already know that PAX3 and SOX10 are required very early in development. The discrete nature of the spotting, with well-defined boundaries, suggests that PAX3 and SOX10 are not required for later steps in pigment-cell development (as discussed below relative to the second migratory phase of pigment-cell development). This phenotype is consistent with melanoblast deficiency at the time of specification or during the first migratory phase of development (Nishikawa et al. 1991).

In humans *PAX3* defects can cause Waardenburg syndrome type I, defined as defects in ear, eye, and facial development with white spotting, as well as spina bifida (Read & Newton 1997).

PAX3 defects are lethal in the homozygous condition, reflecting its important functions before and during neurulation: PAX3 is required for normal development of the neural tube, the melanocyte lineage, and the embryonic paraxial mesoderm. Heterozygous $Pax3^{Sp}/+$ mice exhibit white spotting and more or less severe neural-tube defects. However, they are well covered with pigment cells, depending somewhat on strain background, except for a large white belly spot and extension of the unpigmented areas of the extremities.

Pax3 influences melanocytic proliferation, resistance to apoptosis, migration, and lineage specificity and differentiation. It performs this multitude of tasks via multiple interactions with cofactors and other proteins (Kubic et al. 2008) that vary in different tissues and different stages of the life cycle.

Dominant megacolon (Dom, Sox10, SRY-related box 10)

JAX lists one spontaneous phenotypic allele.
Availability: CMMR, HAR, JAX, MMRRC, and ORNL.

SOX10 is another transcription factor which regulates genes including *Mitf*, *Kit*, and *Dct*, synergistically with PAX3 and/or other factors. In mice, normal SOX10 is required prior to (Box 3.1) and after specification in melanoblasts and the other neural-crest cell lineages (Southard-Smith et al. 1998; Potterf et al. 2001; Mollaaghababa & Pavan 2003). In all three lineages SOX10 interacts with the products of the *Piebald/Ednrb* and *Lethal spotting/Edn3* loci, which function later (days 10–12.5; Shin et al. 1999; Hakami et al. 2006) in the migration/replication of neural, glial, and pigment-cell lineages. Mutant phenotypes at any of these loci include white spotting and megacolon. Megacolon is a congenital failure of peristalsis in the distal gut that results from inadequate innervation and results in intestinal obstruction. Thus, all of these genes are required for normal development of pigmentary and autonomic nervous systems. *Ednrb* and *Edn3* are required during the first migratory phase of melanoblast development, and are therefore discussed below. Mutations of *Dom* are dominant. Heterozygous $Sox10^{Dom}/+$ mice usually die of megacolon before maturity. *Sox10* defects are lethal *in utero* in the homozygous condition.

SOX10 functions to maintain pluripotency of early migrating neural-crest cells, after they have left the immediate vicinity of the neural tube (Hosada et al. 1994; Kim et al. 2003; Lee et al. 2003; Wegner 2005). In *Xenopus*, SOX10 is required for the earliest development of neural crest; it is induced by both FGF and WNT signals (Aoki et al. 2003; Honoré et al. 2003) and is responsible for the upregulation of the endothelin receptor B in enteric neural-crest cells (Zhu et al. 2004). In mice, SOX10 is required prior to and after specification in all the truncal neural-crest cell lineages.

SOX10 has multiple functions in vertebrate development because of its ability to partner with different transcription factors (Southard-Smith et al. 1998; Potterf et al. 2001; Mollaaghababa & Pavan 2003). It is required during the first migratory stage of development of cells of all three neural-crest cell lineages; based on its phenotype, it may not be required after the early stages of melanoblast migration.

Human syndromes that involve SOX10 include Waardenburg syndrome type IV, which involves megacolon as well as the symptoms of Waardenburg syndrome type I, and Hirschsprung disease (aganglionic megacolon).

3.4.2 Genes required for the initial melanoblast migration

The first stage of melanoblast migration runs from the initial delamination of melanoblasts from the neural tube to the time that they move into the developing epidermis. Expression of KIT and MITF (the melanocyte isoform, MITF-M) is the earliest indicator of specification to the melanoblast lineage, as described above. Upon specification, a population of neural-crest cells expressing KIT and/or MITF become apparent mid-dorsally in the neural tube (Opdecamp et al. 1997; Nakayama et al. 1998; fig. 2 in Wilson et al. 2004). They delaminate from the neural tube and make their way to the MSA (Weston 1991). The MSA forms a narrow wedge between the neural tube, the overlying ectoderm, and the dermatome of the developing somite, as described in section 3.2. In mice, ventrally migrating cells specified to the neural and glial lineages are reported to emerge earlier, and from a more ventral part of the neural tube than those that will migrate dorsolaterally (Wilson et al. 2004). In the trunk, the dorsal neural tube is the earliest reported site of cells expressing KIT (Wilson et al. 2004) or MITF (Nakayama et al. 1998).

Melanoblasts may pause, while actively replicating, for up to half a day or so in the MSA (Wehrle-Haller & Weston 1995), after which they continue migrating dorsolaterally, over the lip of the

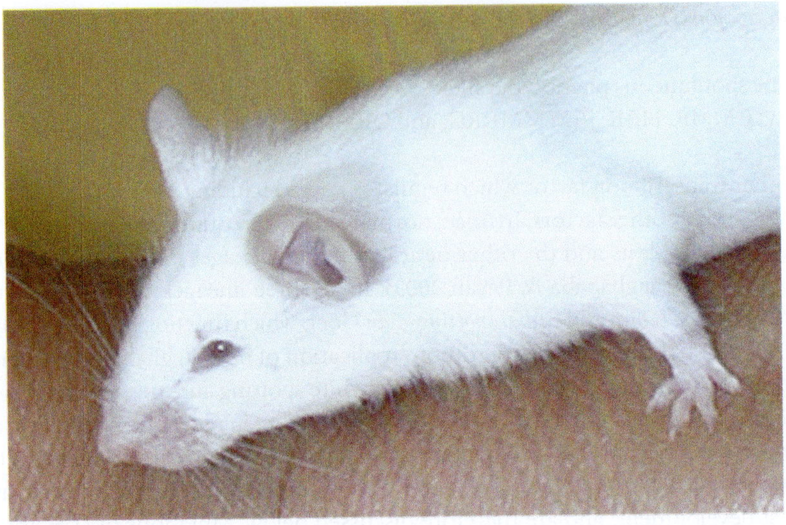

Figure 3.17 Microphthalmic mouse. This mouse is of the genotype C57BL/6J-*Mitf*$^{Mi\text{-}B}$/*Mitf*$^{Mi\text{-}B}$. This is the *Brownish* allele of the *Microphthalmia* (formerly *Mi*) locus, so named because the heterozygote is more pale than wild type, and appeared brownish on the agouti background upon which it was found. A nonagouti black mouse that is heterozygous at this locus is somewhat reduced in pigment intensity. This homozygous mutant mouse displays absence of pigment cells from the skin and hair as well as microphthalmia, illustrating the requirement of MITF for development of the eye and the pigment cells.

developing somite and along its surface toward the ventrum, again while replicating rapidly. For the melanoblasts to accomplish this initial migration successfully, they require five of the classical white-spotting loci. These are: *Mitf* (*Mi*, *Microphthalmia*, *Microphthalmia-associated transcription factor*; Fig. 3.17), *Kit* (*W*, *Dominant white spotting*), *Kitl* (formerly *Steel*, *Sl*), *Ednrb* (*Piebald*), and *Edn3* (*Lethal spotting*).

MITF is a transcription factor that plays a central regulatory role in pigment cells throughout their lifetime. MITF is also required for the development of mast cells (Morii et al. 1996), osteoclasts, and natural killer cells, and is required for normal development of the heart (Tshori et al. 2006). It is encoded at the *Mitf* locus. KIT is an RTK cell-surface receptor encoded at the *Kit* locus, and required for the development of melanoblasts during the migratory phases (Satomi Nishikawa et al. 1991), at least until they reach their destinations in the hair follicles and other locations, and subsequently during each hair cycle. KIT functions by binding KITL encoded at the *Kitl* locus. In mice, EDNRB functions in melanoblasts primarily by binding EDN3.

Microphthalmia (*Mi, Mitf, Microphthalmia-associated transcription factor*)

JAX lists 13 naturally occurring phenotypic alleles.
Availability: CMMR, JAX, MMRRC, ORNL, RBRC, and TIGM.

MITF defines a discrete transcription factor family (Hemesath et al. 1994; Hershey & Fisher 2004). The broad impact of MITF throughout melanoblast development means that phenotypes of mice mutant at *Mitf* vary greatly, depending upon the mutant allele, the effect of the lesion upon the function of the MITF molecule, and the stage(s) of melanoblast development being impacted (Steingrímsson 2008; Steingrímsson et al. 1994, 2003, 2004; Moore 1995). When we created mice heterozygous for $Mitf^{mi\text{-}sp}$ with other *Mitf* mutant alleles, we observed that the white-spotting and fading-with-age phenotypes fall into classes that may prove useful in evaluating the various functions of MITF in relation to different developmental times. For example, mice heterozygous for null alleles, combined with $Mitf^{mi\text{-}sp}$, all appear very similar. Phenotypes of mice mutant at a sampling of these alleles are represented in Figures 3.18 and 3.19. A more extensive survey of phenotypes and photographs of mice mutant at *Mitf* is given in Steingrímsson et al. (2003), where interallelic complementation with Mi^{wh} is discussed. Many questions about this fascinating gene locus remain to be answered, and most of the mice are available, congenic with C57BL/6J, at MMRRC (see Appendix in Chapter 1 for details about repositories).

MITF-M, the pigment-cell-specific spliced form of MITF, functions in the nucleus of the pigment cell to respond to environmental cues imposed by the signaling molecules, and to activate pigment-cell-specific genes during cell differentiation. MITF-M is necessary for pigment-cell survival at all developmental stages. Other isoforms of MITF function in several other cell types; they have alternative first exons (Steingrímsson 2008), are transcribed from their own specific promoters, and activate other genes appropriate to those cell types (Udono et al. 2000; Shibahara et al. 2001; Steingrímsson et al. 2004; Levy et al. 2006).

MITF-M binds to a set of DNA sequences known as M-boxes (or as a subset of E-boxes) in gene promoters, comprising (T)CATGTG(A), sometimes with additional conserved flanking sequences. The genes known to be activated by MITF-M have been reviewed by Vance and Goding (2004),

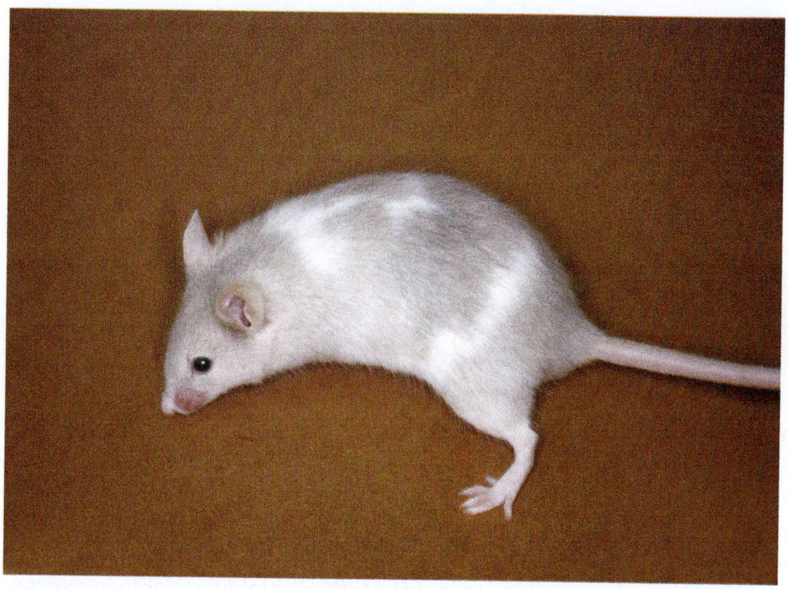

Figure 3.18 White-spotted mouse C57BL/6J-$Mitf^{Mi\text{-}wh}/Mitf^{Mi\text{-}sp}$. Note the evidence of eumelanism dorsally. This tends to become somewhat more pronounced in older mice of this genotype. Another very similar phenotype is seen in mice of the genotype $Mitf^{Mi\text{-}wh}/Mitf^{mi}$, but in that case the dorsal tendency to eumelanism becomes less pronounced with age.

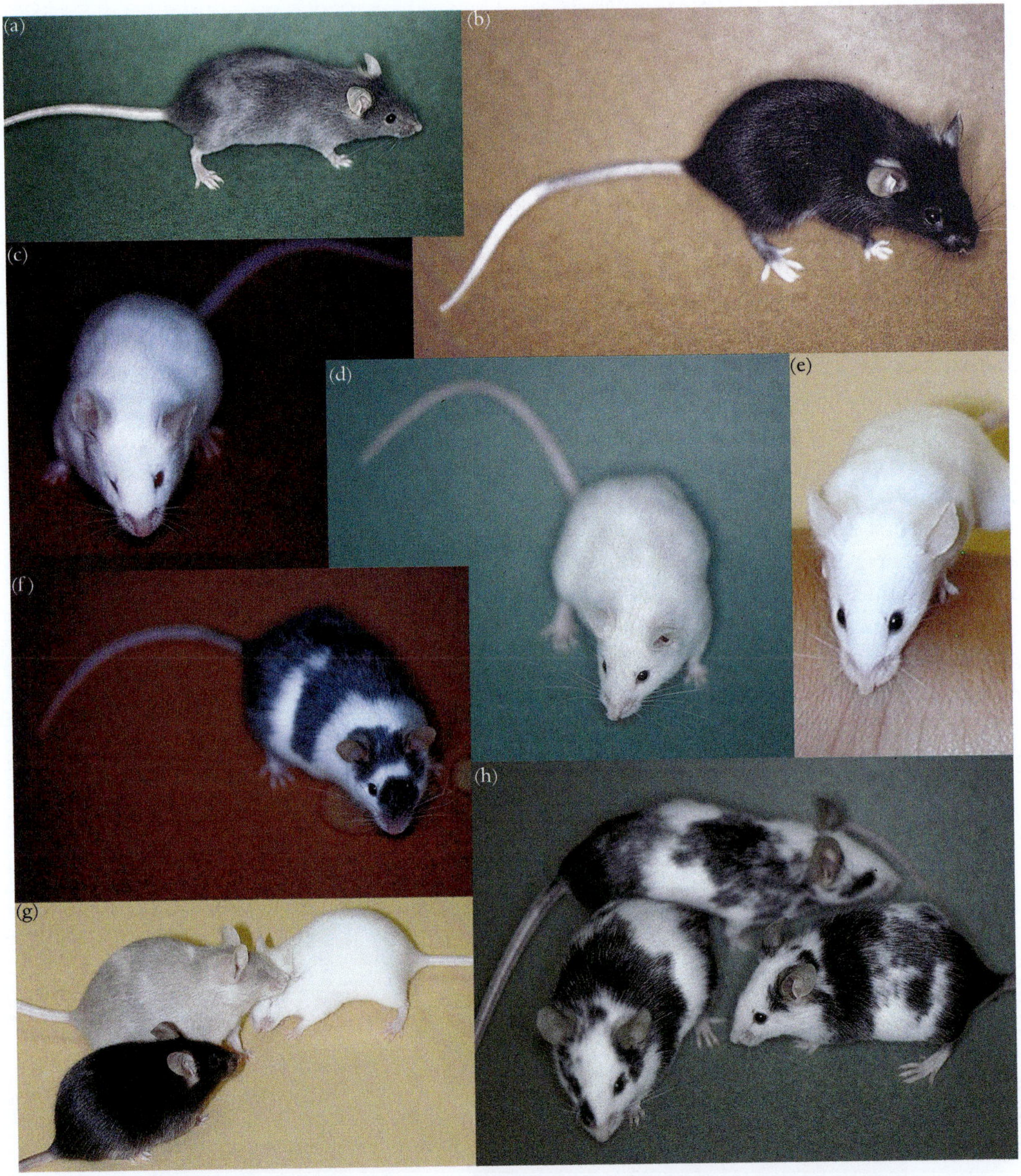

Figure 3.19 The phenotypes of mice that are heterozygous for mutations of *Mitf* are likely to be informative of the varied functions associated with the locus. These are all of the same inbred strain, C57BL/6J. Unless otherwise stated, mutant alleles are in heterozygous combinations with $Mitf^{mi\text{-}sp}$, which when homozygous gives a black mouse appearing similar to wild type. (a) $Mitf^{Mi\text{-}B}$. (b) $Mitf^{mi\text{-}rw}$. (c) $Mitf^{mi\text{-}rw}/Mitf^{mi\text{-}vga9}$, showing a bilateral difference in expression of microphthalmia. (d) $Mitf^{Mi\text{-}Or}$, highly variable in expression of microphthalmia. (e) $Mitf^{mi\text{-}bw}/Mitf^{mi\text{-}bw}$ (black-eyed white). (f) $Mitf^{mi\text{-}vit}/Mitf^{mi\text{-}ew}$. (g) The black mouse is homozygous for $Mitf^{mi\text{-}sp}$, the white, microphthalmic mouse is homozygous for $Mitf^{Mi\text{-}wh}$, and the third mouse is heterozygous for both: in this case microphthalmic in one eye but usually the eyes are grossly normal. (h) The spotted phenotype of $Mitf^{mi\text{-}sp}$ in combination with null alleles, $Mitf^{mi}$, $Mitf^{mi\text{-}ew}$, or $Mitf^{mi\text{-}vga9}$.

Steingrímsson et al. (2004), and Levy et al. (2006). Together with some more recent additions (Fig. 3.9), they include:

- genes involved in melanogenesis – *Dct, Mlana (Mart1), Oca2 (P), Slc45a2 (Aim1, Matp, uw), Si, Tyr,* and *Tyrp1* – and in activating melanogenesis, *Mc1r*;
- genes involved in cell proliferation and motility: *Cdk2, Kit, Met, Snai2/Slug,* and *Tbx2*;
- genes involved in cell survival: *Bcl2* and *Hif1a*;
- genes involved in inhibiting cell proliferation (perhaps only at high levels of MITF, as these are not normally expressed by melanocytes): *Cdkn1a/p21* and *Cdkn2a/p16*.

There are clearly further levels of complexity involved, for example, in generating the different levels of gene expression seen in melanoblasts and melanocytes. Thus, MITF-M is sensitive to many inputs and regulatory controls (Watanabe et al. 1998, 2002; Wu et al. 1999b; Goding 2000; Vance & Goding 2004), including importantly the products of two other classical white-spotting genes, *Pax3* and *Sox10*. Both encode activating transcription factors that were described above as being necessary for specification. They act directly and synergistically with other cellular proteins, to regulate transcription of *Mitf* (Bondurand et al. 2000). The other transcription factors implicated to date in regulation of the MITF-M promoter include CREB, LEF1, and ONECUT2 (Fig. 3.9), with each factor conferring on this promoter the capacity to respond to specific signals. β-Catenin and LEF1 function in the WNT signaling pathway (section 3.2), and both have been shown to impact pigmentation *in vitro* and in genetically manipulated mice (Saito et al. 2003). CREB is a transcription factor that is responsive to cAMP. Pigment cells are sensitive to cAMP level, which (with other factors) is involved in regulating both melanocyte differentiation (Bennett 1989; Buscá & Ballotti 2000) and the pheomelanin/eumelanin switch mechanism (Chapter 5).

MITF is the central organizing molecule of the pigment cell, the 'hub' of the network of melanogenic functions, if you will, and it functions similarly in some other tissues – eye, ear, and heart – using alternative promoters (Steingrímsson 2008). As a transcription factor that interacts differently with multiple other proteins and is required for pigmentation, MITF presumably has different functions during specification, the first migratory phase, the second migratory phase of pigment-cell development, and in the maintenance of stem cells and their cyclic generation of melanocytes in the hair follicle.

We are just beginning to understand in what ways MITF may be responsible for directing or coordinating development of the pigment cell (Steingrímsson et al. 2004). The multiplicity of different phenotypes resulting from mutation at *Mitf* presumably reflects its multiple functions (Steingrímsson et al. 2003) and ranges from absence of pigment cells because of failure at the earliest stages of melanoblast development to, at the end of development, graying with age that reflects defective repopulation of hair follicles following the molt cycle.

3.4.3 Primary controls over the initial melanoblast migration: what does white spotting tell us?

Before continuing the discussion of genetic mutations that interfere with normal development of neural-crest cells, it is important to emphasize their normal behavior (that is, *in vivo*). Normal neural-crest cells have a very great capacity to replicate and migrate away from each other in a permissive environment (in melanoblasts, such an environment would include KITL) unless they are inhibited by contact with other neural-crest cells. In *Xenopus*, this capacity is expressed by the cells on the periphery but not those that are surrounded by other neural-crest cells, and it requires WNT signaling (Carmona-Fontaine et al. 2008). This normal capacity is required to explain the normal and abnormal pigmentary phenotypes of mice that are generated in their first and second migratory expansions.

In the mouse (Wehrle-Haller & Weston 1995), but not in chick (Harris & Erickson 2007), the relationship between KIT expression on the melanoblast and KITL expression in the initial migratory pathway is essential to survival, replication, and migration of melanoblasts from the melanoblast staging area on to and along the dorsolateral pathway (Wehrle-Haller & Weston 1995; Wehrle-Haller 2003). Soluble KITL is required for replication of the specified cells as they make their way through the MSA (Box 3.1e). Then (in mice) the availability of KITL produced on the cellular substrate through which the melanoblasts must migrate determines the tissues available to melanoblast occupation and the migratory pathway of the pigment-cell lineage (Wehrle-Haller & Weston 1995). During the first migratory event EDNRB and its ligand EDN3 are also required.

KIT and KITL are required for melanoblast survival and replication during migration, until the cells arrive at the hair follicle (Yoshida et al. 1996). The phenotypes of mice mutant at *Kit* or *Kitl* are especially useful for visualizing the initial migratory events, because the effects on these early steps are not obscured by normal function at later stages in mice mutant at these loci, as will be shown.

Kit (*c-Kit, W, Dominant white spotting*)

JAX lists 66 spontaneous phenotypic alleles.
Availability: CMMR, HAR, JAX, MMRC, ORNL, RBRC, and TIGM.

KIT (encoded at the *Kit* locus, formerly *W* or *Dominant white spotting*) is an RTK cell-surface receptor that floats in the lipid surface membrane of the pigment cell. KIT functions as a dimer when it binds its ligand. Its ligand is KITL (encoded at *Kitl, Kit ligand*, formerly the *Steel, Sl* locus: see next section). The binding of KITL to KIT activates replication, migration, and survival of the melanoblast and melanocyte.

In mice with mutations at *Kit* the pigment cells are unable to respond normally to KITL. Therefore, the phenotype of homozygous *Kit* or *Kitl* mutant mice is usually black-eyed white with deafness, because the neural-crest-derived melanoblasts fail to survive beyond the earliest stages in the MSA (Wehrle-Haller & Weston 1995). By contrast, the RPE of the eyes is not derived from neural crest, and its cells are not KITL-dependent; therefore the eyes are pigmented.

In heterozygotes, however, where a mutant protein is present as in the $Kit^{W\text{-}2J}/+$ mouse (Fig. 3.20), and because the KIT receptor functions as a dimer, 25% of receptors will be normal dimers, permitting survival and migration of some of the melanoblasts. Null alleles such as Kit^{W} give a milder phenotype when heterozygous, because there is no abnormal protein product. Thus although only half the normal amount of KIT protein is produced, all of it is capable of forming normal dimers. The phenotypes of mice mutant at *Kit* reflect these differences by the amount of pigmentation coverage over the surface of the body.

The heterozygous mice in Figures 3.20 and 3.31 (see below) exhibit the same type of ragged-edged spotting caused by the same mutant allele, with more or less white and pigmented areas resulting from differences in the background genomes. This distinctive phenotype differs from the phenotypes of mice that are capable of a normal secondary migration event (for example, Fig. 3.24, see below) and is characteristic of mice mutant at the structural portion of *Kit*, if they are pigmented.

Exceptions to this phenotype are caused by several phenotypic mutations (*Patch, Rumpwhite, Sash*, and *Banded* are the classical examples) that impact the regulatory region of the *Kit* locus. These are discussed below.

Figure 3.20 C57BL/6J *Kit*$^{W\text{-}2J}$/+ mouse.

While KIT and its ligand may not be required for the entire duration of melanoblast development, they appear to be required for survival, migration, and proliferation of melanoblasts in the dermal layer during the first migratory phase (Matsui et al. 1990; Murphy et al. 1992; Steel et al. 1992; Morrison-Graham & Weston 1993; Reid et al. 1995; Wehrle-Haller & Weston 1995; Langtimm-Sedlak et al. 1996; Cable et al. 1995), and for their migration into the epidermis and thence to the hair follicles (Yoshida et al. 1996) (Box 3.1e–g); in other words, during the times when the melanoblasts are migrating and replicating rapidly.

With regard to migration, it is also interesting to note that yellow *Kit*$^{W\text{-}2J}$/+ mice are almost completely pigmented, with only a small belly spot and without the ragged-edged phenotype of the spotting (Lamoreux & Russell 1979) compared with congenic nonyellow littermates. When the pheomelanic phenotype is caused by mutation at the *Mc1r* locus (formerly known as the *Extension* or the *Recessive yellow* locus) the reduction of the white spotted area is somewhat less than in pheomelanic mice of the genotype A^y/a, mutant at the *Agouti* locus (M.L. Lamoreux, personal observation). Thus mice of the genotype C57BL/6J-A^y/a *Kit*$^{W\text{-}2J}$/+ have only a small belly spot, compared with the C57BL/6J-*a*/*a* *Kit*$^{W\text{-}2J}$/+ mouse shown in Figure 3.20. Because these mice are not pigmented during their embryonic development, this suggests that some factor of the pathway involved in generating yellowness provides a survival value for cells that are mutant at *Kit*.

Pigment-cell survival is also influenced by location on the body of the mouse. For example, the heads and rumps of the *Kit*$^{W\text{-}2J}$/+ mice in Figures 3.20 and 3.31 are more pigmented than the middle. This is characteristic of C57BL/6 mice, and will be further discussed below. The significance here is to point out that the normal survival of the pigment cell does require a normal

Kit genotype, but also requires other factors that depend on the location of the cell on the mouse. That at least some of these are factors of the environment, rather than the pigment cell itself or the *Kit* genotype, is shown by the fact that they vary with strain background (Lamoreux 1999).

KIT is also required for development of mast cells, red blood cells, germ cells, interstitial cells of Cajal, and others. Thus, the pleiotropic effects of *Kit* mutants have informed all of these systems (Geissler et al. 1981; Geissler & Russell 1983a, 1983b; Yasuda et al. 1993; Galli et al. 1993; 1994; Kimura et al. 2004).

And once again, because the patterns of migration of pigment cells from their origin in the neural crest to their destination in the hair follicle are primarily determined by KIT and its ligand, mutants of *Kit* and manipulations of KIT in mice have been used to analyze these patterns. Nishikawa, Yoshida and coworkers (Nishikawa et al. 1991; Yoshida et al. 1996) injected an antibody, anti-c-KIT (ACK), which inhibits KIT, into pregnant mice with normal KITL (thus normal areas that are available to pigment-cell immigration), Based on distribution of pigmentation in the offspring, they described three primary phases of melanoblast replication/migration that require functional KIT during pigment-cell development. The phases have been used as divisions in this chapter. They are as follows.

1 An initial migratory stage of specified melanoblasts on to the dorsolateral migratory pathway and their replication/migration on and in the dermatomes, which express KITL while breaking up into dermis (Box 3.1e & f).
2 A secondary migratory phase of replication/migration after the melanoblasts arrive in the epidermis (Box 3.1g). Beginning on about embryonic day 14.5, expression of KITL in the dermis ceases, expression of KITL is initiated in the basal layer of the epidermis, and the melanoblasts follow the KITL into the basal layer of the epidermis (Wehrle-Haller 2003). In the epidermis, the melanoblasts undergo a second phase of KIT-dependent replication (Yoshida et al. 1996). In a normal embryo at 14.5 dpc, the secondary migration results in the distribution of melanoblasts 'evenly over the entire body surface except for the umbilical, digital and tail tip regions' (Yoshida et al. 1996). (This distribution of pigmentation is typical of normal newborn mice of the C57BL/6J strain and related strains.)
3 Thirdly is the subsequent movement of melanoblasts into the hair follicles.

Steel (Sl, Kitl, Kit ligand, SLF, Steel factor; MGF, Mast cell growth factor; SCF, Stem cell factor)

JAX lists 18 spontaneous alleles.
Availability: HAR, JAX, NMICE, MMRRC, ORNL, RBRC, and TIGM.

The *Kitl* locus encodes KITL, a transmembrane growth factor which is translated from two alternatively spliced transcripts. The larger form contains exon 6, which codes for an extracellular protease cleavage site. Cell-surface proteases readily cut the protein at this site, releasing a soluble form of KITL into the extracellular fluids. The smaller splice variant does not contain exon 6 and so remains intact in the cell membrane (Wehrle-Haller & Weston 1995).

KITL functions on cell surfaces in the tissues through which the melanoblasts must migrate and in the extracellular fluid in the case of soluble KITL. Although other modifying factors influence the pattern of migration of the melanoblasts, KITL is the primary determinant of the locations that are available to occupation by normal melanoblasts. Normal KITL defines the location, and of course interacts with other factors of the tissue environment, some of which are not white-spotting gene loci (Wehrle-Haller & Weston 1995; Lamoreux 1999; Rhim et al. 2000). The ability of the melanoblasts to respond to KITL and survive in those locations is determined at the *Kit* locus and other white-spotting loci that are autonomous to the melanoblast.

The earliest normal melanoblasts delaminate into the MSA, as described above, expressing the cell-surface receptor KIT. These earliest specified melanoblasts respond to the presence of soluble KITL, released from the cells of the nearby dermatome (Box 3.1b). They are thus able to survive and replicate in the MSA and move along the concentration gradient of KITL toward the lateral migratory pathway. Figure 2 in Wehrle-Haller and Weston (1995) suggests that, by the time the melanoblasts leave the MSA, they have replicated sufficiently to fully occupy the MSA.

The availability of soluble KITL is sufficient to enable replication and survival of the initial melanoblasts, and their progression from the MSA to the lateral migratory pathway. For their continued migration through the tissues, as well as survival and replication, the melanoblasts require the membrane-bound form of KITL on the cells of the substrate since this stage of migration fails in $Kitl^{Sl\text{-}d}/Kitl^{Sl\text{-}d}$ mice, which lack transmembrane KITL (Wehrle-Haller & Weston 1995). KIT and KITL are also required for development of other systems in the mouse, including, significantly, hemopoietic and reproductive systems. Mice lacking either protein often die of anemia shortly after birth. Mice that lack functional KITL are black-eyed white in phenotype, phenocopies of mice that lack KIT, and for similar reasons.

The pigment phenotypes of mice that are heterozygous for mutations at *Kitl* differ dramatically from the phenotypes of Kit^{W} mutant heterozygotes (compare Figs 3.20 and 3.21), and the difference reflects the tissues in which the proteins are expressed. KIT is on the migrating cells while KITL expression is required on and around the cells of the tissues through which the melanoblasts migrate (Mayer & Green 1968; Mayer 1973a, 1979; Zsebo et al. 1990; Rottapel et al. 1991; Brannan et al. 1991; Bedell et al. 1996; Mizoguchi 2004).

Figure 3.21 Steel mice, JU/CtLm-$Kitl^{Sl\text{-}d}/+$, flanking a normal control JU/CtLm-+/+ mouse. Heterozygous mice are very pale ventrally.

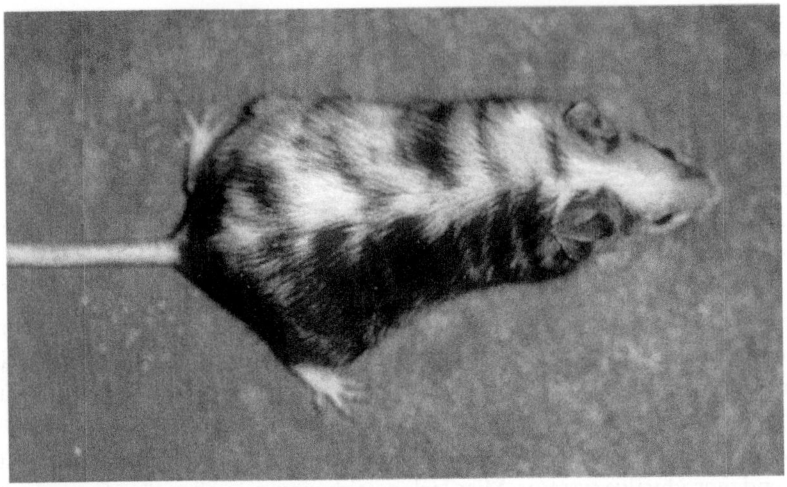

Figure 3.22 VIB (Very Important Baby); parents were both heterozygous *Steel-Dickie*, C57BL/6J-$Kitl^{Sl-d}$/+, as in Figure 3.21.

Because KITL functions as a dimer, some normal growth-factor dimers are available in Steel heterozygotes to support melanoblast development, but a finely scattered array of areas lack melanoblasts, resulting in the overall grayish phenotype seen in Figure 3.21. Illustrative of this phenotype is the apparently unique mouse illustrated in Figure 3.22, which we considered a very important baby mouse, hence his name, VIB. This mouse was born to $Kitl^{Sl\text{-}d}$/+ parents but appeared to be a chimera, having $Kitl^{Sl\text{-}d}$/$Kitl^{Sl\text{-}d}$ stripes which were white and +/+ (or perhaps $Kitl^{Sl\text{-}d}$/+) stripes which were pigmented (M.L. Lamoreux, personal observation). To compare with this anecdotal data is the similarly unique dog pictured in Fig. 3.25, see below). The dog is a presumed chimeric of a phenotype that is autonomous to the pigment cells, while the mouse is a presumed chimeric of a pigment phenotype that is imposed by the tissues surrounding the pigment cell.

$Kitl^{Sl}$/+ mice often have white belly spots, which we have interpreted as an indication of insufficient numbers of migrating melanoblasts (Fig. 3.16) or subsequent death. In addition to the variable belly spot, the $Kitl^{Sl}$/+ mouse always has a very pale ventrum. The demarcation between the darker dorsal and lighter ventral pigmentation is similar in its location to that seen in black-and-tan mice, which are discussed in Chapter 5.

Piebald (*S, Ednrb, Endothelin receptor type B*) and *Lethal spotting* (*Ls, Edn3, Endothelin 3*)

Piebald: JAX lists three spontaneous phenotypic alleles. One of these alleles is known as *Piebald lethal*, not to be confused with *Lethal spotting*.
Availability: ORNL, CARD, JAX, MMRRC, and RBRC.
Lethal spotting: JAX lists one spontaneous phenotypic allele.
Availability: MMRRC and JAX.

In humans, mutations in either of these genes result in Waardenburg–Shah-type syndrome, characterized by pigmentation defects, deafness, and megacolon. All of these symptoms reflect failure of migration of neural-crest cells. *Piebald* and *Lethal spotting* are distinct white-spotting loci (Fig. 3.23). *Piebald lethal*, though sometimes confused with $Edn3^{ls}$ (*Lethal spotting*), is actually an allele at the

Figure 3.23 White-spotted pony, bay in color. The important difference between this pony and the chimeric animals to be discussed in section 3.4.4 is that the chimeric animals (unless of course they are *also* white-spotted) are fully pigmented. The white areas on a white-spotted animal represent areas that have no pigment cells and therefore are not pigmented. Wild type in the horse as in most other animals involves both eumelanin and pheomelanin, distributed over the body or along the shafts of individual hairs in species-specific patterns that are genetically controlled (discussed in Chapter 5). In the horse, wild type is bay, including variations that are sometimes known as brown if they are darker than this pony or blood bay or bright bay if the body is less eumelanic. Bay equines have black mane, tail and points (lower legs) unless other modifying genes are present. This pony also has white spotting, which is not related to the color of the pony.

Piebald (*Ednrb*) locus. Piebald encodes a G-protein-coupled cell-surface receptor, EDNRB, which is capable of binding more than one ligand, including endothelins 1 and 3 (EDN1, EDN3) (Hosada et al. 1994), and in the mouse interacts with EDN1 in development of the head (see section 3.5). However, in murine development of truncal pigmentation, and of the enteric nervous system, EDN3 is the primary ligand of EDNRB (Baynash et al. 1994; Pla & Larue 2003).

EDN3 is an extracellular ligand of EDNRB, encoded at the *Lethal spotting* locus of mice and expressed in tissues surrounding the neural-crest cells (see section 3.3). The binding of EDN3 to its receptor EDNRB is required for the development of both the neural-crest-derived neuronal and the pigment-cell lineages (Mayer 1965; Pavan & Tilghman 1994; Shin et al. 1997) in mouse embryos from 10 dpc until 12.5 dpc. It is not required for the second stage of migration and rapid replication in the truncal epidermis (Shin et al. 1999), but may be necessary for differentiation of melanocytes (Hou et al. 2004). EDN3 is apparently not expressed in the head region of mice, where the ligand for EDNRB is EDN1, an observation that probably explains some pigmentary differences between the head and trunk. From 10 to 12.5 dpc, the melanoblasts and the enteric neuroblasts are migrating and replicating. The similarities and differences of the two processes are interesting. At the level of cellular messaging and at the level of tissue interactions the major signals are similar, with defining exceptions, and illustrate once again a situation in which a pigmentary mutation led to our greater understanding of other systems of the body.

Briefly, in the trunk, neural-crest-derived glial and neuronal precursor cells populate the developing colon primarily from the anterior end but to some extent from the posterior end forward, by a complex process of replication and migration, associated with contact inhibition. This is coordinated by EDN3, EDNRB, and SOX10 (discussed above, encoded at the *Dom* locus). The probable function of the EDNRB / EDN3 coupling is to maintain a sufficient pool of undifferentiated migrating cells to accomplish these tasks (Hearn et al. 1998; Wu et al. 1999a). Both SOX10 and EDNRB are expressed in the migrating enteric nervous system cell (Zhu et al. 2004). An enhancer of *Ednrb* is activated in the enteric-nervous-system precursors as they approach the developing colon. The enhancer contains binding sites for SOX10 (Zhu et al. 2004). Failure of this process explains the megacolon of piebald or lethal spotted mice, as well as $Sox10^{Dom}/+$ mice, as discussed above.

At the same developmental time, melanoblasts are leaving the MSA, entering their own migratory pathway, and migrating ventrad to complete their first migratory event.

Failure to complete this migration results in the characteristic white spotting of *Piebald* or *Lethal spotting* mice, which are very similar to each other. Figure 3.24 shows a *Lethal spotting* mouse with

Figure 3.24 *Lethal spotting* mouse showing white spotting and megacolon phenotypes, C57BL/6J-$Edn3^{ls}/Edn3^{ls}$. Phenotypes of *Piebald* mice are very similar to those of *Lethal spotting* mice. In both cases, transfer to a JU background significantly alters the location of spotting toward the anterior of the mouse, and the variability of phenotypes produced in the process of the backcrossing suggest that both *Lethal spotting* and *Piebald* interact with more than a few 'modifying genes' of the background genome (L. Lamoreux, personal observation). Primary among them may be *Kitl* (*Steel*) (Rhim et al. 2000).

white spotting and megacolon. The unpigmented areas are consistently located in the trunk region of the mice; larger or smaller depending upon the mutant allele and located more or less anteriorly depending upon other modifying loci (Lamoreux 1999; Rhim et al. 2000). The spots are discrete, with well-defined boundaries, quite different from the quality of spotting in mice mutant at *Kit* (compare with Fig. 3.20), and are usually more extensive than those of *Splotch* or *Dominant megacolon* heterozygotes, although with the same discrete character of their boundaries.

We propose that the sharp outlines to the white spotting are created by clonal expansion during the second migration/replication event, after the cells have arrived in the epidermis where EDNRB or EDN3 are not required for their further development.

Summary of what white-spotting phenotypes have told us about the first migratory phase of pigment-cell development

Phenotypes and gene functions of mice mutant at the major white-spotting loci discussed above have informed us that initiation of the pigment-cell lineage is very complex, involving two classical white-spotting gene loci [*Splotch* (*Pax3*) and *Dom* (*Sox10*)] and many other loci to specify the melanoblast lineage, which is defined by continuing expression of KIT and MITF. From that point, KIT and KITL are required to direct the melanoblasts through every phase of replication/migration, while EDNRB and EDN3 are required during the first major replication/migration event. During the first migratory event, the melanoblasts replicate rapidly while migrating from the MSA, through the dermatome and dermis toward the ventrum of the mouse. Further information about this first phase of melanoblast development, from specification until the end of the dorsoventral migration, has been elucidated using chimeric mice that are fully pigmented and chimeric mice that are white spotted, as discussed in the next section.

3.4.4 Chimeric phenotypes of fully pigmented mice inform the normal processes of specification and the first phase of melanoblast migration

The all-white phenotypes of mice that are homozygous for mutations of *Mitf*, *Kit*, or *Kitl* tell us what happens when specification fails or the specified cells fail to survive; that is, there are no pigment cells. However, because of the long migratory pathway of pigment cells subsequent to specification, the information gained from phenotypes representing early melanoblast death may not be directly relevant to normal specification and early migration. For more information on that subject we turn to a phenotype of mice and other mammals whose melanoblasts survive in normal numbers. They have no white spotting, but their melanoblast lineages are formed from chimeric tissues of contrasting colors. We use the word chimera here not only for an animal comprising tissues of two or more different genotypes, but also for an animal expressing more than one color phenotype from the same genotype; for example, following X inactivation, as will be explained.

Tortoiseshell phenotype

One type of chimeric phenotype is known as 'tortoiseshell,' a term taken from the tortoiseshell cat which is chimeric as a result of a cell-autonomous X-linked gene that determines whether pigment cells will produce yellow or nonyellow pigment (Lyon 1961). More broadly, 'tortoiseshell' will refer here to any phenotype of melanoblast-autonomous pigmentary chimerism in the absence of white spotting. The phenotype itself may result from any process that generates chimerism of viable melanoblasts. These processes include X-linkage, as in the tortoiseshell cat (Fig. 3.26) and Cattanach's translocation in the mouse (Fig. 3.28, Fig. 3.29). In females, one X chromosome is

Figure 3.25 This highly unusual Labrador retriever sired yellow pups and nonyellow pups but no chimeric pups, as would be expected if he is the result of an accidental embryo fusion event (Sponenberg & Bigelow 1987). Photograph courtesy Philip Sponenberg.

inactivated randomly in each cell in the early embryo, in the mouse at the time of implantation; the cell's progeny then all retain that specific inactivation, forming a contiguous patch once cell mixing stops in a given tissue. The processes also include embryo aggregation, natural or artificial, as represented by allophenic mice (see below), and as we guess to be represented by the dog in Figure 3.25; early somatic reversion, as in the *Pink-eyed unstable* mouse (Whitney & Lamoreux 1982) (Fig. 3.29); autosomal genotypes such as *Japanese* in the rabbit (Fig. 3.27); and the tortoiseshell guinea pig (Fig. 3.14); and radiation-induced chimerism (Russell 1978).

Allophenic is the term used by Beatrice Mintz (1967, 1971a, 1971b) to refer to mice that are produced by mixing together genetically different morula-stage embryos and allowing them to complete development as one animal. Other methods of artificially creating chimeric mice produce similar phenotypic results (for example, Jaenisch 1985; Huszar et al. 1991). The result, if the genetic alternatives regulate a pigmentary phenotype that is autonomous to the pigment cell (usually alleles of melanogenic enzymes such as TYR or tyrosinase-related protein 1 (TYRP1) as discussed in Chapter 4, or *Recessive yellow*/MC1R as discussed in Chapter 5), is chimeric mice of a 'tortoiseshell' phenotypic pattern.

One of the advantages of the pigmentation model for study of gene functions and relationships is that the phenotype is visually evident and so can be easily used to generate and evaluate hypotheses relative to genetic functions. Tortoiseshell is a perfect example. The tortoiseshell cat is a well-established chimera, as are the artificially created tortoiseshell animals.

Tortoiseshell phenotypes played a significant role in interpreting chimerism (allophenism) in the theories of Mintz (1967, 1971a, 1971b) regarding clonal development of the melanoblast lineage. The core element of her theory of pigment-cell development is that the earliest specified melanoblasts give rise to clonal descendants that migrate laterally, while replicating, to generate normal mammalian pigmentation. The fact that such chimeras can be created so that they express two different genotypes of cell-autonomous pigment colors permits us to visualize the end result of the early melanoblast migration. This theory has continued to inform our studies of melanogenesis and

Figure 3.26 This is a tortoiseshell cat. Tortoiseshell cats are chimeric because of heterozygosity at a gene located on the X chromosome that determines whether pigmentation will be pheomelanic or normal. Therefore tortoiseshell cats are always females (except for the occasional XXY male, in which one of the X chromosomes is inactivated in a fashion similar to what happens normally in females). The X-linked yellow/nonyellow gene in the cat is known as *Orange* (the orange, or yellow, phenotype in the cat is commonly known as red, or ginger). The genotype of this cat is O/O^{+}.

appears to be accurate in its essence. The lateral migration has been confirmed as discussed above. Thus the tortoiseshell phenotype, as defined herein, is a pigment pattern that reflects the clonal descent and the first migratory pathway of early melanoblasts in the absence of white spotting.

The necessary conditions for a tortoiseshell phenotype are three, as follows.

1 Chimerism for a cell-autonomous trait that affects pigment color. This chimerism presumably does not change the developmental processes, but conveniently visualizes them via the color of the pigment cells.
2 Clonal origin of the genetically different populations of cells from which melanoblasts arise. The coincidence of phenotype with other sorts of evidence discussed in sections 3.2 and 3.3 and summarized above supports a clonal origin of the cells that give rise to the pigmentary lineage. Of course, all cells of an embryo arise from a clonal origin, but there is a question about when these specific clones originate. Based upon Lyonization and embryo aggregation, it could be as early as the eight-cell stage. Based upon the phenotypes of pigmentary chimeric mammals it is not later than melanoblast specification. Mintz's hypothesis of clonal origin of melanoblasts suggests that the clones arise as primordial melanoblasts that are rather widely spaced along each side of the neural tube. Alternatively, the clonal event need not coincide with specification of the pigment-cell lineage. The actual clonal segregation might occur earlier in the ectoderm or the developing neural tube (Bennett 1993), so long as the cells are not mixing randomly after that time. Observations in cell culture (see review by Bennett 1993) and in embryos support early plasticity of neural-crest cells. Goding (2000) proposed that the initial specification might be a stochastic event, which could account for widely spaced specification of melanoblasts. There may also be a component of migration of the earliest specified cells in the MSA (similar to that involved with

Figure 3.27 Bonnie Bunny is a yellow and black tortoiseshell *Japanese* color variant of rabbit. As described with Figure 3.14, autosomal chimerism also occurs in the guinea pig. It has been speculated that these autosomal chimerisms involve gene duplication.

Figure 3.28 A female mouse carrying Cattanach's translocation on one X chromosome. The translocation contains the wild-type (*P*) allele at the *Pink-eyed dilution* (*Oca2*) locus. The mouse is autosomally *p*/*p*. The result of heterozygosity at the X chromosome results in Lyonization and a mouse that is chimeric for pink-eyed dilution pigmentation alternating with wild type (wild type is not pink-eyed dilution, which on this background is nonagouti black). The normal pink-eyed dilution mouse phenotype is shown in Figure 3.29. Note the symmetry of markings is not normal but is a coincidence of development in this individual mouse.

Figure 3.29 Two mice of the genotype p^{un}/p^{un} (*Pink-eyed unstable*, new name *Oca2*$^{p\text{-}un}$). The mouse in the background is normally pigmented for mice that are mutant at the *P* locus. The mouse in the foreground is chimeric because of an early somatic reversion event. More commonly, the reversion events occur later in development and result in a small spot or short stripe or half a stripe. These are very informative with regard to the migratory patterns of pigment cells in mice that are fully pigmented.

innervation of the developing gut) as suggested by the work of Huzsar et al. (1991), and also Wilson et al. (2004) and Jordan and Jackson (2000). However, it is not necessary that the specification event and the clonal event coincide. The clonal derivation might occur before melanoblast specification, for example in the ectodermal tissues that give rise to neural crest, if that is the time when clones of different genotypes or 'epigenotypes' stop mixing. In that case it would not be necessary to invoke any process of individual selection of specified cells; melanoblasts might already be within clonal patches when first specified mid-dorsally before proceeding together to the MSA and thence on to the first migratory pathway. This possibility is supported by figure 5 of Wilson et al. (2004). In either case, specification does generate a clone or subclone of genetically uniform melanoblasts that participate in lateral and ventrad migration, and these processes result in the patterns exhibited by tortoiseshell mammals. The process is clearly complex and we do not yet understand exactly in what way the clonal origin of the tortoiseshell pigmentary stripes is established.

3 Melanoblast migration, dorsoventrally in the trunk and tail and anteriorly and ventrally in the head, with minimal mixing of adjacent phenotypic clones. Clonal origin of melanoblasts, followed by random mixing, of course could not result in phenotypic stripes. The actual tortoiseshell stripes appear more or less mixed along their borders (for example, Kim et al. 1995), but they are visible as stripes, therefore supporting the ventrad migration of the expanding clonal groups of melanoblasts. The explanation for lack of mixing at the borders of each clone could be compartmentalization (genetic specification of particular territories that then do not mix). Possibly, the migratory pattern of the cells might be confined to the somite of origin during the first migratory phase. However, while there are differences in migration in different sections of the embryo (cranial, truncal, and tail) the high variability of the striping phenotypes of tortoiseshell animals – recalling that the clones of a tortoiseshell phenotype represent genetic alternatives expressed autonomously in the pigment cell – suggest against compartmentalization of individual melanoblast clones. An alternative explanation is competition among the melanoblasts for the available territories into which they can migrate. In support of this concept, contact inhibition of

neural-crest cells has been reported (Rovasio et al. 1983; Erickson 1985; Carmona-Fontaine et al. 2008). A high rate of replication and competition among the pigment cells for the available territories into which they can migrate is sufficient to explain clonal solidarity. The 'available territories' are generally defined by the tissue environment, primarily by KITL, as discussed above. It is the task of the replicating cells to fill the available spaces. In this first phase of migration the spaces available to pigment-cell migration are primarily lateral, as these processes proceed essentially concurrent relative to adjacent clones of pigment cells. The presence of equally competent clones of melanoblasts with vigorous replication potential, migrating at the same time on parallel pathways should be sufficient to maintain the observed integrity of the striping phenotype. Variability of these factors – differences in the relative vigor of the two clones as for example in combinations of two different inbred strains, or any factor that would leave any supportive area 'empty' of existing melanoblasts – have been studied, and readily account for observed differences in the integrity of tortoiseshell stripes (for example Kim et al. 1995). If the two initial embryos are genetically quite similar to each other, allophenics made from embryos of the same inbred strain for example, then the phenotypes of the resulting mice most often clearly exhibit a moderate number of stripes of the two sorts of pigment cells. Thus the most obvious information gained from allophenic/tortoiseshell phenotypes bears on the clonal origin of melanoblasts.

Chimerism that is not tortoiseshell

Not all chimeric pigment patterns are formed by chimerism for clones of pigment cells carrying alternative versions of genes that they express autonomously. If the gene product affecting pigmentation is expressed by cells other than pigment cells (as with KITL or EDN3), then the resulting chimeric pattern will not reflect the clonal migration of pigment cells. Traits that are controlled by the genotype of the dermis (Fig. 3.30), including *Agouti*-locus phenotypes, discussed in Chapter 5; tabby stripes (and spots) in the cat; coat patterns in the zebra, leopard, and giraffe; probably brindle in the dog and cow, and Blaschko stripes in humans, are not included in the above discussion of tortoiseshell phenotypes.

Figure 3.30 Mouse of the genotype A^{vy}/a. The chimerism is not autonomous to the pigment cell, but of the dermal (somitic) tissues. This mouse also has unrelated *Piebald* white spotting. Note that the presence of white spotting does not change the pattern of the chimerism as *Piebald* spotting does in mice that have chimerism of pigment-cell-specific genes. This mouse is YS/Wf-A^{vy}/a, $Ednrb^{s}/Ednrb^{s}$, and is available at MMRRC.

3.4.5 Genes required for the secondary migratory phase: what white spotting tells us when combined with tortoiseshell chimerism

When Satomi Nishikawa et al. (1991) inhibited KIT during the second stage of migration/replication but not the first, by injecting ACK antibody into normal pregnant female mice, the pattern of pigmentation was similar to that seen in mice mutant at *Kit*, with ragged edges to the spotting as shown in Figure 3.31, and in several other figures in this chapter.

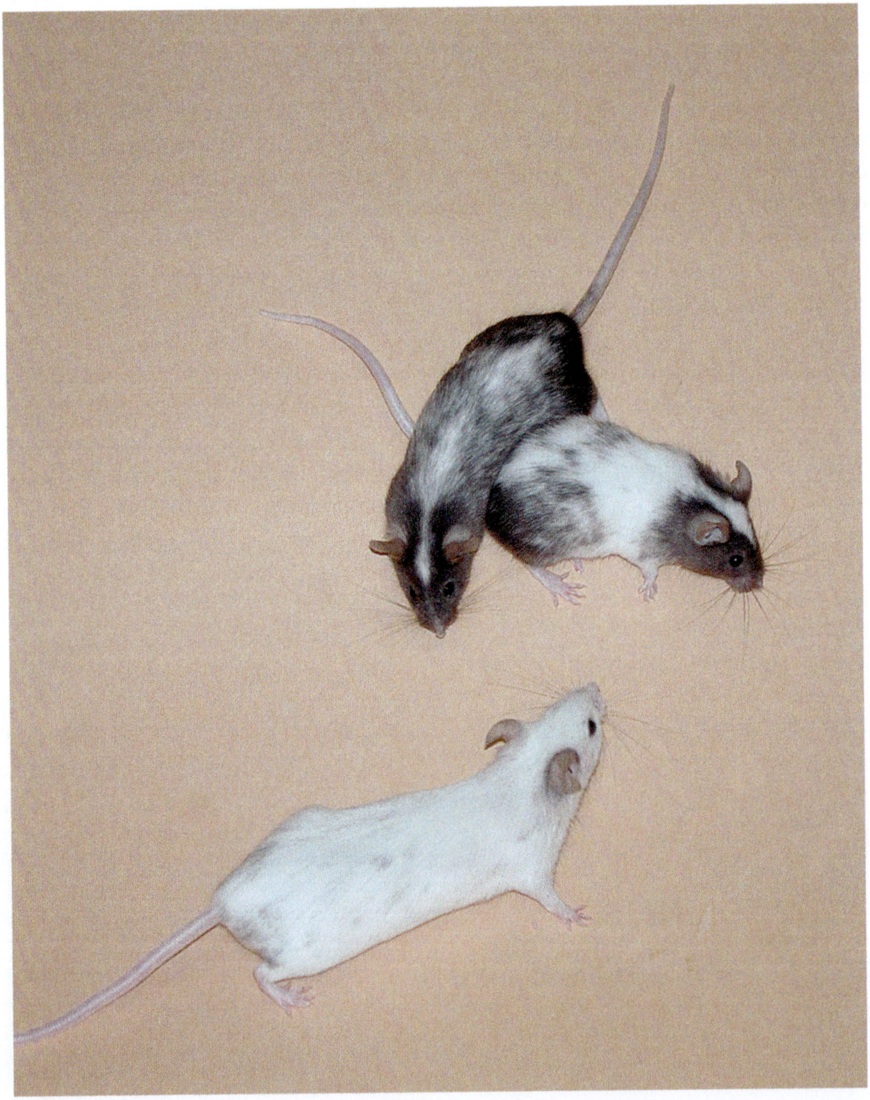

Figure 3.31 *Kit* mutants: three mice of the same mutant genotype, $Kit^{W\text{-}2J}/+$. The lower mouse is of the JU/CtLm inbred strain. Both mice at the top are C57BL/6J. The one on the left is of the usual phenotype. Within this stock occurred a mutation that affects the amount of white; the mouse on the right is of that substrain. These mice are available at MMRRC.

Figure 3.32 Mice of the genotype C57BL/6J-*Ph*/+ and JU/CtLm-*Ph*/+ (with the head spot). *Patch* and *Patch-extended* (Ph^e) are alleles that each represent a deletional mutation encompassing the PDGFRA subunit (*Pdgfra*). However, deletion of *Pdgfra* alone does not result in white spotting, and the *Patch* mutations apparently also cause defective expression or function of KIT (Stephenson et al. 1998). The *Patch* phenotype is very different from the usual phenotype of *Kit* mutant mice. However, at least two *Kit* mutations generate a similar phenotype of fully pigmented, sharp-edged spots. Two of these, *Sash* ($Kit^{W\text{-}sh}$) and *Banded* ($Kit^{W\text{-}bd}$) have been shown to involve regulatory regions of the *Kit* gene.

When Yoshida et al. (1996) inhibited the first but not later stages of development, the pattern of pigmentation appeared similar to the mutant phenotype of *Piebald* ($Ednrb^s/Ednrb^s$) (Fig. 3.2) or *Lethal spotted* ($Edn3^{ls}/Edn3^{ls}$) mice (Fig. 3.24). That is, the spots were discrete, rounded and sharp-edged, a common white-spotting phenotype not restricted to *Piebald* and *Lethal spotting* but also evident in the *Patch* mice in Figure 3.32 and other white spotted mice.

We know that KIT is required during all the migratory phases of melanoblast development; we also know that EDNRB is not required for normal development during the epidermal (second) phase of migration (Shin et al. 1999). Thus, with regard to the major white-spotting genes that are required for melanoblast migration and survival, the difference in pigment pattern between congenic mice mutant at *Kit* and at *Piebald* (*Ednrb*) or *Lethal spotting* (*Edn3*) appears to reflect the different events that occur in the embryo during the initial dorsoventral migration in the dermis compared with the secondary migration of the melanoblasts in the basal layer of the epidermis. This difference in spotting phenotype can be easily explained if one assumes that normal, non-mutant melanoblasts continue to replicate in the epidermis to fill any unoccupied spaces that are KITL-positive. In normal wild-type mice the pigment cells would fill these areas as they enter the epidermis, because of the large numbers of contiguous melanoblasts that are moving together into the epidermis.

In *Kit*-mutant mice, the melanoblast development of which is restricted at all stages, patterns of spotting most closely reflect events of the initial migration. That is, they reflect a dorsal origin of pigment cells that migrate ventrad in the body and rostrally in the head. When the mutant condition is removed by somatic reversion events (presumably from *Kit* +/− to *Kit* +/+) as shown in Figure 3.33, the areas containing normal pigment cells are sharp-edged, well-rounded, and not necessarily oriented on the dorsoventral pathway, although in this mouse they are. Then we see in one mouse the ragged-edge phenotype that results from suppression of KIT at all migratory stages, and the rounded-edged phenotype of the spots that results when that suppression is released.

Patterns of white spotting may or may not primarily reflect the dorsal/ventral migration that is visualized by the tortoiseshell phenotype. Except for mutation of the structural *Kit* gene, white-spotting patterns suggest a secondary clonal formation in the epidermis, each new subclone generating melanoblasts that populate the skin radially from wherever the surviving melanoblast is located as it crosses into the epidermis. If for some reason few melanoblasts have survived to enter the epidermis, but they can function well within the epidermis, then the result will be large, rounded clones adjacent to or within areas that are otherwise unpigmented (Fig. 3.32). This view is supported by phenotypes of triple chimeric mice that were created by Huszar et al. (1991). Two genetically different types of melanoblasts were injected into early embryos that were not otherwise pigmented. When melanoblasts were injected into BALB/c albino mice that have their own unpigmented melanoblasts (Huszar et al. 1991), the few pigment cells that were able to participate in embryonic development appeared in the 'tortoiseshell' pattern (relative to the unpigmented melanocytes otherwise present in the resulting mouse (Fig. 3.34).

When the cells were injected into mice that contained no pigment cells of their own, two types of pattern resulted. In the same mouse or in different mice, when the two pigment types of injected pigment cells colonized contiguous regions and migrated alongside each other, the resulting pigment pattern was again a typical 'tortoiseshell' type of striping pattern. When clones were not adjacent, the resulting clones of pigment cells were rounded and sharp-edged and not necessarily oriented in the dorsoventral pathway. The latter is very reminiscent of the 'calico' phenotype of the cat.

Calico phenotype

A calico cat is a chimeric (tortoiseshell) cat that has, in addition to pigmentary chimerism of the sort described above, a greater or lesser degree of white spotting. The sizes of the patches (subclones) of the different-colored pigment cells correlate with the amount of white spotting. Thus, in a severely white-spotted cat, like the one at the top of Figure 3.35, we can believe that few melanoblasts survive the first migratory phase to arrive in the epidermis, compared with a cat that has no white spotting. These few surviving melanoblasts are released from their inhibition and replicate rapidly to fill the KITL-positive spaces, creating big yellow patches and nonyellow patches. The calico

Figure 3.33 Two mice of the genotype JU/CtLm-Kit^{W-2J}/+. The mouse in front is of the usual phenotype for a JU mouse with abnormal KIT receptors on the pigment cells. Because the receptors function as dimers, and the cells are heterozygous, there is the occasional normal receptor that presumably permits survival of the occasional pigment cell. The mouse behind demonstrates the phenotype of somatic reversion events in mice that are not fully pigmented (compare with Fig. 3.29). In this mouse, the cells in the revertant pigment spot presumably have normal KIT receptors, and they are existing in a space with very few pigment cells, and so of course they behave as normal pigment cells do by replicating and filling up all the adjacent spaces that are permissive. The KITL defines permissive spaces. Somatic reversions most often occur on the anterior or posterior portion of the mouse, probably because it is more likely that there will be pigment cells over this portion of the body. Somatic-reversion events in this genotype are rare (and more difficult to observe) on the C57BL/6J background. They are uncommon on the JU/CtLm background, and usually not so extensive as the one shown. Compare with Figure 3.19, which shows pigmented spots on mice mutant at *Mitf*. These appear not to be somatic reversion events, both because of their more common occurrence and predictable location and because the pigmentation often is not of a revertant phenotype.

phenotype, as we can call it more generally, has also been recognized in other species, including guinea pigs (Fig. 3.14) (Searle 1968).

The calico phenotype is also evident in tricolor mice (Fig. 3.36), depending upon the function of the gene that causes the white spotting. *Piebald* and *Piebald lethal* and *Belted* tricolor mice do exhibit the patchy 'calico' phenotype (M.F. Lyon & M.L. Lamoreux, unpublished observations). Similar to calico cats the patchy phenotype is evident in regions adjacent to the white spot. Normal *Ednrb* (*Piebald*) and *Edn3* (*Lethal spotting*) genotypes are required for the first migratory phase of melanoblasts, but are not required for the second. Thus it appears that melanoblast migration, proliferation, and/or survival is inhibited at the first stage in animals that are mutant at these loci, resulting in a paucity of melanoblasts arriving in the epidermis, especially few in the areas adjacent to the white spot. When the pigment cells arrive in the epidermis and are presumably released from the inhibition, they proceed to clonally populate the available areas in the epidermis that are not

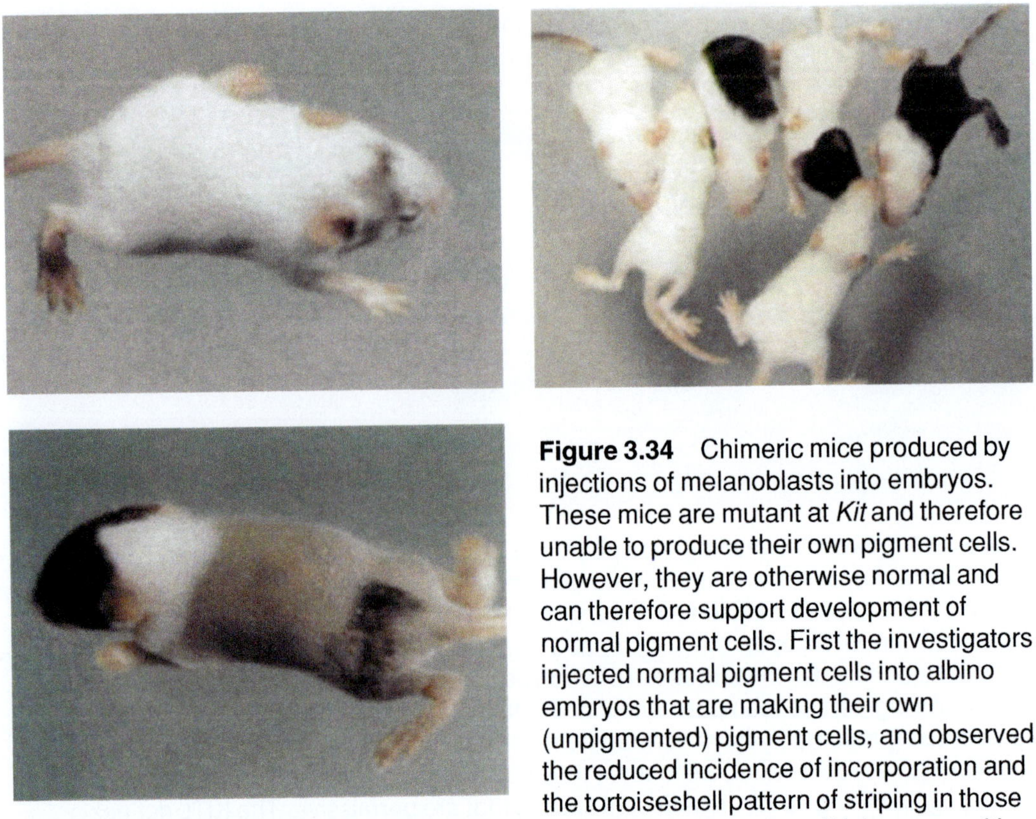

(a) (b) (c)

Figure 3.34 Chimeric mice produced by injections of melanoblasts into embryos. These mice are mutant at *Kit* and therefore unable to produce their own pigment cells. However, they are otherwise normal and can therefore support development of normal pigment cells. First the investigators injected normal pigment cells into albino embryos that are making their own (unpigmented) pigment cells, and observed the reduced incidence of incorporation and the tortoiseshell pattern of striping in those mice that did incorporate some of the alien pigment cells (a). Then they did the same with mice mutant at *Kit* that are not capable of making their own pigment cells and noted the large, sharp-edged clones that resulted as these more frequently became incorporated in the embryo and generated large, discrete clones in the otherwise empty skin (b). These clones did not reflect dorsoventral migration so much as dorsal origin of the pigment cells. In other words, they were more likely to occur near the dorsum but they appear to have expanded radially. Huszar et al. (1991) then injected into developing embryos pigment cells that were capable of making pale pigment typical of pink-eyed dilution mice (see the upper mouse in Fig. 3.29 for phenotype) along with genetically black pigment cells. The result, when cells competed with each other during the first migratory event, was tortoiseshell patterning, clearly reflecting the early migration of adjacent clones of pigment cells toward the ventrum (c). When cells did not compete with adjacent melanoblasts, the result was clear-edged patches that are also shown in (c). Diagram courtesy of Huszar et al. (1991); reproduced with permission of The Company of Biologists.

already occupied, where there are no nearby melanoblasts to discourage their clonal expansion. KIT and KITL are both required throughout melanoblast development. If they are defective as a result of mutation, we therefore would not expect that their melanoblasts would be released from inhibition at any stage of development, including the second migratory period in the epidermis. Therefore we would not expect tortoiseshell mice with a white-spotting mutation at $Kit^{W\text{-}2J}/+$ or $Kitl^{Sl}$ to show the large calico-like patches seen in the tortoiseshell with *Piebald* or *Lethal spotted* or *Belted* mice (Fig. 3.37), and this is the case (M.L. Lamoreux, personal observation). However, the $Kit^{W\text{-}2J}/+$ mice do occasionally exhibit somatic reversion events so that it is possible to observe

Figure 3.35 A calico cat; that is, a tortoiseshell cat with white spotting. The white spotting is not a form of genetic chimerism; all of the pigment cells are of the same genotype with regard to the white-spotting trait that imposes upon them a limited ability to survive, replicate, and migrate. It is not known specifically why they fail to populate the nether regions. Presumably environmental conditions are not so congenial in some portions of the developing embryo as they are in other portions, and the white-spotting genotype confers upon the pigment cells a borderline functionality. The cause of the chimerism is well known as Lyonization (X inactivation). The yellow (*Orange*) gene of the cat is located on the X chromosome. This cat is heterozygous at that locus. The locus functions similarly to the *Extension* locus of the mouse. When mutant at *E* (*Mc1r*), the recessive yellow pigment cells are unable to produce eumelanic pigment; the normal wild-type pigment calls can produce both eumelanin and pheomelanin (depending on the genotypes at other loci). In this cat some of the spots are pheomelanic (yellow, orange) while others are wild-type agouti. The cat below is also a chimeric of the same X-linked genotype as that above (X^O/X^+); that is, she is a chimeric animal with regard to the yellow mutant and the wild-type alleles at the X-linked O locus. Her background genome, however, is quite different from that of the cat above. The background genome is not chimeric and is imposed upon all the pigment cells. First, the cat above is *Agouti*, whereas the cat below is *Nonagouti* (see Chapter 5). Second, the cat above is not genetically 'dilute' (in cats but not mice this locus is *Melanophilin*) whereas the cat below is 'dilute', and is therefore referred to by cat fanciers as a 'blue-cream tortoiseshell' because all the eumelanic and pheomelanic pigmentation is 'diluted' by the mutant *Melanophilin* genotype (see Chapter 4). Importantly for the study of pigment-cell survival during development, the absence of a white-spotting mutation is the most striking difference and defines the difference between the patchy pigmentary clones of the calico cat shown above and the basic dorsoventral patterning of the tortoiseshell shown below.

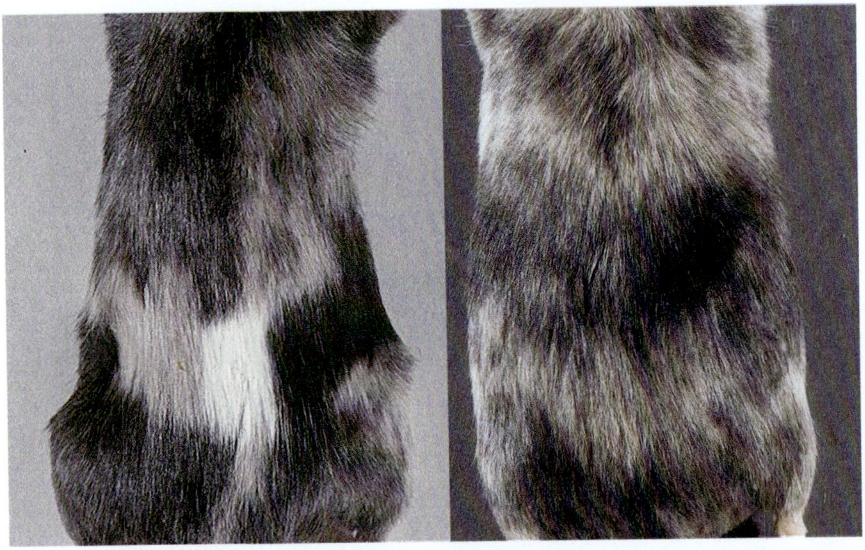

Figure 3.36 The calico pigment pattern can be reproduced in mice. In mice it is easier to evaluate the relationship between the observed phenotype and the known gene functions. On the right is the same mouse as shown in Figure 3.28, which is chimeric because of an X-linked gene. On the left is a mouse with the same X-linked chimerism, but also mutant at the *Belted* locus. Using mice we can demonstrate that the calico pattern can result in chimeric animals from *Piebald* or *Belted* spotting but is not seen in mice that are mutant at *Kit* or *Kitl*.

development of wild-type clones of pigment cells on the same canvas as those that lack normal KIT. The revertant patches are fully pigmented, with rounded edges, as would be expected if the heterozygous cell reverts to homozygous normal. The resulting spots may be large, as in Figure 3.34, or they may be small, in which case they often are not obviously associated with the dorsal midline. In fact, largish patches of pigmentation often occur at the ventral midline, an area prone to white spotting.

All these observations support the idea that melanoblast clonal formation in the epidermis is not primary, but only secondarily reflects the clonal event that occurs in the neural crest. Likely, the epidermal replication event is the normal method of achieving complete coverage of the mouse with pigment cells and probably is the origin of pigmented spots in mice that are mutant at *Piebald* or *Lethal spotting* loci, and very likely other spotting genotypes that produce the sharply delineated, fully pigmented, crisply bordered type of pigmented spots.

Special cases of white-spotting phenotypes: *Patch* (*Ph*), *Rump white* (*Rw*), and some alleles of *Kit* (*W*)

Patch: Two spontaneous phenotypic alleles, both deletions.
Availability: HAR, JAX, and MMRRC.
Rump white: One spontaneous phenotypic allele that represents an inversion.
Availability: HAR and JAX.

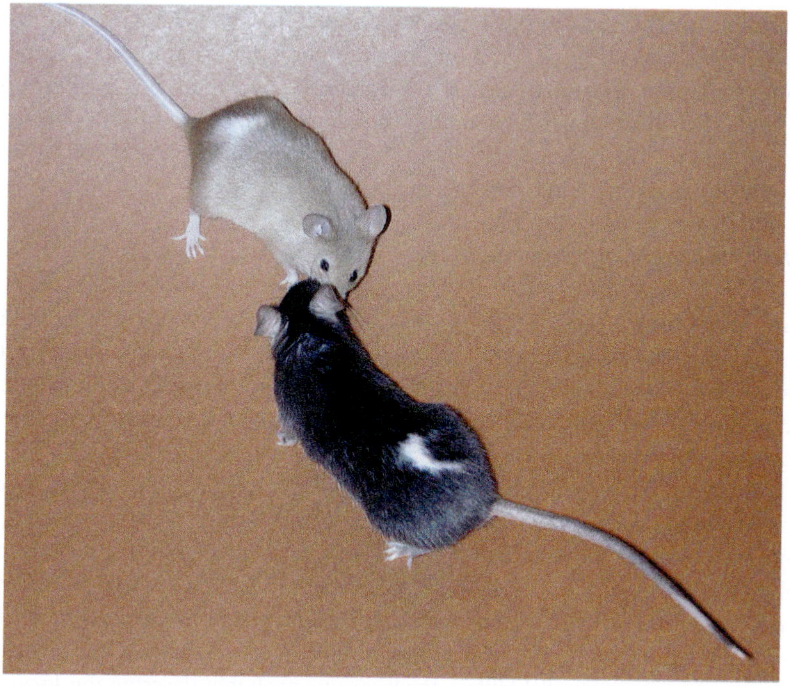

Figure 3.37 *Belted* mice. Left: *Recessive yellow* (C57BL/6J-*a*/*a e*/*e Adamts20*^{bt}/*Adamts20*^{bt}); right: nonyellow (C57BL/6J-*a*/*a E*/– *Adamts20*^{bt}/*Adamts20*^{bt}). The size of the spot varies.

Patch (not to be confused with *Patched*), *Rump white* (not *Rumpwhite*), and four mutations attributed to *Kit* (*Kit*$^{W\text{-}sh}$, *Sash* and *Kit*$^{W\text{-}bd}$, *Banded*, which is the same as *Kit*$^{W\text{-}26H}$), are all phenotypic mutations that represent chromosomal rearrangements near or in the *Kit* locus on chromosome 5 (Nagle et al. 1994; Stephenson et al. 1998; Kluppel et al. 1997; Jordan & Jackson 2000). Mice homozygous for these mutations generally die *in utero* or lack pigment cells in the skin. They have similar phenotypes in the heterozygous mouse; that is, a wide white belt with crisply delimited borders as illustrated for *Sash* in Lyon and Glenister (1982), and *Patch*, on two different backgrounds, in Figure 3.32.

These phenotypes differ from the usual phenotype of pigmented mice that are mutant at *Kit* (Fig. 3.20). All of these mutations have been attributed to rearrangement of the regulatory region of KIT, and they all result in sharp-edged, rounded, pigmented spots. It has been shown that the aberrant phenotype of *Rump white* mice relates to mislocalized ectopic expression of KIT (Hough et al. 1998). Likewise *Sash* embryos express KIT in abnormal locations (Duttlinger et al. 1993, 1995; Hough et al. 1998), and so do *Patch* embryos (Wehrle-Haller et al. 1996).

It is suggested that melanoblast numbers and dermal migration are deficient in the areas abnormally expressing KIT because KITL availability is reduced by its binding to the ectopic KIT on dermal cells, rather than because melanoblasts have reduced KIT activity (Wehrle-Haller et al. 1996). Thus, in these phenotypically unusual *Kit* regulatory mutants, the inhibition is released as they enter the epidermis (Jordan & Jackson 2000) and the melanoblasts can expand clonally from wherever they are when they enter the epidermis.

Most of these secondary patches are found in the dorsal region, indicating an aberrant first migratory event, as demonstrated by the *Kit*$^{W\text{-}2J}$/+ mice shown in several figures in this chapter.

However, heterozygous *Patch* mice (Fig. 3.32), and also heterozygous *Rump white* mice also fairly commonly have ventral islands or peninsulas of pigmentation (M.L. Lamoreux, personal observation), and their pigmentation patterns clearly are influenced by variation in background genome (Lamoreux 1999). The rather dramatic variation in pattern of spotting of *Rw*/+ mice tentatively maps to or near the *Steel* (*Kitl*) locus (M.L. Lamoreux, unpublished results) and is therefore not attributable to the *Rw* mutation, but to background genome.

Summary of first and second migratory stages

In sum, the important contribution of the tortoiseshell phenotype rests in the way it illustrates the first, dorsoventral stage of melanoblast migration. Similarly, the calico phenotype reflects behavior of melanoblasts during the second migratory period of their development. These phenotypes tend to support the observations of Schaible that were formerly difficult to factor in to the prevailing view of melanoblast development.

Schaible (1969) identified what he called 'centers of origin' of pigmentation by observing patterns of pigmentation of large populations of mice segregating for several white-spotting mutations to determine where the pigmented areas are most likely to occur in the least pigmented animals. (Note: Schaible termed these mice 'piebald spotted,' meaning simply white spotted rather than mutant at the *Ednrb* locus.) Schaible also provided a nice summary of much earlier work. Goodale and Morgan (1913), for example, postulated the clonal origin of the patches on tricolor (here referred to as calico) guinea pigs. Schaible's observations probably describe the locations on the mouse where pigment cells are most likely to survive if their viability is reduced by a white-spotting genotype during the first migratory phase, and replicate during the second phase.

Also, in the 1960s and beyond Mayer (Mayer 1970, 1973a, 1973b, 1977, 1979; Mayer & Green 1968), building upon previous work of Rawles (1947), Markert and Silvers (1956), and Silvers and Russell (1955), developed a theory in explanation of white spotting that covered the essence of most of what we have discussed above. To summarize briefly, Mayer agreed that some traits are autonomous to pigment cells. However, the pigment cells live in the tissue environment and depend upon it for developmental cues, and these cues also affect pigment-cell survival, migration, and differentiation. Further, these environmental factors can determine locations of white spots; that is, the areas of the embryo least supportive of melanoblast survival. Mayer viewed the surface of the embryo as a non-uniform field that is more supportive of viable melanoblasts in some areas than in others, as also suggested by Schaible's work. In addition, Mayer described the secondary migration of pigment cells into the epidermis. Overall, in its essence, Mayer's work taken as a whole describes both the initial phase of melanoblast migration and the movement into the epidermis with subsequent expansion, as well as failure of melanoblasts to gain entry into the hair follicles in mice mutant at *Belted* (see below).

In normal fully pigmented mice, according to Mayer, the combination of pigment-cell viability and its environmental support, in all portions of the body, is above a 'survival threshold,' and the body is pigmented over nearly the entire surface. If the total competence of the pigment cells plus environmental requirement is reduced by mutation at any locus, remembering that the environmental component is non-uniform, then we can visualize the least supportive areas of the body by observing where the white spots are most likely to occur. In C57BL/6J mice this is the belt region. Therefore, white spotted mice are usually white in the belt region, regardless of the specific genetic cause of the pigment-cell failure to survive. Today we must also factor in a possible delay of specification in the belt region (Wilson et al. 2004). As the severity of the developmental problem (or genetic lesion) increases, the size of the spot increases, mostly along a gradient, so that the white spot gets bigger and bigger, but eventually adding other regions of the body, such as the forehead, tail, and feet.

Mayer's theory was well supported by his work and also by Rhim et al. (2000), who demonstrated that variations in the location of white spotting on the bodies of mice of differing genetic backgrounds map chromosomally at or near the *Steel* (*Kitl*) locus; KITL of course is expressed in the environment through which the pigment cell must migrate if normal development will result. That region of the genome is not entirely responsible for spot location. Other loci are segregating, and not interacting identically with each white-spotting mutation (Lamoreux 1999), a fact that emphasizes the point that no one factor can be responsible for the behavior of melanoblasts during development. The entire process is a network of signaling interactions working over a continuum of time and space.

There is no conflict between the core theories of Schaible and Mayer, although they were debated, if one recognizes that patterns of pigmentation resulting from the first melanoblast migratory event differ from those imposed by the second. Schaible, working with adult animals, observed the combined effects of both. Mayer worked with embryos that were sometimes in the first migratory stage and sometimes in the second. The primary differences between the observations of Mintz, and those of both Schaible and Mayer results from the fact that Mintz's bicolor chimeric animals are not white spotted and therefore their phenotypes illuminate the first migratory phase of pigment-cell development, but not the later stages. At that time, it was difficult to interpret her work with chimeric combinations that included white spotting, as with the work of Pratt (1982) and others who used different techniques of tissue recombinations to evaluate white spotting, although with present knowledge we can see that the various techniques together provided valid clues to gene functions of white-spotting genes.

Hyperadditive spotting, a topic of considerable discussion in the early days of pigmentation research, takes on a slightly different perspective in the light of the above observations. The numbers of melanoblasts that survive the first migratory phase would of course influence the sizes and numbers of spots that are formed during the second migratory phase, and so an animal that carries two white-spotting mutations is likely to have a larger white area than would result if the spotting mutations had no effect upon each other (Aoki et al. 2005).

While all these interacting factors are sufficient to explain most of the pigment patterns of mice, two sorts of variants remain to be considered. The first involves variable distribution of yellow and nonyellow pigment over the bodies of animals that are not genetically chimeric. This would include dogs, cattle, mice, cats (tabby patterns and spotting), giraffes, and probably zebras. These patterns are not independent of white spotting, as will be considered in Chapter 5.

The second variant that needs to be recognized involves the initial distribution of the earliest specified melanoblasts and their early descendants before the lateral migration begins. Several authors have noted in normal mice that there are more early melanoblasts at the anterior and posterior limits of the trunk region (the areas that will be the shoulders and hips of the animal), and also the ears and head, than in other regions of the embryo (Wilkie et al. 2002; Wilson et al. 2004), and some have specified that the mid-trunk region is populated later, or less (Duttlinger et al. 1993), in development than the extremes of the trunk region (see Fig. 3.31). The method of Huszar et al. (1991) also suggests the possibility of a very early anterioposterior migration/replication component to the ordering of melanoblasts, presumably in the MSA of the trunk region, prior to the lateral clonal migration.

The mice shown in this chapter are mostly of the C57BL/6J or closely related strains. On this strain background, the primary unpigmented area (in mice mutant at any of the major white-spotting loci) is normally just anterior to the hind legs. Wild-type mice often have a tiny belly spot and a little extra white on the tail tip. In JU mice, the area of the primary spot is farther forward, behind the front legs, except in *Rump white* mice where the white area is more posterior (Lamoreux 1999); they have a head spot.

A C3H background also influences size and location of white spots, as well as color phenotype in mice that are yellow because of mutation at the *agouti* locus (see Chapter 5). Clearly, the location of white areas is influenced by genetic background, and this was noticed early in the history of pigmentation research. As early as 1937, Goodale and subsequent coworkers selected a genetically mixed group of mice for size of spot, with great success (Goodale & Morgan 1913; Goodale 1937). In addition, Mayer and Green (1968) demonstrated that the *Steel* (*Kitl*) locus functions in the tissue environment in which the pigment cell is trying to survive. Finally, Rhim et al. (2000) mapped variation of size and location of spotting in piebald mice with different strain backgrounds to the *Steel* region of chromosome 10. Unpublished results of Lamoreux suggest the same relative to *Rump white* spotting. Further work on this subject will no doubt identify additional gene loci that interact with melanoblasts to direct their patterns of migration, replication, and survival.

3.4.6 Homing to hair follicles and maintenance therein

Special cases: *Belted* (*Bt, Adamts20, A disintegrin-like and metallopeptidase* (*reprolysin type*) *with thrombospondin type 1 motif, 20*)

MGI shows 11 naturally occurring phenotypic alleles
Availability: JAX, MMRRC, NMICE, and ORNL.

It is somewhat speculative to include *Belted* under the heading of 'Homing.' However, the preponderance of evidence suggests that the locus functions during melanoblast entry into the developing hair follicle. Mayer (1964), using a method of tissue transplantation, found that the defect in *Belted* mice is different, and occurs later (about at embryonic day (E)13.5; confirmed by Silver et al. (2008), using a more specific technique), than that in *Lethal spotted* mice, and he observed that 'Melanoblasts may be unable to enter the developing hair germ because of some block in migration, or their maturation into melanocytes in the environment of the hair follicle may be prevented' (see also Nishimura et al. 1999).

Belted mice do exhibit 'calico'-type pigment patches (Fig. 3.36), suggesting that, like *Piebald*, *Belted* is not required for the secondary epidermal clonal expansion of melanoblasts. (This does not exclude the possibility mentioned below of a deficiency in crossing the epidermal basement membrane, reducing the number of melanoblasts that reach the epidermis.)

Zhang et al. (2008) found that β-catenin signals the epidermis toward placode and hair shaft fate, at the expense of epidermal differentiation, and activates pigmentation and innervation. ADAMTS20 (A disintegrin-like and metallopeptidase (reprolysin type) with thrombospondin type 1 motif, 20) was among the pigmentation regulators that they identified in transcript-profiling experiments.

Belted is the result of a defect in a novel ADAMTS family member (Somerville et al. 1994; Rao et al. 2003). ADAMTS proteins are capable of processing a variety of extracellular-matrix components and secreted molecules, including the common connective-tissue proteoglycan versican (Silver et al. 2008). ADAMTS20 and ADAMTS9 together make up a subset of the family that has been conserved over long periods of evolution (Somerville et al. 2003). ADAMTS9 is more widespread in the embryo. They both exhibit metalloprotease activity (Llamazares et al. 2003). ADAMTS20 is expressed in the dermal mesenchyme.

Belted is recessive. The homozygous *Belted* mouse differs from other white-spotted mice that have been studied in that the location of the spot does not seem to be influenced by the strain background (Lamoreux 1999). However, Silver et al. (2008) observed that ADAMTS20 enhances the ability of melanoblasts to respond to soluble KITL, at least at late stages of development. Because KITL generally defines the migratory pathways of melanoblasts, this observation lends further support to the concept that ADAMTS20 is required for the melanoblast to find its destination in the hair follicle. On the other hand, they reported that belted embryos also show excessive death of dermal melanoblasts just at the time they normally enter the epidermis, suggesting some restriction of this transition (Silver et al. 2008). Perhaps ADAMTS20 is used in passing though the basement membrane.

Yellow *Belted* mice have smaller average spot size than black *Belted* mice (Hauschka et al. 1968) and they have more 'islands' of pigmented hairs within the white area, suggesting that something about the yellow signaling pathway helps to counteract the lesion generated by mutation at *Belted*. Or, conceivably it intersects with the belted/KIT/KITL interface, as KIT is known to be strongly affected by the yellow signaling pathway. This is true whether the mice be yellow because of mutation at the *Agouti* (*A*) or the *Recessive yellow* (*E, Mc1r*) locus, and there may be an impact of heterozygosity at the *Recessive yellow* locus (M.L. Lamoreux, personal observation).

In normal mice, the numbers of epidermal melanoblasts in the hairy portions of the body begin to decrease from the first day after birth (Yoshida et al. 1996). They reach extremely low levels in the epidermis by about 1 month of age (Hirobe 1984a, 1984b), as the melanoblasts migrate from the basal layer of the epidermis into the hair follicles, again following a pattern defined by the expression of KITL (Kunisada et al. 1998b). KIT is not required for maintenance of pigment stem cells in the niche area of the hair follicle.

3.4.7 The hair follicle

After migrating into the epidermis, and before beginning their entry into the developing hair follicles, the melanoblasts replicate to fill any remaining available spaces. As the hair follicles begin to develop, about at E16.5 in the mouse, the melanoblasts immediately begin to enter and actively proliferate in the follicles (Botchkareva et al. 2003). Again, the most basic explanation for this migration presumes that the migratory pathway is defined by KITL in the surrounding cells and recognized by KIT on the migrating cell. However, as at all stages of migration, little is understood of the specific elements of the environmental milieu that interact to inform the migratory process. For example, work in cell culture suggests that PAX3 can simultaneously function to activate a melanogenic cascade in differentiation of pigment cells, by activating *Mitf*, while simultaneously competing with *Mitf* for activation of *Dct*, thus preventing expression of terminal differentiation markers until external stimuli abolish PAX3-mediated repression (Lang et al. 2005).

As the melanoblasts enter the hair follicle, some migrate directly to the bulb area of the follicle (Krause & Foitzik 2007), where the first generation of hairs will begin to grow, in the mouse about at the time of birth (Chase 1951; Slominski & Paus 1993; Tobin et al. 1998) (Fig. 3.38). This population of differentiating cells requires expression of KIT for their survival (Nishimura et al. 2002) and migration to the bulb (lower) area of the hair follicle. In the bulb region of the hair follicle, they differentiate (Chapter 4) to produce pigmented melanosomes that they then transfer to the keratinocytes of the growing hair. After the hair is formed the lower half of the follicle regresses, which is the catagen stage of the molt cycle. The hair remains attached during the resting (telogen) stage, until the next growth phase is initiated (anagen; Fig. 3.38), a process that is influenced by sonic hedgehog (Sato et al. 1999).

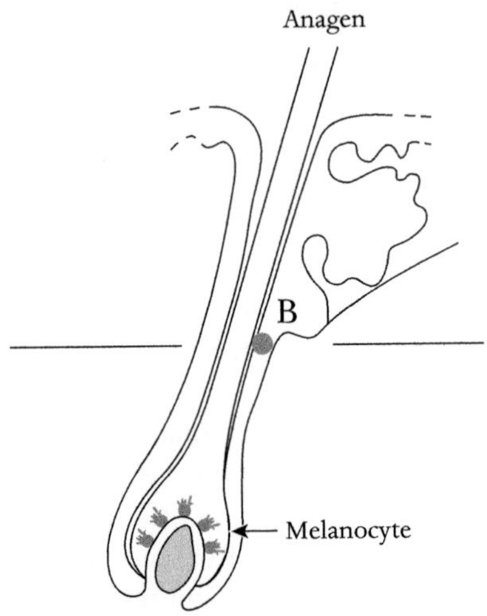

Figure 3.38 A mature hair follicle. B indicates the bulge region of the hair follicle and the gray spot shows where the melanocyte stem cells reside. The horizontal line represents the level to which the follicle regresses after the growth phase is complete. Diagram courtesy of Nishimura et al. (2002).

The *Foxn1* (*Nude*) locus is required for normal hair growth, as illustrated by the nude phenotype of the mice. In nude mice that are pigmented, the disruption of the hair cycle, just as the pigment is forming, results in loss of the hair, thus loss of the pigmentation and stimulation of a new molt cycle. The molt cycles of mice proceed approximately in a wave, from front to tail. In mice homozygous for the *Traveling wave* mutant allele of *Foxn1*, the early loss of hair results also in loss of pigmentation and initiation of a new cycle; thus the mouse is pigmented in stripes that appear to travel from head to tail (Suzuki et al. 2003a).

In normal mice, during the initial migration of the melanoblasts into the hair follicle, some of the cells are retained as quiescent stem cells (Fig. 3.38) (Nishimura et al. 2002; Osawa et al. 2005; Mak et al. 2006; Moore & Lemischka 2006; Moriyama et al. 2006; Nishikawa & Osawa 2007; Yonetani et al. 2007). Stem cells are cells that have the ability to both self renew and generate differentiated progeny. During the initial migration, they are sequestered in the upper (bulge) portion of the hair follicle, known as the niche. This portion of the follicle does not regress during molt cycles. Therefore the stem cells remain available to provide a fresh set of melanocytes to populate the lower bulb region of the hair follicle as it regrows during each successive molt cycle. Melanoblastic stem cells have been shown to survive in the presence of the ACK antibody, which blocks the function of KIT (Nishimura et al. 2002), indicating, quite uniquely among the melanoblasts, they do not require KIT for survival and are presumably not responsive to KITL, a fact that may explain why they stop at the niche area during their migration into the follicle, rather than proceed all the way to the bulb of the growing hair.

The stem cells remain quiescent until the next molt cycle (6–8 weeks later; see section 1.9), and each of the subsequent hair growth cycles (Lavker et al. 1993). Keratinocyte stem cells also are stored in the niche (Oshima et al. 2001). All this is accomplished in a manner so precise and intricate that the process results in exactly the right kinds and colors of hairs in exactly the right locations (Galbraith 1964), and this is repeated at intervals throughout the life of the mouse.

In these cycles of renewal, the hair follicle offers unique access to highly regenerative epithelial-mesenchymal interactions involving numerous stem cells. The cycles are controled by changes in the complex local signaling milieu, and thus they inform important ongoing research aimed at answering questions about the relationships between gene function and phenotypes (Waters et al. 2007). How does the hair follicle 'know' where and how to form? How does the pigment cell find its place in the follicle? What is its place in the follicle? How are the follicular changes coordinated with the melanocyte migration and differentiation? The subject of stem cells (Fuchs et al. 2004) and their environmental controls is a very important topic of research, and a great deal of work has been ongoing to define the characteristics of melanoblastic stem cells, compared with other cells of the melanocyte lineage (Nishikawa & Osawa 2007).

Development of the hair follicle itself is also of great interest (Van Steensel et al. 2000; McElwee & Sinclair 2008), but is outside the scope of this book, except to mention an interesting overlap of functions of several of the signaling pathways (WNT/β-catenin, Delta/Notch, TGF-β, SHH, BMP, HGF/MET, and probably others) that are active in the development of the neural crest and the specification of melanoblasts, and also function in the development of the hair follicle (DasGupta & Fuchs 1999; Huelsken et al. 2001; Alonso & Fuchs 2003a, 2003b; Millar 2003; Blake & Ziman 2005; Jamora et al. 2005; Moriyama et al. 2006; Nishikawa & Osawa 2007). Included among the factors that influence development of the follicle is the HGF/SF↔MET receptor signaling pathway (Lindner et al. 2000). The impact of this signaling pathway on the yellow/nonyellow phenotypes is discussed in Chapter 5. Development of the hair follicle must also involve keratinocyte-derived factors (Hirobe 2004). Thus the white-spotting or graying-with-age phenotypes that may be associated with mutation at these loci must be interpreted with awareness of the multiple functions of many of the same factors both in the embryo and after birth.

We have seen that the white-spotting phenotype is caused by failure of melanoblasts to complete the journey to and become incorporated into the hair follicle. If melanoblasts reach the hair follicles but fail to thrive there over time, the result is an age-related loss of pigmentation that, in the mouse, is not usually localized. Patterned progressive pigment loss, in the mouse, normally occurs only as a secondary result of a molt cycle, because the newer hairs will be less pigmented than the older (Figs 3.3 and 3.4). At the completion of each molt the mouse will be less intensely pigmented, overall, than it was before the molt began and will maintain this state until the next molt cycle. Graying with age does not necessarily mean *old* age, but usually refers to abnormal progressive graying of mice, beginning at the first molt (which is approximately at puberty) (Lerner et al. 1986; Spritz 2008; Boissy & Spritz 2009).

In humans, chickens, and pigs, graying with age is characteristically multifactorial. In inbred mice, mutations that cause abnormal loss of pigmentation with age occur at the *Mitf* and *Bcl2* loci (Moore & Lemischka 2006; Nishikawa & Osawa 2007). BCL2 is a widely expressed mitochondrial protein that inhibits apoptosis (Veis et al. 1993).

The work of Nishimura and colleagues (2005) suggests that graying with age results from aberrations or reduction in number of the melanocyte stem cells of the hair follicle. They implicate *Mitf* and conclude that it is necessary for activating the expression of BCL2 so as to guarantee survival of the stem cells. Recently, the role of BCL2 in melanocyte stem cells has been analyzed carefully by Mak and coworkers (2006), who reported that BCL2 is essential for the emergence of melanocyte stem cells, and graying results from its absence. Notch signaling (Moriyama et al. 2006) or phosphatase and tensin homolog (PTEN) deficiency (Inoue-Narita et al. 2008) also protect against hair graying.

In contrast to Nishimura and colleagues, who suggested that melanocyte stem cells are initially present in the bulge region of *Bcl2* mutant mice and then disappear at 6.5 days after birth, Mak et al. (2006) do not detect any melanocyte stem cells in the bulge region of hair follicles from *Bcl2* mutant embryos. This suggests that BCL2 is required early for melanoblast survival and may not be important for the melanocyte stem cells. Interestingly, Bouillet et al. (2001) have shown that all the effects of BCL2 deficiency are relieved in animals simultaneously deficient for pro-apoptotic factor BIM, indicating that BIM opposes BCL2 in melanocyte stem cells.

Patchwork (*Pwk*)

JAX lists one spontaneous mutant allele.

Patchwork (*Pwk*) is a recessive mutation that results in a unique hair-follicle phenotype in which each follicle consistently produces hairs that are either unpigmented or pigmented (Aubin-Houzelstein et al. 1998; Aubin-Houzelstein & Panthier 1999). The pigmented hair bulbs contain functional melanocytes while the unpigmented hair bulbs contain no melanocytes, and the phenotype of each hair follicle is maintained through successive molt cycles. Thus, beginning about E18.5, the hair is able to function in some ways as an independent unit. There is evidence that *Pwk* functions in an autocrine pathway, in which melanocytes produce a factor that supports their own survival (Aubin-Houzelstein et al. 1998).

The gray horse

One of the classical 'vitiligo' mutant phenotypes as defined in humans is not available in mice, but occurs in the gray horse (Fig. 3.39), not on the body but occasionally on the nose or around the eyes. The gray horse is born with the slightest indication of decreased pigmentation, usually a few white hairs in the eyelashes but otherwise normally pigmented, and then proceeds to become less intensely pigmented with every new hair cycle. If the animal is wild type – that is, bay with a pattern of pheomelanic and eumelanic pigmentation in the form of dapples on its body – then it may develop into a dappled gray horse at about the age of 4 years, and proceed to become more uniformly white at later ages. This demonstrates that the eumelanic portions of the horse retain pigment longer than the pheomelanic ones. If the animal is pheomelanic in pigmentation, then it may not have dapples, its pigmentation loss may be more rapid, and the melanoma that is often a part of the syndrome may be more likely as it approaches old age (Norris 2008; Pielberg et al. 2008). Scattered reports suggest theat pheomelanogenesis, at least in some circumstances, also reduces incidence of melanoma in Sinclair swine, other breeds of pigs, and in mice.

The gray mutation is caused by a duplication that affects both Syntaxin 17 (a locus newly described with this finding) and NR4A3 (Pielberg et al. 2008). Syntaxins are involved in membrane fusion events, so this one may have something to do with protein routing (see also Chapter 4), whereas NR4A3 is a nuclear receptor without an identified ligand, and also a proto-oncogene. In another example of the phenotype informing our understanding of the genotype, Smith et al. (2008) have described an interaction between the NR4A class of molecules and the MC1R, which functions in pigment-type switching (Chapter 5). Thus it was difficult to decide in which chapter to place this gene. However, the gray horse is one of our oldest models of 'vitiligo,' and, in addition to progressive graying, gray horses also often exhibit the patchy loss of skin pigment that is defined in humans as vitiligo. So perhaps it is appropriate that gray horses be placed at the end of this developmental chapter to remind us again that all of development is interconnected. As noted by Norris (2008), the question of the gray horse continues to provide 'another telling example of how the molecular investigation of animal coat colour phenotypic traits and associated diseases can convey new understanding to the functional consequences of genotypic variation.'

Figure 3.39 Gray horses. Judging by pigmentation, the horse on the left may be a year or two older than the one on the right, but they are both quite young, apparently between 4 and 6 years old. In old age, gray horses may become essentially all white, and by that age usually have developed melanoma.

3.5 The head, heart, ears, and eyes

The vertebrate head represents a major and definitive evolutionary specialization that supports the brain with associated sense organs. Many of these specialized tissues are derived from neural crest (Gans & Northcutt 1983; Le Douarin & Kalcheim 1999). Neural-crest progenitor cells are the main contributors to craniofacial cartilage and connective tissue, and of course the melanocytes and peripheral neurons and glia (Barrallo-Gimeno et al. 2003).

The migratory pathways of the neural-crest cells that populate the skin of the head, the inner ear, and choroid and iris of the eye start primarily at the midbrain/hindbrain junction (Baxter & Pavan 2003; Wilson et al. 2004) and to some extent from the vagal neural tube, which area also gives rise to a stream of neural-crest cells that populate the developing heart. The RPE is induced *in situ* in the part of the optic vesicle that contacts the ectoderm. (The two optic vesicles are outgrowths of the neural tube, which form the retinae.)

Migration patterns of the melanoblasts of the head and heart differ from those of the trunk described above (Fig. 3.40), a fact that can be seen in embryos labeled with melanocyte markers (Baxter & Pavan 2003; Wilson et al. 2004) and can be inferred from the phenotypes of chimeric mammals. In addition to a difference in the migration patterns, the pigment patches in the head are also larger, suggesting a considerable clonal expansion during the second migratory phase in the epidermis, and often are quite specifically delineated (see Fig. 3.6), raising the suggestion of

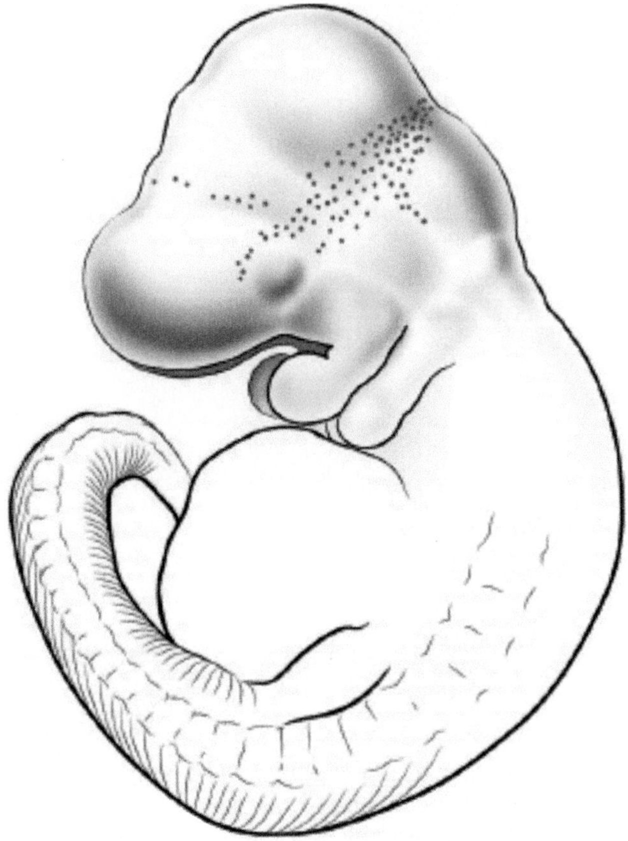

Figure 3.40 Diagram of embryo taken from Baxter and Pavan (2003). This diagram shows migratory patterns of melanoblasts in the head region of the embryo at day 10.5 of development. Cells that hybridized with PMEL17 (silver protein) and not DCT are shown in red. Blue indicates regions where cells hybridized with both proteins.

compartmentalization of melanoblast development in the head, a possibility that has not been studied. It may also involve the physical separation of parts of the developing face, which fuse only later (Bennett 1993).

It is of passing interest that an area in the mid-forehead in many white-spotted mammals is most likely to contain a white spot, and this probability is heritable (Lamoreux 1999; Rhim et al. 2000). This may reflect relatively sparse melanoblast migration into the frontonasal process of the embryo (diagram in Bennett 1993). In the absence of white spotting, an area of pheomelanic or pink-eyed dilute pigmentation may often be found in that location in chimeric animals, suggesting that these pigment types may be better able to survive than wild-type pigmentation in some tissue environments.

In the mouse, endothelin receptor A (EDNRA) is expressed in neural-crest cells of the head region during their delamination from the neural tube. The ligand for EDNRA is EDN1. In the absence of EDNRA, the pharyngeal arches fail to develop (Clouthier et al. 1998; Pla & Larue 2003).

In addition to generating pigment, melanocytes have evolved specialized locations and functions in the sensory organs and in the heart. For the most part, these functions are not well understood.

3.5.1 The ear

The inner ear is composed of several cell types, among them melanocytes which are necessary for its normal function. Deafness is a heterogeneous disorder, with many genes implicated, most of which are not associated with pigmentation. Given the complexity of the inner ear it is not surprising (Steel & Brown 1994; Steel 2002) that the expression of deafness may be strongly influenced by background genome (Anagnostopoulos 2002). The following web page is devoted to identifying gene expression in the developing ear: www.ihr.mrc.ac.uk/legacy/hereditary/genetable/index.html (Steel & Bussoli 1999).

White spotting is one clear cause of deafness. Deol (1970) noted the relationship between abnormalities of pigmentation and of the inner ear. Deafness may be associated with any of the major white-spotting phenotypes (Cable et al. 1992; Motohashi et al. 1994; Steel & Bussoli 1999; Yajima et al. 1998). Intermediate cells in the stria vascularis of the mammalian cochlea are melanocytes, which contain melanin pigments and are capable of synthesizing melanin. These melanocytes are required for normal development of the cochlea, as shown by studies of mutant mice with congenital melanocyte anomalies. Melanocytes are also needed for developed cochleae to function normally, as evidenced by studies of mutant mice with late-onset melanocyte anomaly and humans with acquired melanocyte anomaly. Mutations at *Splotch* (*Pax3*) additionally result in malformations of the inner ear (Steel & Brown 1994). The ear needs melanocytes for both normal development and normal function (Cable et al. 1992; Tachibana et al. 1992; Steel & Brown 1994; Tachibana 1999; Grimm et al. 2007).

The pigment itself, however, is not necessary for normal hearing. Albino mammals, whose melanosomes are not pigmented, ordinarily have normal hearing (although their eyesight is not normal). Pigment might protect against damage to the cochlea from stress, trauma, or chemical insult (Tachibana 1999). But more importantly, the melanocytes themselves apparently provide cation channels that are necessary to the function of the ear (Cable et al. 1992; Cable & Steel 1998; Grimm et al. 2007).

Apparently melanocytes in the stria vascularis, but not in other locations of the body, express *Gsta4*. *Gsta4* encodes one of the cytosolic glutathione-S-transferases (GSTs) (Uehara et al. 2009). This is an interesting observation also in light of the obscure relationship between pigment-type switching (Chapter 5) and glutathione. It would be interesting to know whether or not *Subtle gray* mice (Chapter 5), in which melanocytes are deficient in glutathione, are deaf.

The study of hereditary hearing impairments of course is not limited to the pigment cells, but the study of pigment cells in inbred mutant mice has provided clues to the functions of the inner ear and at the same time demonstrated differential function of MITF in different tissues (Uehara et al. 2009). Given modern genetics techniques, we are now in a position to deal with two objectives simultaneously: first, identification of the causative genes and the underlying pathogenic process in each form of deafness and, second, elucidation of the molecular and cellular mechanisms of hearing (Leibovici et al. 2008).

3.5.2 The heart

In addition to pigment cells of the skin, eye, and inner ear, a population of neural-crest cells exists in the heart (Rosenquist & Finnell 2007). Between E10.5 and 11.5, melanoblasts travel to the heart from the neural crest, between the post-otic area and the third somite, along the anterior and common cardinal veins and branchial arch arteries. By E14.5, large numbers are found in the atrioventricular endocardial cushions and persist in the adult mouse where they are found in the valves (mitral, tricuspid, and aortic) and the septa (ventricular and atrial) (Mjaatvedt et al. 2005; Brito & Kos 2008; Yajima & Larue 2008). These cells are absent in *Piebald-lethal* ($Ednrb^{s\text{-}l}/Ednrb^{s\text{-}l}$), $Kit^{W\text{-}v}/Kit^{W\text{-}v}$, and $Mitf^{mi\text{-}vga9}/Mitf^{mi\text{-}vga9}$ mice. No relationship has been reported between albinism or white spotting and differential cardiac performance in the mouse. Cardiac melanocytes are also found in birds, but not in fish or frogs.

3.5.3 The eye

Pigment cells and pigment are required for the normal development of the vertebrate eye and for its function throughout life (Clark 1986; Arnheiter 1998; Chow & Lang 2001; Bharti et al. 2006; Hu et al. 2008). The vertebrate eye develops from an evagination of the neuroepithelium in the region of the ventral forebrain. The pigmentation of the eye comes from two sources. The pigment cells of the uvea (choroid, iris) are derived from melanoblasts that migrate to the eye from the neural crest as the head is developing. They are capable of making eumelanin or pheomelanin (Wakamatsu et al. 2007b), with eumelanin predominating, especially in the uvea (Hong et al. 2006). The pigment cells of the choroid continue to produce melanosomes throughout life, while the melanization of the RPE is largely completed before birth (Lopes et al. 2007). Conversely, the vertebrate RPE is generated in place, directly from the optic neuroepithelium (Bharti et al. 2006). The RPE is a highly specialized tissue with multiple functions in development and throughout life (Chow & Lang 2001; Martínez-Morales et al. 2004, 2009). RPE cells form a monolayer of cuboidal epithelial cells that are joined together laterally by tight adherens and gap junctions. These cells are required for development of the retina and the normal eye (Raymond & Jackson 1995; Giménez et al. 2005), and of course normal vision. On their basal side, they contact the Bruch's membrane that lies between them and the choroid. On the apical side, they directly contact the outer segments of the photoreceptor cells of the retina (Clark 1986). Study of pigmentation has informed the function of the eye, as it has informed our understanding of many organs and tissues of the body. Detailed analysis of any of these is beyond our scope here.

While the functions of the RPE differ from those of other pigment cells, several of the white-spotting gene loci are critical to its development. Following the pattern with other organs and tissues, development involves many of the same players, but specific interactions are, at least in part, unique (Bharti et al. 2006). Not surprisingly, MITF is again a central element in development of this specialized type of pigment cell. MITF and orthodenticle homolog 2 (OTX2) (Martínez-Morales et al. 2003) are central elements in development of this specialized type of pigment cell (Mochi et al. 1998) interacting with PAX6 and with many of the genes that are required for melanogenesis in neural-crest-derived melanocytes. Other common factors including sonic hedgehog (Perron et al. 2003; Dakubo et al. 2008), FGF and RAS (Zhao et al. 2001), frizzled (Chang et al. 1999; Kameya et al. 2002), KITL/KIT, and EDN3/EDNRB function in the neural-crest-derived tissues, choroid and iris, but not in the RPE, a fact that explains the classic black-eyed-white phenotype.

The mutant alleles of MITF are numerous and the phenotypes resulting from the various genotypes influence pigmentation of the organs in complex ways (Fig. 3.19). This should make mice mutant at *Mitf* an excellent model to study similarities and differences that relate to eye RPE compared with neural-crest-derived pigmentation.

Albino mammals also have profound retinal abnormalities, including photoreceptor deficits and misrouted neural pathways from the eye to the brain. Patterns of cell production are distorted and delayed. At maturity, the central region is underdeveloped, rod photoreceptor numbers are reduced, and pathways at the optic chiasm are disrupted, such that many retinal axons tend to connect to the opposite brain hemisphere at the expense of the same side, reducing binocular vision

(Jeffery 2001a, 2001b; Giménez et al. 2004). Humans differ somewhat from mice in this respect (Neveu et al. 2006). These retinal abnormalities result in significant visual impairment (Kinnear et al. 1985; King et al. 2003a). Lavado et al. (2006) demonstrated, using transgenic mice, that the presence of L-dopa will repair the abnormal optic development that is caused by albinism. Lopez et al. (2008) showed that L-dopa is the ligand for GPR143, the product of *Oa1* (*Ocular albinism 1*; see section 4.4), and Young et al. (2008) found evidence that this receptor signals through $G_{\alpha}i3$. Thus, the *Oa1* gene product functions as an intracellular receptor that responds to its ligand (L-dopa, generated by melanogenesis) by activating a signaling pathway that is necessary (probably indirectly) for normal development of the optic nerve.

The fact that there are differential controls over pigmentation of the eye and skin is further emphasized by the black-eyed-white phenotypes of mice that are mutant at *Kit* or *Kitl*, which impact the neural-crest melanocytes but not the RPE cells (Mintz & Klein-Szanto 1992), as do some *Albino*-locus alleles and some alleles of *Mitf*. TYR is differentially regulated in the two pigment-cell lineages (Porter & Meyer 1994; Camacho-Hübner & Beermann 2001). The complexity of regulatory controls over the *Tyr* locus and the MITF protein and locus (Bharti et al. 2008) suggest a wealth of information yet to be mined by study of the genes that interactively and differentially affect the RPE and neural-crest melanocytes.

Figure 3.41 Calico cats. These pictures are not intended to show the usual but indeed a most unusual manifestation of the combination of Lyonization plus white spotting to produce calicos with perfect stripes. The genotype of the cat is otherwise *Nonagouti black* and *Nondilute* (*a/a B/B D/–*). It is rare because of the beautiful arrangement of the stripes/patches; the cat was used as the poster-feline for an animal welfare organization.

4

'Albinism' and the Failure of Normal Melanosome Maturation

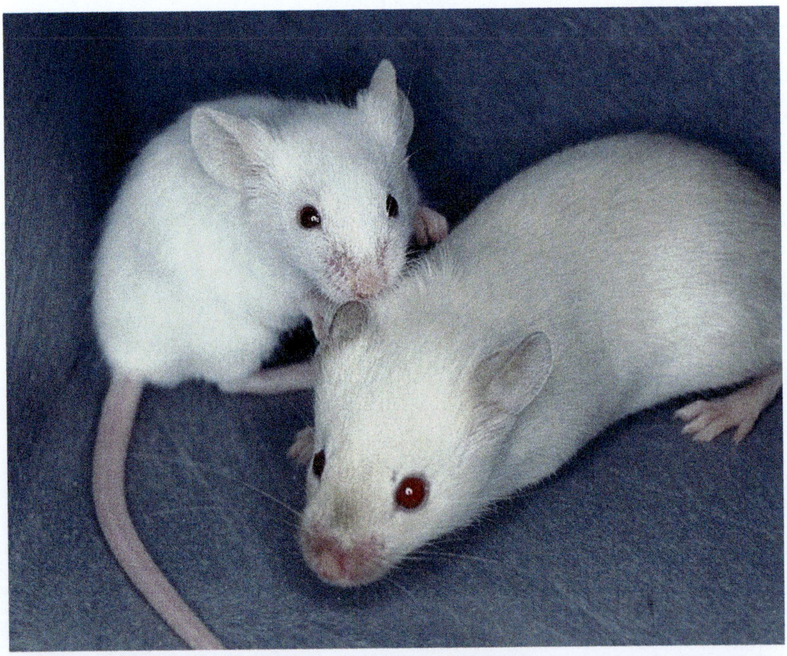

Figure 4.1 Mice mutant at the *Albino* (*C*, *Tyr*, *Tyrosinase*) locus. The mouse at the back is $Tyr^{c\text{-}44H}/Tyr^{c\text{-}44H}$ (*Dark-eyed albino*) and the mouse at the front is $Tyr^{c\text{-}h}/Tyr^{c\text{-}h}$ (*Himalayan*). For a discussion of alleles at the *Albino* locus see Beermann et al. (2004).

The Colors of Mice: A Model Genetic Network, 1st edition. By M. Lynn Lamoreux, Véronique Delmas, Lionel Larue, and Dorothy C. Bennett. Published 2010 by Blackwell Publishing Ltd.

4.1 Background

In this chapter we discuss the process of creation of the differentiated product of the pigment cell, the melanosome, which is fulfilling expectation as an excellent model for studying basic principles of organelle biogenesis, protein processing, sorting, and trafficking, and a number of human pigment-related diseases such as albinism, Hermansky–Pudlak syndrome, Chediak–Higashi syndrome, and Griscelli syndrome (Yasumoto et al. 2004; Raposo & Marks 2007).

In mice, the development (Chapter 3) and maturation of melanocytes (pigment cells) to their differentiated phenotype is complete soon after birth, when most of the pigment cells of the skin have come to reside in the hair follicles, some in the bulge region where they remain as melanocyte stem cells, and others in the bulb, where they differentiate to become pigmented melanocytes that deposit pigmented organelles (pigment granules, melanosomes), in elegantly controlled patterns, into the growing hair. In every subsequent molt cycle some of the melanocyte stem cells move down into the regenerating bulb to repeat the cycle of growth and differentiation. In humans, many melanosomes are also deposited into the basal layer of the epidermis, whereas adult mice have few melanocytes in the epidermis.

Processes required in the construction and function of a normal melanosome include:

- production and aggregation of the proteins that make up the primary matrix of the melanosome (section 4.2);
- production of the melanogenic enzymes (section 4.3);
- controlling the internal environment required by the melanosomal enzymes as they are transported to and maintained within the melanosome (section 4.4);
- processing and routing the melanosomal enzymes to the melanosome (section 4.5);
- transporting the melanosome from the area around the nucleus where it is produced to the periphery of the cell, where it is exported, usually to a keratinocyte of the skin or hair (section 4.6).

Defects in the genetic code that organizes these functions, preserved in mutant inbred mice and manipulated cell cultures, have elucidated several human diseases, including cancers, and have contributed impressively to our understanding of how cells function.

Failure of normal pigmentation within melanosomes defines albinism; however, the terminology in the literature is somewhat confusing and has evolved over time. In earlier days, the term was used differently by mouse geneticists and by physicians. Mouse geneticists generally restricted the term 'albino' or 'albinism' to apply to mutation at the *Albino* (*C*, now termed *Tyrosinase*, *Tyr*) locus; physicians often used the term albinism to refer to any sort of hypopigmentation. Today we can usually be more specific. Physicians recognize that the difference between albinism and white spotting is significant and the words 'partial albinism' (to refer to white spotting) and hypopigmentation (unless the cause is unknown) are falling out of favor. The geneticists' outlook has been broadened to include any defect in the processes involved in generating a normally pigmented melanosome within a viable pigment cell.

The *Albino* (*C*, *Tyr*, *Tyrosinase*) locus encodes the rate-limiting enzyme necessary for the production of melanin pigment within the melanosome (Fig. 4.2). However, many other loci with various distinct functions contribute to normal melanization within the pigment cell; defects at several of these loci can result in a generally similar 'albino' phenotype. Therefore, we will use the term albinism to apply to any failure of differentiation of viable melanocytes. This usage is not entirely logical with regard to melanogenesis, but is now standard.

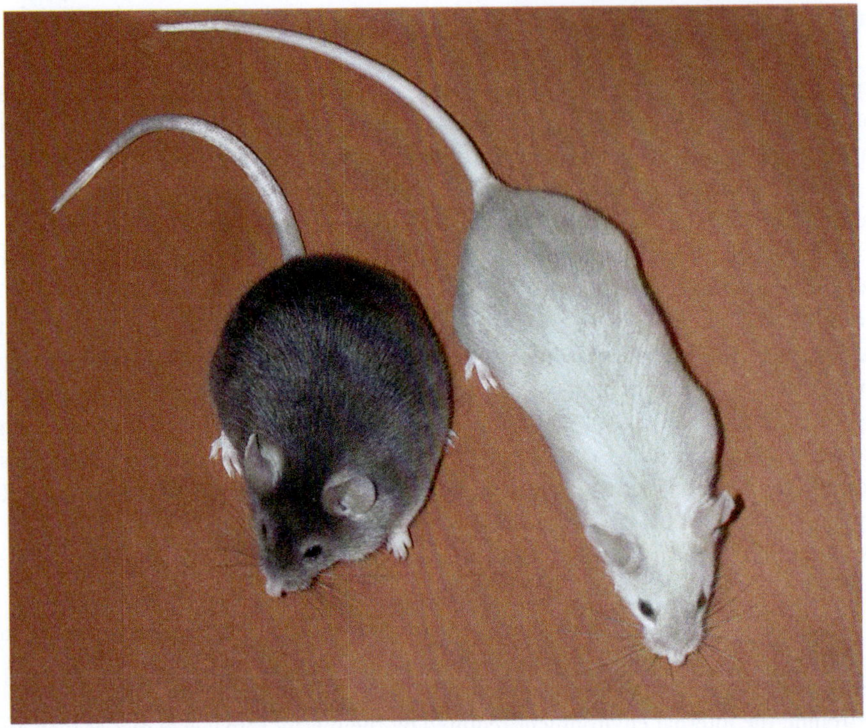

Figure 4.2 Mice mutant at the *Albino* locus. Both alleles generate a mottled (chimeric) phenotype, apparently the result of mutation at an upstream regulatory region. On the left is $Tyr^{c\text{-}m}/Tyr^{c\text{-}m}$ (*Chinchilla mottled*) and on the right is $Tyr^{c\text{-}em}/Tyr^{c\text{-}em}$ (*Extreme-dilution-mottled*).

Albino mammals have healthy pigment cells, but are incapable of making normally pigmented melanosomes. Completely albino mice are phenotypically white (lacking pigment in the skin and hairs) with pink eyes (pink because there is no pigment to mask the color of the blood circulating in the eyes). These white mice, unlike white-spotted mice, do have melanocytes, but their melanosomes are unpigmented. Melanogenesis is necessary for normal development and function of the eyes. Moreover, the retinal melanin behind the rods and cones reduces light-scatter, thus increasing the acuity of vision. Melanogenesis, interestingly, is also required for normal development of the optic chiasma. Therefore, the syndrome of 'albinism' also includes visual impairment, regardless of which genetic lesion causes reduced melanogenesis. Oculocutaneous albinism is a leading cause of visual anomalies in humans and other mammals. More than 10% of admissions worldwide to institutions for the visually impaired are due to some form of albinism (Manga et al. 2001).

A complex array of genetically controlled processes is required to generate normal melanosomes, and many genes regulate these processes, and therefore the overall syndrome of reduced pigmentation associated with visual defect is similarly complex. Complete albinism caused by failure of the rate-limiting enzyme tyrosinase (TYR) is referred to as oculocutaneous albinism type IA (OCA1A). Albino mammals with lesser melanogenic defects have lesser ocular and cutaneous defects (OCA1B, OCA2, OCA3, OCA4), which are defined by the genetic lesion as will be discussed below. Oculocutaneous albinism generally refers to a defect of mechanisms specific to melanogenesis and so specific to the structure and function of melanosomes.

Figure 4.3 Two of these mice have routing defects. Front to rear: *Cappuccino* (*Cno*^{cno}/*Cno*^{cno}), *Ruby-eye 2* (*Hps5*^{ru2}/*Hps5*^{ru2}), and the control black mouse. *Ruby-eye 2* (*Hps5*^{ru2}/*Hps5*^{ru2}) has been renamed for the human disease, *Hermansky–Pudlak Syndrome type 5*. *Ruby-eye* and *Ruby-eye 2* mice look very similar to each other, as do other mice whose mutations affect the various proteins of BLOC2 (biogenesis of lysosome-related organelle complex-2). These similarities were used in identifying the components of the BLOCs. The different appearance of *Cappuccino* (*cno*/*cno*) mice reflects their relationship with BLOC1. The black mouse is a control representing the fact that all three of the mice are *Nonagouti*, *Black*, but the control is normal at both the *Ruby-eye 2* and *Cappuccino* loci.

The melanosome is a member of the secretory lysosome-related family of organelles, which includes melanosomes and insect pigment granules, as well as platelet dense granules, lytic granules, MHC class II compartments, basophil granules, and azurophil granules. Hermansky–Pudlak syndromes are a set of related disorders of this lysosomal class of organelles (Fig. 4.3). The symptoms of Hermansky–Pudlak syndrome accordingly include those of oculocutaneous albinism, as well as a bleeding disorder and, in some patients, ceroid storage and progressive lung disease. The pigmentation problems associated with Hermansky–Pudlak syndrome result from defects of the proteins whose function is to route the melanosomal proteins from the endoplasmic reticulum (ER) to the melanosome. Other members of the secretory lysosome organelle family are similarly affected by failure of correct routing of their specific proteins (Huizing et al. 2008).

Chediak–Higashi syndrome, also a routing defect, is a rare, autosomal recessive disorder characterized by reduced pigmentation, severe immunologic deficiency with neutropenia and lack of natural killer cells, a bleeding tendency, and neurologic abnormalities. Griscelli syndrome reflects a defect in the normal transport and transfer of melanosomes themselves from the pigment cells into the keratinocytes of the skin and/or hair. Mice mutant at loci that are responsible for all this activity

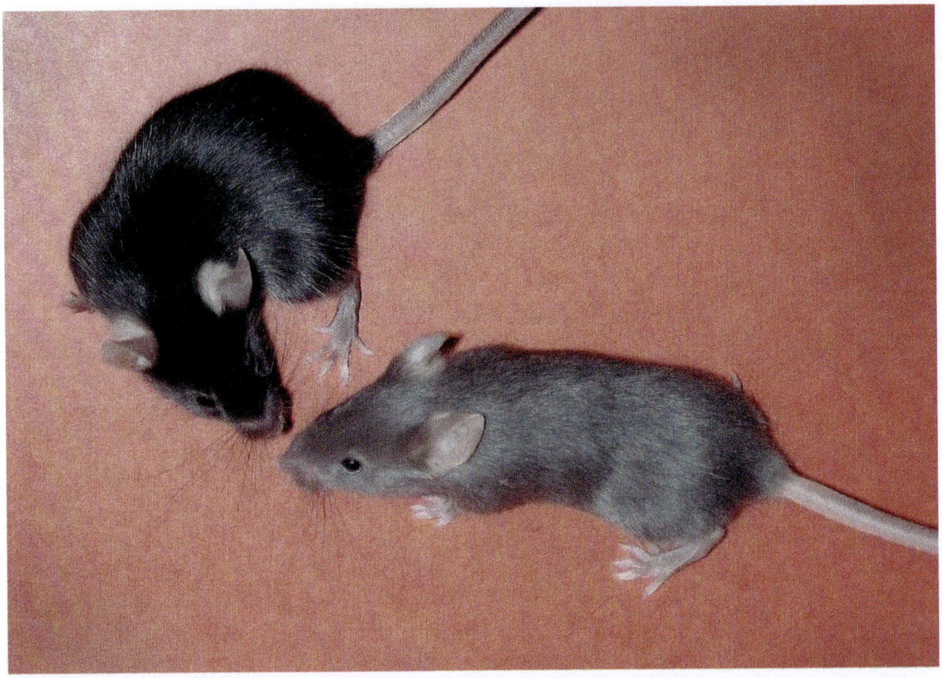

Figure 4.4 The *Nonagouti, Black* control mouse on the left contrasts with the *Nonagouti, Black, Dilute* ($Myo5a^{d}/Myo5a^{d}$) mouse on the right. Both are C57BL/6J. The dilute phenotype is caused by failure of the melanosomes to be properly transported to the periphery of the cell for deposition into the growing hair. The pigment itself is not 'diluted,' nor very much reduced, but its deposition into the hairs is irregular so that light is scattered. This phenotype is characteristic of mice and other vertebrates that are defective in melanosome transport.

have played a significant role in describing both the comparable human diseases and the cellular and molecular defects responsible for them (Barral & Seabra 2004).

The melanocyte is capable of producing two major types of melanosomes containing the two major classes of melanin pigment: eumelanin is dark black or brown (Fig. 4.4), while pheomelanin is variously referred to as red, yellow, or orange, depending upon species (Fig. 4.6), and in mice as yellow. Eumelanins and pheomelanins have been analyzed chemically and quite well described (Ozeki et al. 1997a, 1997b; Napolitano et al. 2000; Prota 2000; Lamoreux et al. 2001; Ito & Wakamatsu 2006).

Pigment-type switching, from eumelanin to pheomelanin or the reverse, is a complex process that involves rapid and reversible changes in a coordinated array of processes and proteins that function in producing melanosomes. Here we discuss eumelanogenesis and the melanogenic defects that have illuminated the array of processes necessary for creation of eumelanosomes (Fig. 4.7). From time to time we will comment on differences between eumelanic and pheomelanic mice/cells when the information is available. However, pigment-type switching itself will be described in Chapter 5.

Domestic animals have contributed significantly to our understanding of pigmentation, and especially of melanogenesis, and so a cautionary note regarding terminology may be useful here. Because artificial selection of livestock and pet species began hundreds of years before the science of

Figure 4.5 Agouti 'Silver' Tabby cat. The *Tabby* locus in the domestic cat is unlike murine loci phenotypically, and apparently genetically (Lyons et al. 2006). The cat is an agouti, in that respect similar to the cat in Fig. 1.7 and to wild-type animals of most species, including mice. Agouti animals have hairs with pheomelanic and eumelanic bands. The feline *Tabby* locus imposes a spatial pattern to the banding (Robinson 1971). Hairs in the yellow stripes on the animal are primarily pheomelanic; hairs in the nonyellow stripes are tipped with eumelanin and most often banded with pheomelanin. This cat also carries a gene that causes the pheomelanic pigmentation to be drastically reduced in the same way (grossly) that *Albino*-locus mutations do in mice. Cat fanciers refer to this dominant gene as the *Inhibitor* (*I*) gene (Turner & Robinson 1980) or more commonly as the silver phenotype, not to be confused with the *Silver* gene locus in other species (Menotti-Raymond et al. 2009). This picture was chosen to illustrate the differential effect of some mutant alleles on pheomelanin or eumelanin (a normal tabby cat would have ruddy yellow stripes between the black (or brown) stripes) and also to interpret some of the breeder terminology.

genetics, the names of breeds and colors are often much more artistic than they are scientific. A 'breed' of domestic animal is the result of artificial selection according to human desires, more lately to a 'standard of excellence' that is written and interpreted by breeders and fanciers. Individual breeds have been selected (mostly unawares) for specific collections of genotypes. However, a breed of domestic animal is not an inbred strain as defined in mice. A breed is genetically highly variable; an inbred strain is genetically homogeneous as a result of 20 or more generations of sibling mating or the equivalent. Similarly, the breeders' descriptions of colors are not the technical names of phenotypes as used in genetics. The technical name of a phenotype is (or should be) as specific as the nomenclature of the gene that generates it. For example, the term 'silver' applied to breeds or color variants of livestock or pets may (or may not) reflect a variant at the *Silver* locus. The cat in Fig. 4.5, though labeled by cat fanciers a 'silver tabby,' is actually the result of mutation at the *Inhibitor* locus (Turner & Robinson 1980), so named because of phenotypic inhibition of pheomelanic pigmentation. It is not our purpose to unravel breeder terminology for multiple species. We will use phenotype terms that apply specifically to murine mutant loci, but from time to time we will also mention phenotype names of other species that sometimes cause a bit of confusion.

Figure 4.6 A red (orange) cat. The eumelanin/pheomelanin alternative of the cat and the Syrian hamster are regulated by the genotype of an X-linked gene; referred to in the cat as *Orange* and in the hamster as *Sex-linked yellow*. Similar to the *Recessive yellow* locus of the mouse, the phenotype in the cat (and the hamster) is cell autonomous, as demonstrated by the patchiness of the coat pattern. The result, in cats (and hamsters) that are heterozygous at this locus, is the chimerism that we discussed in Chapter 3. The *Sly* gene of the Syrian hamster has been identified as a 'melanocortin pathway component that acts similarly to, but is genetically distinct from, *Mc1r*' (Alizadeh et al. 2009).

Figure 4.7 Nonagouti, brown, eumelanic, nontabby (a/a b/b O^+/O^+ t/t) Persian breed of cat. This color cat is known by cat fanciers as chocolate, not to be confused with the mouse *Chocolate* locus. The gene(s) that determine hair (and nose) length are independent of the color genes.

4.1.1 How to make a melanosome

The complex job of the differentiated pigment cell in the skin, hair, eyes, and a few other organs is to generate the proteins required to create a melanosome (pigment granule), and assemble them appropriately. Assembly also has to be done carefully, because melanogenesis is a toxic process.

The forming melanosome buds off from the ER of the melanocyte as a multivesicular body, a stage I melanosome (Fig. 4.8a). The protein product(s) of the *Silver* locus, SI/PMEL17/GP100 (GP100 is the human protein; it is known as GP87 in mice), is then delivered to the multivesicular body. When sequestered within the multivesicular body, the SI protein molecules are clipped into two portions, of which one self-assembles to form the highly structured, striated internal lattice that defines the stage II eumelanosome and produces its elliptical outline (Fig. 4.8b; see below).

In biogenesis of the eumelanosome, proteins including the three catalytic enzymes are transported through the cytoplasm in small membrane-bound transport vesicles to the maturing melanosome. When the transport vesicles arrive at the melanosome, they and the melanosomal membrane fuse, and the melanogenic enzymes, which are transmembrane proteins with the active site now inside the organelle, become active. The enzymes then begin to make pigment and eumelanins polymerize on the intraluminal matrix. Deposition of visible pigment, while striations are still visible, defines the stage III melanosome and melanogenesis (Fig. 4.8c).

At stage IV (Fig. 4.8d) the maturation process is complete. Melanin may incorporate some nonessential or harmful molecules that happen to be present, such as nicotine. Thus the specific structure of a melanin is somewhat variable, but the basic structure has been described (Ito & Wakamatsu 2003, 2008). As the stage IV melanosome becomes fully infused with melanin pigment, the pigment cell transports the pigment granule along microtubules to the ends of its dendrites, which contact neighboring keratinocytes of the skin or hair base. In humans, much of the skin color

(a) (b)

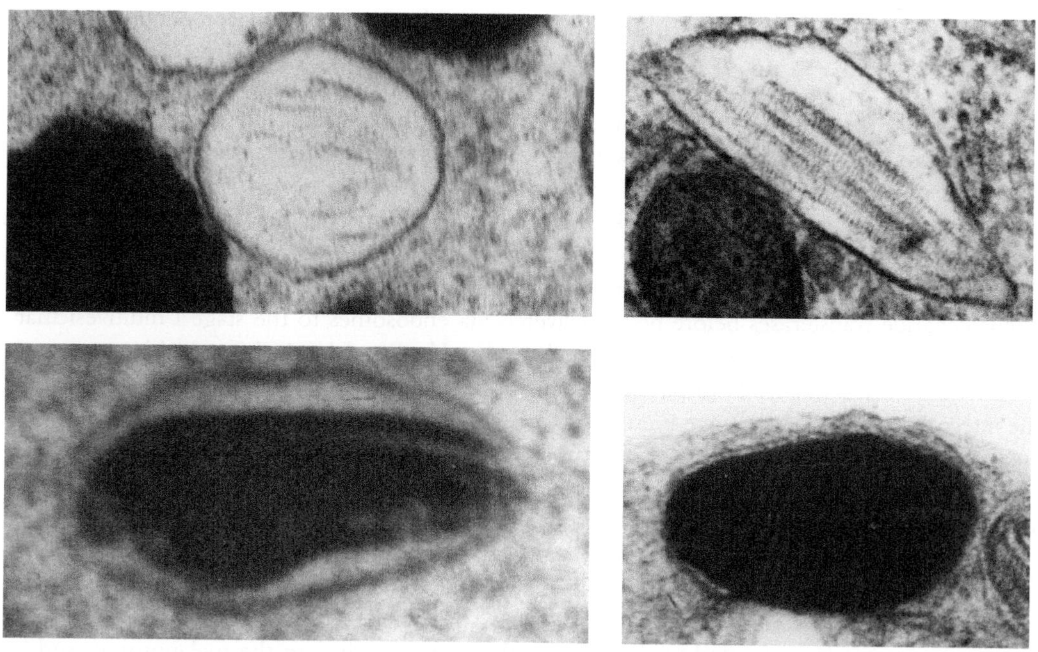

(c) (d)

Figure 4.8 Eumelanosomes: (a) stage I; (b) stage II; (c) stage III; (d) stage IV.

(apart from the pink, which is due to hemoglobin) results from deposition of melanosomes into the basal layer of the epidermis. In mice, most of the melanocytes reside in the hair follicles, where they deposit melanosomes into the hairs during the anagen (growing) phase of the hair cycle.

Pheomelanosomes of mice lack any structured matrix because they lack SI protein; the products of the *Pink-eyed dilution* and *Underwhite* loci are also absent. The catalytic enzymes that contribute to the structure of the melanosome are also affected. Tyrosinase-related protein 1 (TYRP1) and dopachrome tautomerase (DCT) are essentially absent from pheomelanosomes, whereas TYR activity is reduced (Kobayashi et al. 1995; Lamoreux et al. 1995). Thus a pheomelanosome is both structurally and functionally quite different from a eumelanosome. Pheomelanogenesis is discussed in Chapter 5.

In sum, the construction of a melanosome requires several classes of proteins: (1) the primary matrix protein, the silver protein, SI/PMEL17/GP100; (2) the enzymes that catalyze melanogenesis; (3) membrane proteins that regulate the interior milieu of the melanosome, and (4) proteins that are involved with the processing and routing of proteins and transport of the melanosome within the cell. And of course each of these proteins is encoded at a different locus. Thus, a large number of pigmentary genes are involved with the construction and delivery of melanosomes. A defect in any of the involved processes can result in 'albinism.' We will discuss these aspects of pigmentation under the five categories of structure: the matrix (section 4.2), melanogenic enzymes (section 4.3), regulators (section 4.4), the routing proteins that sort and deliver these to the maturing melanosome (section 4.5), and transport of melanosomes within the pigment cell (section 4.6).

4.2 The melanosomal matrix

Silver (*Si*, *Pmel17*, *Gp87*)

MGI lists one spontaneous phenotypic allele.

Availability: CMMR, JAX, and MMRRC (see Appendix in Chapter 1 for details about repositories).

Si encodes a structural protein responsible for converting the melanosome from the immature, relatively amorphous membrane-enclosed vesicle to the typical striated structure of the stage II eumelanosome. The protein encoded at the *Silver* (*Si*) locus (SI, GP87, PMEL17, PMEL; also known in humans as antigens gp100, NKIbeteb) is a melanocyte-specific protein required for the formation of the matrix upon which melanins are deposited in eumelanosomes (Kwon et al. 1994, 1995; Theos et al. 2006). Synthesized as a 70-kDa transmembrane protein, it is modified by Golgi-associated oligosaccharide transferases before being delivered via endosomes to the stage I multivesicular melanosomes (Harper et al. 2008). There it is cleaved and becomes luminal, and highly structured in eumelanosomes (Raposo & Marks 2007). The 'fibrils,' which resemble amyloid, form much of the internal structure of the stage II eumelanosome and confer its elongated shape (Theos et al. 2006; Harper et al. 2008). Once this structural reorganization occurs, eumelanogenic enzymes TYR, TYRP1, and DCT are separately targeted to the melanosomes, and become competent to synthesize eumelanin pigment (Yasumoto et al. 2004).

The murine *Silver* (Si^{si}) mutation is the result of a G-to-A substitution in the cytoplasmic tail of the SI protein that gives a truncated protein lacking targeting motifs in its cytoplasmic tail (Martínez-Esparza et al. 1999; Solano et al. 2000; Theos et al. 2005a). Because of this defect, SI protein tends to be translocated to the cell membrane rather than to the melanosome, and is

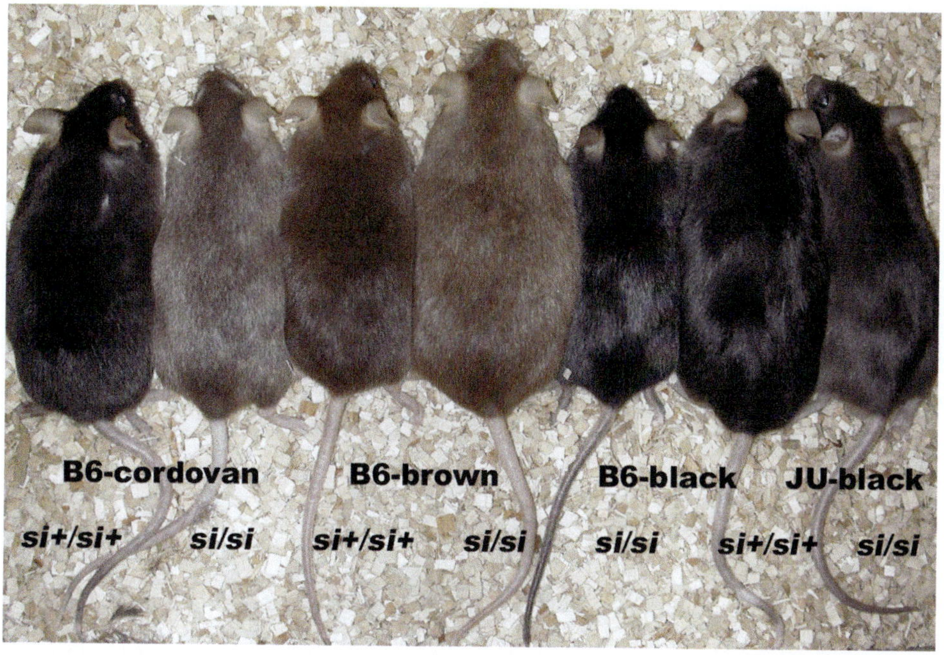

Figure 4.9 Silver (*Si*^{si}/*Si*^{si}, labeled '*si*/*si*') and non-silver ('*si*+/*si*+') mice that are *Cordovan, Brown*, or *Black* (b^c, *b*, or *B*) at the locus (*Tyrp1*) that determines whether black or brown pigment will be deposited on the melanosome in eumelanic mice. Silvering is more apparent in homozygous *Silver* mice that are mutant at *Tyrp1*. *Silver, Black* mice of the C57BL/6J and JUCtLm strains show very little silvering but there is a slight difference in their appearance compared with mice that are wild type at the *Silver* locus. In *Cordovan* mice, however, the difference between silver and nonsilver is easy to see. Do not confuse the light reflections with silvering. Strain background also influences expression of *Silver*.

inefficiently exported from the ER, resulting in overall larger melanosomes with a round morphology and few or no striations (Theos et al. 2006). The general phenotype in animals with mutations in *Silver* is a reduction in the amount of pigmentation; as a potential explanation, there is evidence in mice for a reduced number of melanocytes (Chase & Rauch 1950), supported by the poor viability of cultured *Silver* melanocytes, which is counteracted by an inhibitor of TYR (Spanakis et al. 1992; D.C. Bennett, unpublished results). These melanocytes are not hypopigmented, suggesting that the pigmentary defect *in vivo* may likewise be one of cell death, perhaps due to failure of SI-deficient melanosomes to bind and sequester toxic melanin intermediates. In mice the reduction in coat pigmentation is modest (see Silvers 1979) and minimal in C57BL/6J mice (M.L. Lamoreux, personal observation; and see Figs 4.9 and 4.11). In some other animals, null mutations lead to nearly complete absence of pigment. Thus, phenotypes of vertebrates mutant at *Silver* are somewhat variable, but generally semi-dominant.

The mouse melanocyte apparently expresses only one type of silver protein (Martínez-Esparza et al. 1999). The *Si* gene promoter in mice contains an M-box, indicating probable regulation by microphthalmia-related transcription factor (MITF; among other transcription factors). In the zebrafish, mutation of the *Si* homolog *Fdv* (*Fading vision*) locus results in a pale body color (Schonthaler et al. 2005). The vision defect is caused by developmental aberrations and reinforces the fact that pigmentation is necessary for normal development of the eyes.

Figure 4.10 Merle dogs. Ripple (rear) is a *Nonagouti, Black* (a/a, $Tyrp1^B/Tyrp1^B$) dog that is heterozygous for the dominant *Merle* mutation at the *Silver* locus. Puppy (front) is similar, but *Black-and-tan* (a^t/a^t) at the *Agouti* locus, and also carries the dominant modifying gene that causes the patches of different colors to be larger and more defined (Clark et al. 2008). He also has white spotting. See also Fig. 2.1.

Dominant white, as seen in the ubiquitous White Leghorn egg-production breed of chicken, is a pigment defect associated with the homolog of the *Si* locus, as are the *Smokey* and *Dun* mutant alleles of the chicken, whose effects are less extreme (Kerje et al. 2004).

Silver mutations found in dogs, horses, cattle, and mice have been useful in clarifying the relationship between phenotypes and their genetic cause. 'Merle' and 'harlequin' dogs of many breeds (Fig. 4.10) are heterozygous for a transposable element that essentially inactivates the *Silver* gene (Clark et al. 2006, 2008). The patchy patterns of pigmentation in these heterozygous dogs reflects excision of the element, plus the influence of a dominant modifier gene that affects the size and character of the resulting clone of normally pigmented hairs (Whitney & Lamoreux 1982; Sponenberg & Lamoreux 1985). The mutant modifier gene is lethal in the homozygous state (Clark et al. 2008). Dogs that are homozygous for the *Merle* mutation are very pale, nearly white in color, with occasional pale spots that presumably represent reversion to the heterozygous condition.

In some breeds of ponies and horses, a color variant that is usually known as 'silver-dapple' typically causes very pale mane, tail and legs, with silvered, brownish looking dapples on the body. This phenotype is caused by a mutation at the *Silver* locus (Brunberg et al. 2006). Many variations in phenotypic expression of this mutant depend upon the basic color of the pony or horse resulting from segregation of alleles at other loci (for a discussion of the relationship between genotypes and colors of horses see Sponenberg (2009) and the University of California at Davis website, www.vgl.ucdavis.edu/services/coatcolorhorse.php). The typical 'silver dapple' pony or horse (Fig. 4.12) is heterozygous for a mutant silver protein and is also bay in color. In horses, bay is the

Figure 4.11 Two brown mice with and without silvering. A C57BL/6J-*Tyrp1*^{b}/*Tyrp1*^{b}, *si*/*si* (brown, silver) mouse is shown at the rear, to the right, and a control mouse, brown without silver, C57BL/6J-*Tyrp1*^{b}/*Tyrp1*^{b}, *Si*/*Si*, is at the front, to the left. Do not confuse the light reflections on both mice with the silvering on the right-hand mouse.

wild-type color, consisting of patterns of eumelanin and pheomelanin that typically are not agouti, as in mice, but are distributed over the body (Fig. 1.17). Bay horses often have a more or less obvious pattern of dapples on the body that are formed by irregular distribution of eumelanin and pheomelanin pigments. We have referred to this pattern as it is enhanced by mutation at the *Gray* locus of the horse in Chapter 3. Bay horses always have eumelanic 'points;' that is, mane, tail and legs. In such a pony or horse that would otherwise have a black mane, tail, and legs and some dark hairs on the body, the eumelanic (black in this example) pigmentation is reduced by heterozygosity at the *Silver* locus, but any pheomelanic pigmentation is not appreciably reduced. The result is a 'silver dapple' phenotype with reduced pigmentation of the eumelanic dapples of the body and nearly white mane, tail, and legs.

The *Silver* locus has been implicated, by mapping, as a cause of the dilution of the Charolais breed of cattle, which are genetically pheomelanic (Gutiérrez-Gil et al. 2007) but a creamy off-white in color. *Silver* mutant phenotypes are reported in both pheomelanic and eumelanic Highland cattle (Berryere et al. 2007) where the heterozygous pheomelanic animals appear to be more intensely pigmented than the eumelanic, and the homozygotes, both black and yellow, are approximately of the shade of Charolais breed of cattle, as shown in Figure 4.13. Figures 2.6 and 5.11 show cattle that carry the *Silver* gene from a Charolais origin.

In all the examples given above, *Silver* mutations of other species inform the condition in mice, where the silver phenotype is not so strongly expressed on many inbred backgrounds. In each other species, the mutation reduces the eumelanic (black / brown) pigmentation more than it reduces

Figure 4.12 Silver dapple color, Rocky Mountain breed of horse. Photograph courtesy of Philip Sponenberg (Sponenberg 2009).

Figure 4.13 Charolais breed of bull, believed to be homozygous for a mutant *Silver* gene. Courtesy of the Britten Trust.

pheomelanic (yellow/red) pigmentation on the same animal or similar animals. The differential reduction of eumelanic compared with pheomelanic pigmentation is not surprising, given that pheomelanic pigment cells do not contain silver protein, at least not in mice (Chiu et al. 1993).

In *Silver* mutant mice, a reduction in yellow pigmentation has been discussed by Silvers (1979) and was originally reported by Dunn and Thigpen (1930). Those mice were not inbred. Lamoreux (unpublished results) created three lines of C57BL/6J *Silver* mice. In the process of backcrossing on to the C57BL/6J inbred strain it was evident that at least one and probably two modifying factor(s) were lost as the backcrossing progressed. Reduced pigmentation was not seen on a yellow background, but was seen on a eumelanic background (Figs 4.9 and 4.11). The eumelanic C57BL/6J-*si*/*si* mice were much less affected by the mutant genotype than the original stock that had been selected for maximum expression of the silvering. These mice are available at MMRRC.

Silver phenotypes of mice are influenced by the genotype at the *Brown* (*Tyrp1*) locus, as described in Silvers (1979), so that the silvering is more evident in mice that are heterozygous than homozygous for the $Tyrp1^{b}$ mutation. A similar relationship is evident in the C57BL/6J-silver lines. Cordovan silver (B6-$Tyrp1^{b\text{-}c}$/$Tyrp1^{b\text{-}c}$, Si^{si}/Si^{si}) mice are more silvered than black silver (B6-$Tyrp1^{+}$/$Tyrp1^{+}$, Si^{si}/Si^{si}) mice (Fig. 4.9), and mice that are heterozygous for brown ($Tyrp1^{+}$/$Tyrp1^{b}$ Si^{si}/Si^{si}) are more silvered than either homozygote (these mice are available at MMRRC). The reduction of pigmentation is more obvious at the base of the hairs than superficially, and on the B6 background did not increase with age after the first molt, as described by Dunn and Thigpen (1930). These interactions suggest a physical relationship between the SI protein and TYRP1, as proposed by Silvers. Given that the SI is a major structural protein of the melanosome and TYRP1 is an enzyme that forms an integral part of the melanosome, it is not surprising that their functions might interact.

The 'white-based hair' phenotype in mice has been attributed to progressive loss of pigment cells from the hair follicle during hair growth, so that the tip of the hair is pigmented and the base is not. This phenotype has been reported in mice mutant at the *Si*, *Tyrp1*, and *Dct* loci. In the case of $Dct^{Slt\text{-}Lt}$, Budd and Jackson (1995) demonstrated that *Albino* (Tyr^{c}/Tyr^{c}) mice of the mutant genotype did not experience the progressive melanocyte loss that is seen in the hairs of pigmented mice. They proposed that the melanocyte death was the result of toxic compounds of the melanogenic pathway leaking from the melanosome because of its aberrant structure. This explanation may also apply to white-based hair phenotypes of mice mutant at *Tyrp1* and *Si*, to the high death rate of cultured *Silver* melanocytes (Spanakis et al. 1992), and to the more severe pigmentary defects in *Silver* mutants of other species.

The availability of inbred mice mutant at *Si* and at *Tyrp1*, as well as several vertebrate species that carry a variety of naturally occurring mutations of the *Si* locus, opens a number of avenues to enhance our understanding of the cell biology of the melanocyte and of other cells that may be more important to us. It is worth noting that Harper (2008) found the SI protein to fulfill several criteria as an amyloid. The ways in which the pigment cell sequesters melanin safely in the melanosome, and disposes of defective melanosomes or their components, may therefore be potentially useful in the study of several human disorders that involve amyloids, including Alzheimer's disease. In Chapter 3, we briefly described *Notch* as a factor in differentiation of melanoblasts. *Notch* and presenilins are components of the signaling pathway involved in processing amyloid precursor protein, and they function in intracellular protein transport, including transport of TYR (Wang et al. 2006; De Strooper & Annaert 2001) and proteins defective in familial Alzheimer's. *Notch* is also required for normal pigmentation, apparently acting at the stage of differentiation of the melanocyte or melanosome (Moriyama et al. 2006; Schouwey et al. 2007; Aubin-Houzelstein et al. 2008). These relationships further expand the number of useful questions that can be addressed using mutations associated with pigmentation.

Clearly, the phenotypic expression at the *Silver* locus is responsive to several other factors in the cell. It is tempting to wonder whether one of these may be *Gpnmb*, discussed below.

Gpnmb (Glycoprotein transmembrane NMB)

MGI lists seven mutant alleles at this locus, of which one ($Gpnmb^{ipd}$; which has been changed to $Gpnmb^{R150X}$) is spontaneous.
Availability: JAX.

GPNMB is a highly glycosylated type I transmembrane protein that shares significant sequence homology with several melanosomal proteins, including 30% homology with SI. GPNMB is expressed in melanocytes and melanoblasts and its expression is dependent on MITF (Loftus et al. 2008). In early development *Gpnmb* mRNA is detected at high levels in the retinal pigment epithelium. Sequence homology to known melanosomal proteins, mRNA expression, and subcellular localization suggest that GPNMB is an intracellular, endosome-/melanosome-specific protein important for melanin biosynthesis and the development of the retinal pigment epithelium and iris. GPNMB is involved in the syndrome of defects that cause glaucoma in DBA/2J mice, which are mutant at the *Dilute*, *Brown*, and *Agouti* loci (Chang et al. 1999; Anderson et al. 2002; Bachner et al. 2002). Anderson and coworkers (2002) identified the spontaneous mutant of murine *Gpnmb* that contributes to iris pigment dispersal, and they also found that an allele of *Tyrp1* contributes to iris stromal atrophy in the DBA/2J murine model for glaucoma. They suggest that each of the proteins is required for normal melanosome structure. Further, they suggest that the glaucoma syndrome results from leakage of toxic substances from the melanosome. This conclusion is based in the observation that the syndrome is absent from mice that lack melanogenesis because of absence of functional TYR. This is reminiscent of the similar proposals to explain the white-based hair phenotype in $Tyrp1^{B\text{-}Lt}/-$ mice, $Dct^{Slt\text{-}Lt}/-$ mice, and *Silver* mice. Furthermore, Anderson and coworkers (2006) report differing expression of the iris stromal atrophy phenotype in congenic inbred strains that carry different combinations of alleles. This concept of interacting factors driving the phenotype correlates with our observations of polygenic interactions between SI and TYRP1 driving the phenotypes of the hair coat (see above), including the requirement of additional modifying genes. Perhaps one of these is *Gpmnb*. Further study is needed.

4.3 The enzymes that catalyze melanogenesis

The deposition of eumelanin pigment upon and within the framework of the stage II melanosome defines the beginning of stage III, as the organelle matures into a fully pigmented stage IV eumelanosome. At least three enzymes, the tyrosinase-related proteins, namely TYR, TYRP1, and DCT, are involved in eumelanogenesis (Spritz & Hearing 1994; Winder et al. 1994). During pheomelanogenesis, TYR activity is reduced, while TYRP1 and DCT are not expressed (Fig. 4.14).

Albino (C, Tyr, Tyrosinase)

Mutations at the human TYR locus cause OCA1A and OCA1B.
MGI lists 105 phenotypic alleles including 26 spontaneous ones, many of which are repeat mutations to the null condition.
Availability: HAR, JAX, MMRRC, ORNL, and RBRC.

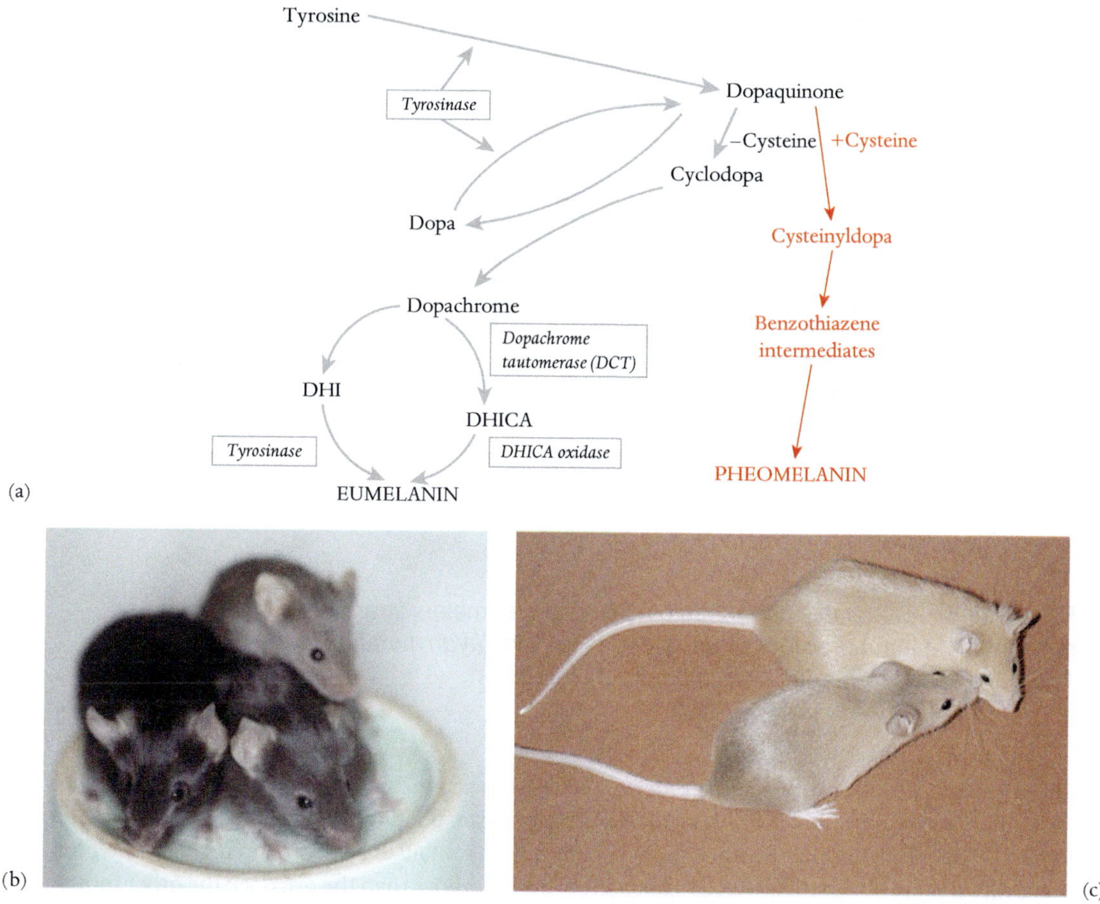

Figure 4.14 (a) The melanogenic pathway (simplified from Ito & Wakamatsu 2008; reproduced with permission of Wiley-Blackwell). The melanogenic enzymes, tyrosinase, 5,6-dihydroxyindole-2-carboxylic acid (DHICA) oxidase encoded at *Tyrp1*, and dopachrome tautomerase (DCT) are shown in boxes and are discussed in this chapter. (b) In nonagouti mice that are otherwise wild type at pigment loci, mutation in *Dct* results in slaty mice (b, lower right mouse). Mutation in the DHICA oxidase gene (*Tyrp1*) results in brown mice (b, rear). The mouse on the left in this picture is wild type. The eumelanin/pheomelanin switch mechanism is discussed in Chapter 5. (c) The two pheomelanic mice are (front) *Recessive yellow*, in which the MC1 receptor is mutant, and (back) *Lethal yellow*, in which the *Agouti* locus is mutant.

The mouse *Albino* locus encodes the rate-limiting enzyme required for melanogenesis, TYR (Beermann et al. 2004). The function of TYR is to catalyze formation of melanin pigment in the melanosome, and the result of correct gene function is pigmented melanosomes. At the phenotypic level, therefore, the function of the *Albino* locus is to determine whether or not, and how much, melanin will be produced. Oculocutaneous albinism type IA (OCA1A) is the human syndrome that results from complete absence of TYR activity. If melanogenesis is only diminished, because of a defect at this locus, the defect is known as OCA1B. The defects in naturally occurring OCA1A- and

Figure 4.15 Eumelanic nonagouti black cat and eumelanic nonagouti black siamese cat (mutant at the *Albino* locus, homologous to *Himalayan* mutation of the mouse), c^h/c^h. The siamese cat illustrates an ocular effect of albinism.

OCA1B-like mice are encoded at the murine *Albino* locus and include temperature-sensitive alleles (similar to 'siamese' in the cat of Fig. 4.15 and the himalayan phenotype in rabbits, or the mouse in Fig. 4.1). The mottled, chimeric mice shown in Fig. 4.2 are mutant at an upstream regulatory region of the *Albino* locus. Some *Albino* locus alleles affect the hair more than the eyes (see the other mouse in Fig. 4.1). One allele (*Platinum*) lacks the normal response to pheomelanogenesis.

Except in *Platinum* ($Tyr^{c\text{-}p}/Tyr^{c\text{-}p}$) mice (Orlow et al. 1993; Beermann 1995, 2004), pheomelanin pigmentation is more reduced in non-null *Tyr* mutant mice than eumelanin (Lamoreux et al. 1986; Lamoreux 1986; Lamoreux & Pendergast 1987). This same relationship is evident in *Agouti*, *Chinchilla* ($A/-\ Tyr^{c\text{-}ch}/Tyr^{c\text{-}ch}$) mice, but the best way to observe this difference in mice is to compare inbred mice that differ at the *Agouti* and *Tyrosinase* loci. A black (C57BL/6J-$a/a\ Tyr^{c\text{-}ch}/Tyr^{c\text{-}ch}$) *Chinchilla* mouse will be nearly indistinguishable from an ordinary black C57BL/6J-a/a mouse, whereas the phenotype of the *Chinchilla* yellow mouse will be dramatically paler than that of a C57BL/6J-A^y/a yellow mouse.

If TYR is absent or nonfunctional, the melanocyte will contain unpigmented melanosomes and the animal will be an albino, lacking melanin pigment in the skin, hair and eyes. The phenotype of albino animals includes abnormal development of the optic nerves (it is not always so obvious as in the cat on the right in Fig. 4.15). The optic-nerve abnormality is common to animals with reduced melanogenesis, but is not generally shared by animals with white spotting. White-spotted animals, but not albino animals, very often are deaf. Based on these phenotypes, it has been inferred that pigment cells, not necessarily pigmentation, are required for normal function of the inner ear; while melanogenesis itself is necessary for normal formation of the optic chiasma. These observations led to studies that clarified the mechanisms by which pigment cells in the cochlea are necessary for normal hearing (Steel & Brown 1994; Steel & Bussoli 1999), explaining the auditory effect of white spotting. And it has recently been shown that the optic nerves, for their proper development, require the normal function of OA1, described below, a G-protein-coupled receptor that uses an intermediate product of melanogenesis as ligand (Lavado et al. 2006; Lopez et al. 2008).

TYR is a copper-containing type I transmembrane glycoprotein that is the rate-limiting enzyme necessary to catalyze the initial step in melanogenesis, conversion of tyrosine to dopaquinone, and participates in other steps, including oxidation of DOPA (see Fig. 4.14). The intermediate compounds produced during eumelanogenesis can be toxic to the cell, a fact that may explain the intricate control that the cell exerts over the processing and trafficking of TYR. Defects in the enzyme itself – or in any of a number of proteins involved in its processing – result in phenotypic albinism.

TYR undergoes a complex process of post-translational processing before reaching the melanosome. This processing involves N-glycosylation in several sites, including one located in the CuB copper-binding site (Olivares et al. 2003), as the enzyme moves from the ER to the Golgi, and is sorted to the melanosome (Watabe et al. 2004). Sorting to the melanosome requires a dileucine motif in the cytoplasmic tail (Höning et al. 1998). It is interesting that in melanosomal proteins the dileucine is preceded by a highly conserved proline (PLL), strongly suggesting functionality, but this is not yet understood. The dileucine motif is recognized by adaptor proteins 3 and 1 (AP-3 and AP-1), which are thought to function in partially redundant pathways to transfer TYR from distinct endosomal subdomains to melanosomes (Theos et al. 2005b). Adaptor proteins are components of transport vesicles and enable them to be targeted to the correct cellular compartment.

Processing and folding of TYR in the ER requires the chaperone molecule calnexin (Toyofuku et al. 2001). Phenotypic and other evidence strongly suggest that TYR and TYRP1 (encoded at the locus that determines whether eumelanic pigment will be black or brown, and discussed below) interact physically. On a C57BL/6J background, black eumelanic mice that are homozygous for the *Chinchilla* allele (*Tyr*$^{c\text{-}ch}$/*Tyr*$^{c\text{-}ch}$) are somewhat reduced in intensity of color. The pigmentation appears less 'saturated,' and while the difference is small it has been sufficient to identify mutant mice. Congenic *Brown* mice are not affected in that way by the *Chinchilla* mutation. These phenotypes suggest that TYR and TYRP1 may interact physically, perhaps in the ER (Toyofuku et al. 2001); however, they probably are routed separately from the ER to the melanosome, and they clearly function together in the melanosome. The interaction in the melanosome may be necessary for normal eumelanogenesis (Kobayashi & Hearing 2007). In living tissues, TYR is not normally active outside the melanosome, and in fact all of the melanogenic enzymes are carefully controlled by the cell from the time of their production to their arrival in the melanosome. Many gene products are required for the maturation and correct routing and functioning of TYR, as discussed in section 4.5. TYRP1 (below) is also required for full TYR activity.

Albino mice have been widely used for many biological applications; a panel of deletion mutations caused by irradiation has been used for evaluation of albinism, mosaicism, and the functions of nearby loci (Russell 1979; Russell & Raymer 1979; Russell et al. 1979, 1982; Rinchik 1993a). The common murine *Albino* (OCA1A-like) defects are caused by *Tyr* mutations that interfere with the processing or catalytic function of the protein. Proteins that are not correctly processed and sorted through the ER are translocated to the cytosol and degraded by proteasomes (Svedine et al. 2004; Watabe et al. 2004). Other naturally occurring mutant alleles illuminate additional details about the functions of this locus (Beermann et al. 2004).

Mice homozygous for the alleles *Chinchilla-mottled* (*Tyr*$^{c\text{-}m}$) or *Extreme-dilution-mottled* (*Tyr*$^{c\text{-}em}$) are chimeric, with stripes of pale and paler pigmentation. The c^{em} mutation arose within the c^{m} stock; thus the mottling of both are of the same origin. The mutation modifies the expression of the *Tyr* gene via a chromosomal rearrangement that impacts a DNase I-hypersensitive region 15 kb upstream of the gene (Porter et al. 1991). In mice that are homozygous for one of these mottled *Albino*-locus mutations and also for *Piebald* (*Ednrb*^{s}), the calico pattern of patchiness (as described in Chapter 3) is present in pigmented areas near the white spotting (M.L. Lamoreux, personal observation), an observation that further supports the clonal nature of the chimerism (see Chapter 3) (Porter et al. 1991; Porter & Meyer 1994). The paler phenotype of the $Tyr^{c\text{-}em}/Tyr^{c\text{-}em}$ mice, compared with the $Tyr^{c\text{-}m}/Tyr^{c\text{-}m}$, is explained by an additional mutation that abolishes an N-glycosylation site associated with the second metal-ion-binding site of the enzyme (Lavado et al. 2005). Thus the problem of these mice, with regard to making pigment in the eyes and hair follicles, is caused by a combination of coding and noncoding genomic alterations resulting in several abnormalities that include suboptimal gene expression, abnormal protein processing, and reduced enzymatic activity (Lavado et al. 2005). Furthermore, the eyes of these mice tend to be darker than expected relative to the coat pigmentation.

Mice of the genotype $Tyr^{c\text{-}44H}/Tyr^{c\text{-}44H}$ (also called *Dark-eyed albinism*) exhibit a dramatic difference between body pigmentation and eye pigmentation. Their fur is nearly white, yet the eyes nearly fully pigmented. Schmidt and Beermann (1994) sequenced the cDNA to determine whether the mutation affected the structural gene or the promoter. They determined that this allele is the result of a single base-pair change in the structural gene, suggesting that the darker color in the eye compared with hair results from a difference in protein processing.

Mice homozygous for the *Platinum* allele (Tyr^{p}) are very pale in their pigmentation, in eyes and hair, in spite of the fact that the cell lysates of the eyes and skin exhibit high TYR activity (Townsend et al. 1981; Lamoreux & Pendergast 1987; Orlow et al. 1993, 1994). This mutant TYR lacks the normal cytosolic tail, as the result of a point mutation (Beermann et al. 1995). Normally, the cytosolic tail of TYR binds to AP-3 (Höning et al. 1998) for transport to the melanosome. Most of the TYR in platinum mice does not arrive at the melanosome, but takes the default routing pathway to the plasma membrane.

The few alleles discussed above, and their multiple interactions with other pigment-related proteins, illustrate some of the ways in which the *Albino*-locus mutant alleles have been used to illuminate basic functions of the cell. More examples are found in the remainder of this chapter. It is not surprising that such extensive controls over the processing and transport of TYR are necessary in the cell. As mentioned, TYR is the rate-limiting enzyme for melanogenesis, and its substrate, tyrosine, is a ubiquitous amino acid. In the presence of its substrate, TYR can generate a melanin-like product even without other enzymes. Thus, cell survival also depends upon careful control of the catalytic activity of TYR.

Tyrp1 (*Brown, B, Tyrp1, Tyrosinase-related protein 1*)

Mutations at the 'black/brown' locus can cause OCA3 in humans and can cause glaucoma in mice, especially in combination with other susceptibility genes.

MGI lists 57 phenotypic alleles, nine of spontaneous occurrence, which are probably all re-occurrences of the three original mutant alleles: *Brown*, *Cordovan*, and *Light*.

Availability: JAX, MMRRC, ORNL, and RBRC.

This locus encodes TYRP1, another transmembrane melanosomal enzyme that determines whether eumelanic pigment will be black (when active) or brown (when inactive or absent) (Bennett et al. 1990; Zdarsky et al. 1990; see the biochemical pathway in Fig. 4.14). Melanin pigment in *Brown* ($Tyrp1^{b}/Tyrp1^{b}$) mice (Fig. 4.16) or cats, dogs, or sheep (Fig. 4.17) is chocolate-colored, rather than black. *Cordovan* ($Tyrp1^{b\text{-}c}/Tyrp1^{b\text{-}c}$) mice are darker, more akin to charcoal-brown. *Light* ($Tyrp1^{B\text{-}lt}/-$; sometimes referred to as *B-Light*) mice have an unusual pigment defect, in that the base of the hairs lacks pigment, while the tips are of a color resembling that of *Cordovan* mice (Johnson &

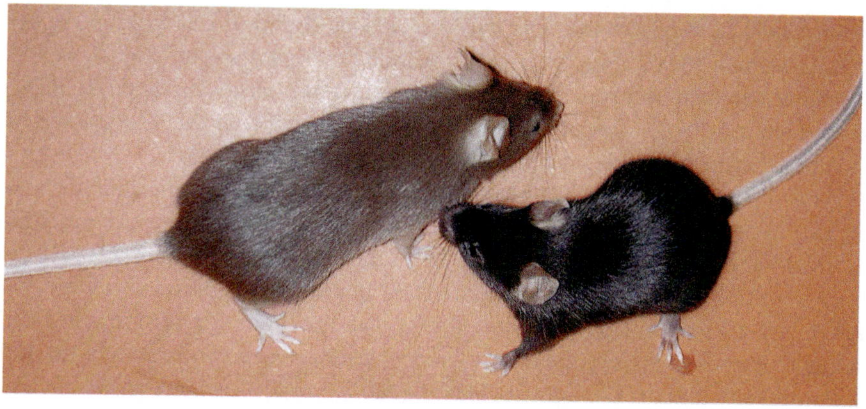

Figure 4.16 A C57BL/6J-*Tyrp1*^{b}/*Tyrp1*^{b} (eumelanic brown) mouse is shown on the left and a eumelanic black control mouse, C57BL/6J-*Tyrp1*^{B}/*Tyrp1*^{B}, is shown on the right.

Figure 4.17 Brown Shetland breed of sheep.

Jackson 1992). *B-Light* is a semidominant defect; the others are recessive. The cause of the B-light phenotype may be progressive loss of melanocytes from the hair follicle during each cycle of hair growth. The loss may be caused by leakage of toxic melanogenic intermediates from the mutant melanosomes, resulting in melanocyte death, as suggested by Budd and Jackson (1995) with regard to slaty-light mice. These have a similar white-based hair phenotype and the *Slaty* product, DCT, also forms part of the structure of the melanosome. OCA3 in humans is the result of mutation at the *TYRP1* locus (Boissy et al. 1996; Manga et al. 1997).

Various catalytic functions have historically been attributed to the *TYRP1* by different reports or work done under different conditions (Murray et al. 1983; Jiménez et al. 1991; Winder et al. 1993, 1994; Kobayashi et al. 1994; Orlow et al. 1994; Spritz & Hearing 1994; Vijayasaradhi et al. 1995; Toyofuku et al. 2001). Some of the conflicting results are attributable to an apparent difference in the function of TYRP1 between human and murine cells (Boissy et al. 1998); other differences in experimental results may reflect the tight physical and biochemical relationship between TYR and TYRP1. As technologies improved, it was concluded that the catalytic function of TYRP1 in mice (though not detected in humans) is 5,6-dihydroxyindole-2-carboxylic acid (DHICA) oxidase (Kobayashi & Hearing 2007). Other functions include stabilization of the TYR protein and contribution to the structure of the melanosome (Sarangarajan & Boissy 2001).

Brown ($Tyrp^b/Tyrp^b$) mice (Fig. 4.16) produce only slightly less eumelanin than do *Black* ($Tyrp1^B/-$) mice. The difference in color of pigment arising from the crucial point mutation in this allele apparently has a three-fold causation. First, the DHICA oxidase function of TYRP1 is absent in these mice, resulting in incorporation of more DHICA and less 5,6-dihydroxyindole (DHI) into the melanin (DHICA-based melanin is brown) (Ozeki et al. 1995). Second, the intimate relationship of TYRP1 and TYR in the melanosome may influence the enzymatic functions of both (Kobayashi et al. 1998; Toyofuku et al. 2001; Kobayashi & Hearing 2007). Third, brown melanosomes are smaller than black ones and spherical rather than elongated (Russell 1946, 1949; Hearing et al. 1973), potentially influencing the way in which the spectrum of white light is reflected to the eye. This suggests a function of TYRP1 as a structural component of the melanosome, or alternatively a requirement for normal TYRP1 for correct incorporation or aggregation of matrix protein SI into the melanosome and its matrix, since SI-deficient eumelanosomes are also spherical. Silvers (1979) and the above section on SI both describe phenotypes that suggest a direct interaction of SI with TYRP1.

TYRP1, along with several other melanogenic proteins, is not expressed in melanocytes that are making pheomelanin (Lamoreux et al. 1995; Kobayashi et al. 1995), as discussed in Chapter 5. As a result, a yellow pigmentary phenotype is epistatic to nonyellow (eumelanic) whether it be black or brown. Although TYRP1 is not concurrently expressed in cells that are making pheomelanin, nevertheless there are a few eumelanosomes in yellow mice, especially in the tips of the dorsal hairs. For this reason, there is a slightly warmer tone to the phenotypes of yellow animals that are also homozygous for $Tyrp1^b$ (*Brown*) compared with those that are yellow and genetically $Tyrp1^B/-$ (*Black*).

Melanocytes (identified by a silver-based stain technique) are more numerous and more dopa-reactive in the epidermis of 3-day-old newborn inbred C57BL/10J *Brown* mice than in congenic *Black* mice (Hirobe 1984b). This difference appears to be related to a difference in the rate of differentiation.

It has been reported that homozygosity for a *Brown* allele ($Tyrp1^{isa}$) causes glaucoma in the DBA/2 inbred strain of mice, which is mutant at *Dilute*, *Brown*, and *Nonagouti*. Because DBA/2J carries known mutant alleles at two additional pigmentation loci, and at *Gpnmb* (see above), it is not a convincing model for a one-gene hypothesis, but it does suggest interaction of the melanosomal membrane with TYRP1 (Anderson et al. 2002). So, while the DBA/2 glaucoma model is useful (Inman et al. 2006) and does seem to involve a mutant TYRP1, it does not convincingly attribute specific phenotypic effects to the *Tyrp1* locus.

Slaty (*Slt*, *Dct*, *Dopachrome tautomerase*, formerly *Trp2*, *Tyrosinase-related protein 2*)

MGI lists three spontaneous phenotypic alleles: *Slaty* (twice) and *Slaty-light*.
Availability: JAX, MMRRC, and RBRC.

The *Slaty* locus on chromosome 14 was originally termed *Trp2* because it encodes a member of the tyrosinase family of enzymes with TYR and TYRP1 (Cassady & Sturm 1994), which is regulated in part by MITF (Schwahn et al. 2005). *Slaty* encodes the melanosomal enzyme dopachrome tautomerase (Körner & Pawelek 1980; Hearing et al. 1992; Jackson et al. 1992; Tsukamoto et al.

Figure 4.18 All these mice are nonagouti, eumelanic black. From left to right, at the *Slaty* locus: *Slaty* (Dct^{slt}/Dct^{slt}), *Slaty-light* ($Dct^{Slt\text{-}Lt}/Dct^{Slt\text{-}Lt}$) and wild type.

1992; Budd & Jackson 1995) (see Figs 4.14 and 4.18). Mutation at the *Dct* locus of mice is responsible for a significant decrease in dopachrome-tautomerase activity and reduced intensity of eumelanic pigmentation (Jackson et al. 1992; Budd & Jackson 1995).

DCT is a type I transmembrane protein that functions as an enzyme in the eumelanosome. Eumelanic *Slaty* (Dct^{slt}/Dct^{slt}) mice are somewhat reduced in intensity of pigmentation compared with the wild type. *Slaty-light* ($Dct^{Slt\text{-}lt}$) is a semi-dominant allele that reduces the intensity of pigmentation yet more. *Slaty-light* hairs lack pigment at their bases in a fashion similar to the hairs of *B-Light* mice, described above. The white-based hair phenotype, and the work of Budd and Jackson (1995), suggest that the structural integrity of the melanosome may depend upon DCT (and other melanosomal proteins), and when aberrant may lead to progressive melanocyte death during each hair growth cycle as a result of leakage of toxic intermediates of melanogenesis. DCT is not expressed during pheomelanogenesis (Kobayashi et al. 1995; Lamoreux et al. 1995), as discussed in Chapter 5.

In the mutant *Slaty* ($Dct^{slt}/-$) mice, DCT is delivered normally to the stage II melanosome. There is no evidence of interaction with TYR or TYRP1 during processing (Costin et al. 2005). There seem to be no phenotypically evident interactions with *Silver*, *Brown*, or *Albino*, or with *Agouti*-locus alleles, but this possibility has not been investigated thoroughly.

4.4 Membrane proteins that regulate the internal milieu of the melanosome

Ocular albinism 1 (*Oa1*, *Gpr143*, *G protein-coupled receptor 143*)

MGI lists two phenotypic alleles, neither of which are spontaneous.
Availability: TAC.

Ocular albinism type 1 (*Oa1*, now *Gpr143*) encodes a seven-pass integral transmembrane protein, a G-protein-coupled receptor (Sone & Orlow 2007; Young et al. 2008), which has therefore now been renamed G-protein-coupled receptor 143 (GPR143). GPR143/OA1 differs from all other known G-protein-coupled receptors in being intracellular rather than on the plasma membrane, targeted to late endosomes/lysosomes and melanosomes (Schiaffino & Tacchetti 2005).

It has long been deduced from murine phenotypes that the process of pigmentation (as opposed to the presence of pigment cells) is required for normal development of the optic chiasma, and this was confirmed for OA1 by Incerti et al. (2000). Lavado et al. (2006) reported that L-dopa (a product of melanogenesis) is required for the appropriate development and innervation of the retina. Subsequently Lopez et al. (2008) showed that L-dopa is the ligand for GPR143, and Young et al. (2008) found evidence that it signals through $G_{\alpha}i3$. Thus, OA1 functions as an intracellular receptor that responds to its ligand (L-dopa, generated by melanogenesis) by activating a signaling pathway that is necessary (probably indirectly) for normal development of the optic nerve.

In humans, oculocutaneous albinism type I is also known as Nettleship-Falls oculocutaneous albinism. The effect of this condition on the eye is typical of the albinism syndrome: hypopigmentation of the retina, nystagmus, strabismus, foveal hypoplasia, deficient crossing of the optic fibers, and reduced visual acuity (Oetting 2002). The syndrome is characterized additionally by the presence of abnormally large melanosomes. The reduced pigmentation is not caused by absence of pigment cells, but by reduction in numbers of melanosomes within the cells. Analysis of an *Oa1*-knockout mouse suggests that the macromelanosomes are formed by abnormal growth of a single melanosome, and not by the aggregation and fusion of several smaller melanosomes (Incerti et al. 2000). It thus appears that the size of each melanosome is normally limited by $G_{\alpha}i3$ signaling. This makes sense because $G_{\alpha}i3$ is a member of the $G_{\alpha}i$ or G_i protein family, which inhibit cAMP synthesis. Thus it could mediate negative feedback to prevent too-rapid or too much melanogenesis, through CREB and MITF (Fig. 3.10). The exact mechanism is unknown, but since macromelanosomes are not found in *Oa1*-null mice that are also *Albino* ($Tyr^{c\text{-}2J}/Tyr^{c\text{-}2J}$ or $Slc45a2^{uw}/Slc45a2^{uw}$, *Underwhite*), it appears that melanin synthesis is involved in the overgrowth of the melanosomes. The structure of the eye is not affected by absence of pigment in albino mice (Cortese et al. 2005) or by the pheomelanin/eumelanin alternative (Newton et al. 1996).

Palmisano et al. (2008) reported defective regulation of organelle transport in *Oa1*-null melanocytes, the melanosomes aggregating at the cell periphery. This suggests a potential third function for GPR143.

Mottled, Mosaic, Pewter, Atp7a (*ATPase, Cu^{2+}-transporting, α polypeptide, MNK, Menkes protein*)

Twenty-three spontaneous phenotypic alleles of 37 phenotypic alleles are listed on JAX.
Availability: ORNL, JAX, MMRRC, HAR, TIGM, and RBRC.

The *Mottled* (*Atp7a*) locus of the mouse is X-linked and includes 37 naturally occurring alleles (*Mottled, Blotchy, Dappled, Brindled, Viable brindled, Tortoiseshell*, and *Mosaic* are the best studied), some of which are homozygous (hemizygous) lethal *in utero* at about embryonic day (E)12.5. This is the mouse equivalent of Menkes syndrome in humans (Mercer 1998). Mutations in this gene affect copper metabolism and, depending on the allele, result in abnormal pigmentation, with vibrissae, hair, and skeleton abnormalities. Behavior may be abnormal and defects of collagen and elastin fibers are reported (Monaco & Chelly 1995; Mercer & Llanos 2003).

The phenotypes of this series were analyzed extensively during the 1960s and 1970s in an effort to deduce the nature of the lesion. These studies are described in detail in Silvers (1979), available online at www.informatics.jax.org/wksilvers/, which provides an excellent example of the application of phenotypic studies to the understanding of gene function. Menkes et al. (1962) and Goka et al. (1976) suggested that the mottled syndrome is caused by a primary defect in copper transport, a conclusion that was confirmed subsequently using cellular and molecular methods (Reed & Boyd 1997; Levinson et al. 1994; Mercer et al. 1994; Cecchi et al. 1997; Cunliffe et al. 2001; Llanos et al. 2006; DeBie et al. 2007; Setty et al. 2008). Hunt (1974), reasoning from the aggregate defects of coat color and texture, neurological disturbances, aortic aneurysm, and skeletal abnormalities, suggested the locus as an animal model for Menkes disease.

Briefly, the phenotype of heterozygous females appears as a chimerism of pale and dark intensity of the basic color that is determined at other loci. The hair is also a complex of strands of fine and normal texture. The average percentage of dark and light color responds to selection that apparently relates to X-linked modifiers (Falconer & Isaacson 1972). *Brindled* and *Viable brindled* males suffer from a slight tremor and uncoordinated gait, as well as disturbances of the circulatory system (Sidman et al. 1965; Hunt & Johnson 1972a, 1972b), and *Brindled*, *Viable brindled*, and *Dappled* females may have aortic lesions (Rowe et al. 1974). When the animals are raised up by the tail, the hind legs grasp each other (Falconer 1956) and their feet may be abnormal (Phillips 1961). Some dappled (Phillips 1961) or mottled (Lyon 1960) females may develop calcified lumps in the periosteum on the vertebral column in the lumbar and thoracic regions. Rib and limb bones of hemizygous dappled males may be distorted around the time of their death. Mice bearing the $Atp7a^{To}$ (*Tortoiseshell*), $Atp7a^{Blo}$ (*Blotchy*), *or* $Atp7a^{Mo\text{-}vbr}$ (*Viable brindled*) alleles each display a defect in the cross-linking of both collagen and elastin.

The hemizygous males of course have only one X chromosome. If that X carries a mutation at the *Mottled* locus, in surviving males the hairs are short and wavy and very pale, while pigmentation is found primarily in skin of the ears, tail, and scrotum and the eyes, which are normally pigmented (Phillips 1961).

In combination with either *Piebald* ($Ednrb^{s}/Ednrb^{s}$) or *Belted* ($Adamts20^{bt}/Adamts20^{bt}$) white spotting, in the homozygous condition, dappled phenotypes change from what we have described as the tortoiseshell pattern (section 3.4.4; the tortoiseshell phenotype, not related to the tortoiseshell allele) to what we have described as the calico pattern of pigmentation (section 3.4.5; the calico phenotype). This fact strongly suggests an influence of *Atp7a* over pigmentary development, but does not rule out its requirement also for development of skin. Interestingly, the calico phenotype was not expressed in combinations of piebald with heterozygous *Tortoiseshell* ($Atp7a^{To}/Atp7a^{to}$) females (Dickie 1954). Mercer (1998) discussed the possibility that control over copper metabolism, via the *Mottled* (*Atp7a*) locus, may be influenced differently in different tissues, a possibility that would explain the observation.

The availability of copper as a cofactor is essential for many cellular processes, including pigmentation. TYR, the rate-limiting enzyme that initiates pigmentation in the melanosome, requires copper ions as a component of its three-dimensional structure. Furthermore, copper is required for the oxidation reaction itself. To fill this need, ATP7A located in the melanosomal membrane functions to transport copper from the cytosol into the maturing organelle, thus making it possible for TYR to attain its functional state. ATP7A is carried to the developing melanosome by transporter protein complex BLOC1 (biogenesis of lysosome-related organelle complex-1) that is discussed below (Setty et al. 2008). The questions of why TYR is not active until it reaches the melanosome, and how this step of melanogenesis is so exquisitely controlled in living cells have been posed for decades. The answers appear to lie in the complex controls over maturation of TYR, including its delivery to the melanosome (discussed in section 4.5), and the availability of copper ions via ATP7A, which is separate to the melanosome (Branza-Nichita et al. 1999; Olivares et al. 2003; Setty et al. 2007); and perhaps also including its functional interactions with TYRP1.

4.5 Protein processing and routing to the maturing melanosome

TYR, TYRP1, and DCT are the enzymes that catalyze eumelanogenesis. Failure of the function of these enzymes results in reduced pigmentation. This may result from mutation of their structural loci, but may also result from mutation at other loci, responsible for the processing or transport of the enzymes. As stated above, the enzymes, in living tissue, are not functional until they arrive in the melanosome. They are processed and transported by a series of genetically controlled events from their origin in the ER until they are delivered to the stage II eumelanosome, where they become active. This correct sorting and routing is necessary for normal pigmentation and because activation of the toxic melanogenic pathway would be harmful to the cell if it occurred outside the melanosome. Control over these processes is accomplished by successive protein-sorting events. Proteins that are not correctly processed and sorted are destroyed through the proteasome pathway and removed from the cell (Halaban et al. 2001).

Thus, while the number of enzymatic melanosomal proteins is few, the number of genes that must function correctly to process those enzymes, create a melanosome, and transport the enzymes to the melanosome in safety is rather large. Mutations at these other loci also result reduced pigmentation of the melanosome. Similarities and differences among the phenotypes of mice mutant at various loci have helped to reveal the steps in this overall process (Halaban et al. 1988; Raposo & Marks 2007).

4.5.1 Protein processing

For their correct post-translational processing, melanosomal enzymes require glycosylation; to fold up into their correct secondary and tertiary structures, they require specific environmental conditions in the ER and Golgi, and the enzymes must be prevented from initiating melanogenesis until they are incorporated into the stage II melanosome. While the conditions required for these accomplishments are still under study, it is likely that both the pH and the redox potential (affecting disulfide bridging) are important for correct folding of proteins, and for correct melanogenesis in the eumelanosome. Several pigmentation mutations have helped to clarify these processes, while at the same time illuminating additional functional causes of albinism.

Pink-eyed dilution (*P, Oca2, Oculocutaneous albinism type II*)

Malfunction of the pink-eyed dilution protein in humans results in oculocutaneous albinism type II (OCA2).
MGI reports 63 phenotypic alleles, 18 of which are spontaneous. Six of the spontaneous mutant stocks arose as germline-revertant events from *Pink-eyed unstable* (p^{un}/p^{un}) mice.
Availability: JAX.

This locus has given us several important clues to biological processes in general and to the cell biology of pigmentation specifically, but it may be difficult to find all the information because of the nomenclature. One must search for *Pink-eyed* (not *Pinkeye* or *Pink*) or search under the new name *Oca2* (*Oculocutaneous albinism type II*). It is customary to rename mouse loci when the specific affected protein or function has been identified, so we can look forward to another change when this occurs.

The *Pink-eyed dilution* locus (Fig. 4.19) encodes an integral membrane protein with 12 trans-membrane domains (Rinchik et al. 1993b; Rosemblat et al. 1994), and with a variety of alternative

Figure 4.19 The mouse at the back is homozygous for *Pink-eyed dilution*, $Oca2^{p}/Oca2^{p}$. The mouse at the front is homozygous for the *Pink-eyed unstable* allele and would therefore now be referred to as $Oca2^{p\text{-}un}/Oca2^{p\text{-}un}$. Somatic reversion events are evident in the front mouse. The mouse behind illustrates the usual phenotype of mice mutant at *Oca2*. Phenotypes range from completely pale to completely dark, but most mice are pale or have small revertant stripes. Note: the older symbols are used in the text (*p*, p^{un}, etc.).

proposed functions. These include maintaining an acidic pH in the melanosome (Puri et al. 2000; Ancans et al. 2001; Brilliant 2001), or transporting glutathione in the ER to facilitate protein folding (Staleva et al. 2002). Rosemblat et al. (1994) suggested that P functions in tyrosine transport to the melanosomal interior, but Potterf et al. (1998) found that the rate of tyrosine transport is normal in *P*-mutant eumelanosomes. However, TYR is apparently misrouted to the plasma membrane of *P*-null melanocytes, causing abundant extracellular melanin synthesis (black culture medium), and indicating that the P protein plays a critical role both in the processing and sorting of TYR to the melanosomes (Potterf et al. 1998; Manga et al. 2001; Chen et al. 2002; Toyofuku et al. 2002). In culture, the melanosomal enzymes are found in cytosolic vesicles in abnormally large amounts; they are secreted from the cell into the culture medium (Hirobe et al. 2003; Costin et al. 2003) and, in the *p*/*p* mouse, into the serum (Wakamatsu et al. 2007a). The pigmentation of the mutant melanocytes in culture can be partially restored by providing high levels of tyrosine, the substrate for melanogenesis (Silvers 1979; Sviderskaya et al. 1997; Chen et al. 2004). This perhaps compensates for the low level of TYR delivered to melanosomes. Further insight into the function of P protein is provided by recent evidence that the protein is definitely present in melanosomes, and that pigmentation is deficient if mutations prevent P reaching the melanosome (Sitaram et al. 2009). This does not mean that the protein has no function elsewhere (for example in the ER); still, more work is needed for a full understanding of this intriguing protein.

Pheomelanic *p*-mutant mice are just as intense in their yellow pigmentation as yellow mice that are not mutant at the *p* locus (Silvers 1979; Lamoreux et al. 1995; Kobayashi et al. 1995). Actually they are more yellow-looking because the few little specks of eumelanin that are characteristic of yellow mice have been reduced to pale gray. The eyes of mice are never pheomelanic because the pigment-type switching mechanism (discussed in Chapter 5) does not function in the retinal pigment epithelium of mouse eyes. Therefore the eyes of *P*/– mice (whether the mice are yellow or not) are black, whereas the eyes of *p*/*p* mice lack pigment and appear pink.

The lack of a p-mutant phenotype in yellow mice is not surprising, considering that the *P* gene appears not to be expressed in yellow mouse skin (Chiu et al. 1993), including the 'tan' (pheomelanic) part of the black-and-tan mouse (Rinchik et al. 1993b). This indicates that the P protein does not take part in pheomelanogenesis, and it raises the question of how TYR is then correctly routed to pheomelanosomes, if P protein is required for routing. It is possible, however, that routing mechanisms are entirely different in pheomelanocytes.

The phenotype of *Pink-eyed extra dark* (p^x/p^x) mice is darker than the majority of pink-eyed dilution phenotypes, which result, in eumelanic mice, in a soft grey pelage and pink eyes. *Pink-eyed unstable* (p^{un}/p^{un}) mice exhibit somatic reversion events in which clones of wild-type melanocytes develop alongside the mutant melanocytes. The p^{un} allele is the result of a duplication within the wild-type *P* gene (Gondo et al. 1993), inactivating it. The duplication frequently reverts to normal. Revertant events are accompanied by loss of the duplicated section. This can occur early or late in development, and it can occur in the skin or in the eyes. The study of the various phenotypes thus produced have been useful in evaluating hypotheses about the developmental history of melanocytes, as is discussed in section 3.4.

A panel of radiation-induced deletions that were identified by inactivation of the *P* locus has been useful in evaluation of functions of adjacent genes (Russell et al. 1995).

Underwhite (Uw, Matp, Membrane-associated transporter protein, Slc45a2, Solute carrier family 45, member 2)

Human OCA4 is caused by mutation at the *SLC45A2* (*MATP*) locus.
MGI lists 11 phenotypic alleles: five are spontaneous and three of these differ in phenotype.
Availability: JAX.

In humans, oculocutaneous albinism type IV (OCA4) is caused by mutation at the *Underwhite* locus, later termed the *Matp* (membrane-associated transporter protein) locus (Newton et al. 2001; Inagaki et al. 2004) in humans and in mice, because it encodes a membrane-associated protein with homology to solute transporters. And now it is termed *Slc45a2*, solute carrier family 45, member 2. This locus is expressed early in development in the eyes and neural crest cells (Baxter & Pavan 2002) of eumelanic mice (Fig. 4.20). The defect is autonomous to the pigment cell (Lehman et al. 2000).

The *Underwhite* locus was originally described by Dickie (1964) co-isogenic with C57BL/6J. Sweet et al. (1998) reported the origins and phenotypes of three separate mutant alleles – *uw*, uw^d, and Uw^{dbr} – at The Jackson Laboratory. Guénet and Babinet (1982) reported a fourth allele. The presence of both dominant and recessive mutations at a locus strongly implies that the protein functions as a molecular complex with other proteins, possibly as a homo- or hetero-oligomer (Sweet et al. 1998). Despite the proposed transporter role, the specific function of *Underwhite* is not described; however, the protein is necessary for normal routing of melanogenic proteins to the melanosome and shows structural and functional similarities with P (Pink-eyed dilution) protein (Costin et al. 2003) (Fig. 4.21).

SLC45A2 protein has 12 transmembrane regions. It functions, after apparently normal processing of the melanosomal proteins in the Golgi, to direct them into stage II melanosomes. In the absence of SLC45A2 protein, in cultured $Slc45a2^{uw}/Slc45a2^{uw}$ primary melanocytes, the melanosomal proteins are in large part secreted from the melanocytes into the culture medium (Costin et al. 2003). This is similar to the P protein; the function of the P protein involves carrying the TYR to the melanosome, and in the absence of P protein, much of the TYR is secreted from the melanocyte.

Figure 4.20 Homozygous *Underwhite* mouse with black control mouse.

Figure 4.21 Palomino horse and child. The horse is heterozygous for a mutation at *Underwhite* locus, representing one of the early papers on horse genetics (Castle & King 1951), and one of the most recent loci identified (see also Castle & Singleton 1961; Adalsteinsson 1974).

As is true of several other genes involved in melanosome differentiation, *Underwhite* expression appears to be under the direct or indirect control of MITF (Du & Fisher 2002). The *Uw* homolog in medaka fish is the medaka *B* locus (Du & Fisher 2002), which is thus not homologous to *B* (*Tyrp1*) in the mouse.

Mutation at a different locus (*Slc24A5*) causes a pale phenotype in zebrafish and is responsible for a good deal of color variation in the skin of humans (Lamason et al. 2005) but not in mice (Vogel et al. 2008). Probably the difference in phenotype between the mice and the humans/zebrafish relates to the fact that most of the pigmentation of the latter is in the skin, rather than in hairs.

4.5.2 Routing, or trafficking, proteins from their production site to the melanosome

In this chapter we have discussed the role of structural proteins in the formation of the melanosome in the cytosol, and the enzymes that function within the melanosome to generate melanin pigment. TYR is the rate-limiting enzyme of melanogenesis. Its competence is often reduced in albinism by *Tyr* mutation or by mutant proteins that interact with TYR. As mentioned, these interacting proteins include TYRP1, which appears to interact physically with TYR, and modify its native functions, and in mice to catalyze a subsequent step in melanogenesis. DCT also catalyzes a later step in melanogenesis. P protein and Underwhite protein are apparently required for normal transfer of melanogenic enzymes to the melanosome and for its function upon arrival. We have noted that TYR is recognized by AP-1 and AP-3, through the dileucine motif in its cytoplasmic tail, and that AP-1 and AP-3 are required to route TYR to the melanosome.

We have explained why the enzymes must be carefully escorted to the melanosome in an inactive state, to protect the cell from toxic intermediate products of melanogenesis. We have not explained how the enzymes manage to make their way from the Golgi to the melanosome. In this next section we will introduce the mouse mutations that have clarified the processes involved with routing, or trafficking, of the enzymes to the melanosome.

The process of sorting and routing proteins within the cytoplasm, from their place of origin to their various destinations, is referred to as routing, or trafficking. As transmembrane/organellar proteins, the melanosomal proteins are synthesized and initially processed in the ER, then pass through the Golgi. At the *trans*-Golgi network, the melanosomal proteins are sorted and packaged into small, membrane-bound transport vesicles that carry them initially (in some cases) to endosomes, and then to their final destinations, the maturing melanosomes (Fig. 4.22).

As shown in Figure 4.22, emerging data indicate that different melanosomal proteins are routed in different ways. Silver protein makes a brief initial visit to the plasma membrane (Theos et al. 2006), while TYR and TYRP1 both appear to travel initially to a sorting endosome. From these endosomes, Silver protein is routed to early melanosomes, where it enables matrix formation to produce a stage II melanosome. TYR moves directly to stage II melanosomes with the help of either AP-1 or AP-3 and the large routing complex BLOC1, which contains at least eight different component proteins including the products of five mouse color genes (see section BLOC1, below) (Raposo & Marks 2007). TYRP1, however, is routed by BLOC1 and AP-1 apparently only to an intermediate type of endosome, a stage that also requires the chocolate gene product RAB38 (or its relative RAB32). Thus homozygous *Rab38*^{cht} (*Chocolate*; Loftus et al. 2002) mice have a deficiency of TYRP1 in melanosomes and a brown coat color resembling that of homozygous TYRP1-deficient *Brown* mice. TYRP1 then requires the BLOC2 complex (of three color gene products, also detailed below) to reach the melanosome (Raposo & Marks 2007). The function of the remaining BLOC – BLOC3 – remains poorly understood, but one inferred function, shared with LYST, is to prevent aggregation or fusion together of melanosomes and related organelles (Fig. 4.22).

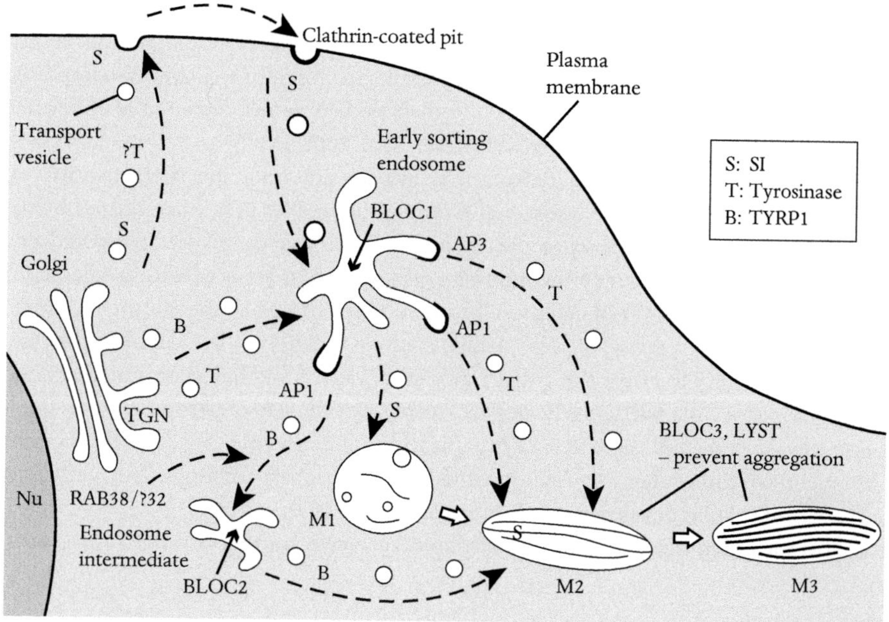

Figure 4.22 Model for routing of proteins to eumelanosomes. Current understanding of routing is undoubtedly incomplete, but it seems that different melanosomal proteins travel by various routes. The figure compiles data for the three most studied: SI (shown as S; Silver protein, PMEL, GP87), TYRP1 (shown as B; or 'Brown' protein), and tyrosinase (shown as T). BLOC1 and BLOC2 are routing complexes found in endosomes as shown, and are required for exit of the proteins being routed. Little is known about the molecular action of BLOC3, but it does appear to prevent melanosomes (and other organelles) from clumping together, as does lysosomal trafficking regulator (LYST), since defects in LYST or BLOC3 components result in giant or aggregated organelles respectively (Suzuki et al. 2002). M1–M3: stage I–III melanosomes; Nu, nucleus; TGN, *trans*-Golgi network. Partly based on figure 4 of Raposo and Marks (2007), with grateful acknowledgement. Other information from Setty *et al.* (2007) and Suzuki *et al.* (2002).

For the most part, the routing proteins are not specific to melanosomes or melanocytes, but function ubiquitously to organize and construct the lysosomal family of organelles. For this reason, the symptoms are similar if one or another of the proteins is aberrant. The common pleiotropic effects involve melanosomes, lysosomes, and platelet dense granules, and thus include prolonged bleeding times, abnormal lysosomal enzyme levels (which may lead to lung or kidney disorders), and reduction in intensity of pigmentation (Li et al. 2004; Raposo & Marks 2007). Mutant mice with these symptoms have been collected over many years, and it was recognized that their symptoms are similar to a group of human diseases known as Hermansky–Pudlak syndrome (sometimes called HPS) (Swank et al. 1998; Li et al. 2004).

Hermansky–Pudlak syndrome is a set of rare autosomal recessive disorders in which oculocutaneous albinism, bleeding, and lysosomal ceroid storage result from defects of multiple cytoplasmic organelles: melanosomes, platelet-dense granules, and lysosomes (Oh et al. 1998). The clinical symptoms of the syndrome in mice and humans include those of oculocutaneous albinism: reduced pigmentation and accompanying loss of visual acuity. In addition, Hermansky–Pudlak syndrome patients experience prolonged bleeding times, associated with defective platelet dense granules, and early death caused by fibrotic lung disease that results from malfunction of lysosomes (Li et al. 2004). Typically, melanosomes in Hermansky–Pudlak syndrome are fewer in number and tend toward a preponderance of early-stage forms (Nguyen et al. 2002).

These symptoms are the result of defects that impact not only the melanosome, but also the other members of the secretory lysosomal family of organelles. They are caused by the cell's inability to route the proteins properly to the organelle. Given that there already existed a panel of mutant mouse models for a similar panel of human diseases, as the tools became available to define the molecular defects at the level of cell biology, our understanding of the routing systems of the eukaryotic cell has quickly expanded. The details of their function at the cellular level have progressed beyond our ability to cover the entire story in this space and have continued at a dramatic pace. This ongoing story is a particularly elegant example of the use of parallel mouse and human abnormalities, evaluated at all the levels of organization – phenotypic, physiological, organismal, cellular, biochemical, molecular – to rapidly elucidate a previously unknown but very significant functional aspect of cellular development (Raposo & Marks 2007).

There follows a quick note about the unrelated recessive mutations *Faded* (*fe*) and *London gray* (*Lgr*).

Faded (*Fe*) and *London gray* (*Lgr*)

Faded availability: MMRRC.
London gray availability: MMRRC, TAC, and TIGM.

These loci have not been defined. Based entirely on phenotype, it is possible that they fit into the routing, or trafficking, category. The two phenotypes differ, however. London gray appears to have a phenotype somewhat more akin to fading with age, but does not fade very much, if at all, after first molt. Further, the London gray defect seems to be less uniform, consisting of very small darker and lighter areas of pigmentation.

Faded (*fe*/*fe*) mice tend to become slightly lighter at first molt. In the experience of one author (M.L. Lamoreux) they do not have skin lesions, as previously reported, when they are maintained in a controlled, parasite-free environment; perhaps they are predisposed to problems in different situations (Figs 4.23 & 4.24).

4.5.3 Adaptor proteins

Some of the Hermansky–Pudlak syndrome-like mutations affect one of the so-called adaptor proteins, specifically adaptor protein 3 (AP-3). These include the *Mocha* and *Pearl* mutations, respectively in *Ap3b1* and *Ap3d*, encoding the B1 (formerly β3A) and D1 (δ) subunits respectively of the tetrameric AP-3 complex, which regulates trafficking of membrane vesicle proteins to and from the *trans*-Golgi and endosomal compartments (Robinson & Bonifacino 2001). Adaptor proteins (which also include AP-1 and AP-2) appear to lack any intrinsic enzymatic activity, but instead mediate specific protein–protein interactions that drive the formation of protein complexes that mediate vesicle trafficking.

For example TYR is mis-sorted in melanocytes of Hermansky–Pudlak syndrome type 2 (HPS2) patients, who have a dysfunctional AP-3 complex due to mutations in the *Ap3b1* gene. This finding

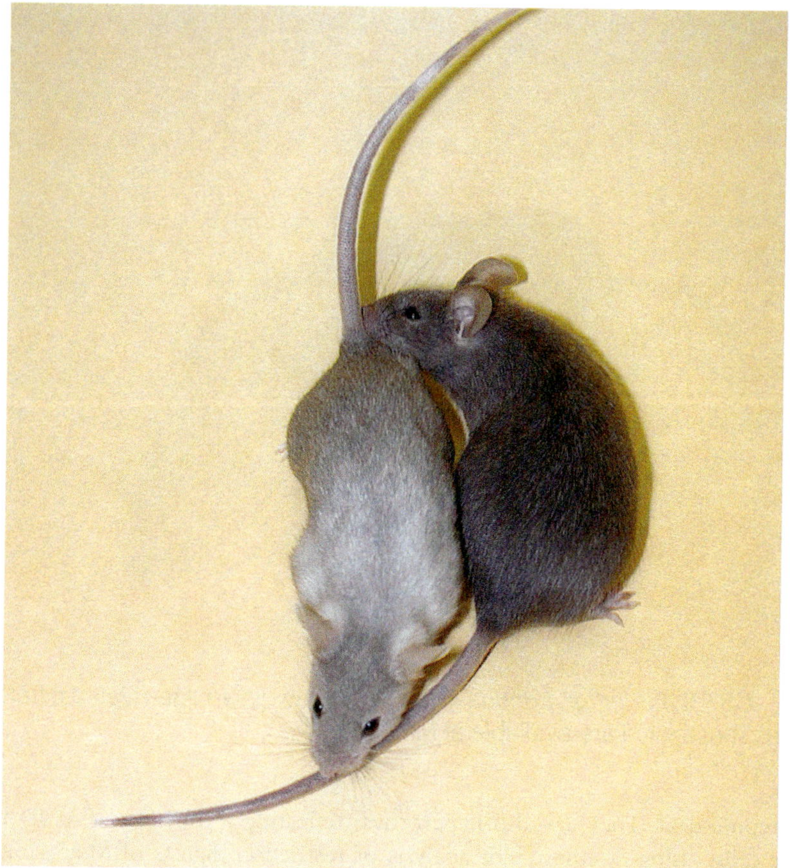

Figure 4.23 C57BL/6J-*fe*/*fe*, *Faded* mice, showing age-related differences in pigmentation. These mice are available at MMRRC.

is consistent with the presence of a dileucine-containing sequence that is found in the C-terminal tail of TYR and several other of the proteins involved in melanogenesis and which is recognized by AP-3.

Mocha (*Mh, Ap3d, Adaptor-related protein complex 3, delta subunit*)

JAX lists four phenotypic alleles, two of which are spontaneous.
Availability: JAX.

AP3D is a subunit of AP-3. The AP3D subunit is implicated in intracellular biogenesis and trafficking to melanosomes, and possibly platelet dense granules and neurotransmitter vesicles. The *Mocha* ($Ap3d^{mh}/Ap3d^{mh}$) mouse mutant shows coat and eye color dilution, reduced levels of renal lysosomal enzymes in urine, and prolonged bleeding due to storage-pool deficiency in the dense granules of platelets. Mocha mice also have balance problems due to otolith defects and

Figure 4.24 C57BL/6J-*lgr/lgr*, *London gray* mice, showing an age-related difference in pigmentation. These mice are available at MMRRC.

eventually become deaf. They are also hyperactive, reflecting a function of AP3D in neurons, among other cells. Using Southern-blot analysis of restriction digests of Mocha mouse DNA, Kantheti et al. (1998) determined that *Mocha* is a null allele of the *Ap3d* gene.

Pearl (*Pe*, *Ap3b1*, *Adaptor-related protein complex 3, β1 subunit*)

The related human condition is Hermansky–Pudlak syndrome type 2.
MGI reports 17 phenotypic alleles, 14 of which were spontaneous.
Availability: JAX, MMRRC, and ORNL.

Homozygous mutant *Pearl* ($Ap3b1^{pe}/Ap3b1^{pe}$) mice exhibit all the symptoms of oculocutaneous albinism, plus elevated kidney levels of lysosomal enzymes and a platelet dense-granule deficiency that results in prolonged bleeding. Feng et al. (1999) reported positional/candidate cloning of the autosomal recessive *Pearl* locus, and presented evidence from mutational analysis that the primary *Pearl* gene defect is in the *Ap3b1* gene. The *Ap3b1* gene encodes the large B1 subunit of AP-3 (Dell'Angelica et al. 1998). Unlike *Mocha* mice (above), *Pearl* mutants have no behavioral symptoms because neuronal AP-3 contains an alternative β subunit.

Zhen et al. (1999) found that the β3A subunit was undetectable in all cells and tissues of the *Pearl* mouse. In addition, expression of other subunit proteins of the AP-3 complex was decreased. The subcellular distribution of the remaining AP-3 subunits in platelets, macrophages and a melanocyte line from *Pearl* mice was changed from the normal punctate, probably endosomal, pattern to a diffuse cytoplasmic pattern. Ultrastructural abnormalities in mutant lysosomes were likewise apparent in mutant kidney and a cultured mutant cell line.

Canine cyclic hematopoiesis is also known as gray collie syndrome because it arose in this breed and affected dogs have reduced coat pigmentation. The disorder resembles human Hermansky–Pudlak syndrome type 2. Benson et al. (2003) showed that homozygous mutation of the *AP3B1* gene causes canine cyclic hematopoiesis. C-terminal processing of neutrophil elastase exposes an AP-3 interaction signal responsible for redirecting neutrophil elastase trafficking from membranes to granules. Disruption of either neutrophil elastase or AP-3 apparently perturbs the intracellular trafficking of neutrophil elastase.

4.5.4 RABs

A RAB is a small GTPase, a member of a family of about 70 proteins in mammals; some are expressed ubiquitously whereas others are cell-type-specific. The RAB family of proteins is a subclass of the RAS superfamily. RAB GTPases regulate many steps of membrane traffic, including vesicle formation, vesicle movement along actin and tubulin networks, and membrane fusion. RAB38 functions predominantly in melanocytes, and is localized to melanosomes. RAB27A functions primarily in melanosome transport to the periphery of the cell, and will be discussed in section 4.6.

Chocolate (Cht, Rab38, RAS oncogene #38)

MGI reports one spontaneous mutant allele, *Rab38*^{cht}.
Availability: JAX and CMMR.

RAB38 is unusually specific in its function, which is apparently to transport TYRP1 from the place of production to the maturing melanosome. The absence of TYRP1 from the melanosome mimics mutation at *Tyrp1*, so that mice mutant at the *Rab38* (*Chocolate*) locus are phenotypically similar to mice mutant at *Tyrp1*. Loftus et al. (2002) clarified and confirmed these relationships and proposed that RAB38 plays a role in sorting TYRP1 to the melanosome. Wasmeier et al. (2006) confirmed this role and demonstrated that RAB38 and the closely related RAB32 are important, functionally redundant regulators of melanosomal protein trafficking and melanocyte pigmentation. In the absence of RAB38 the mouse phenotype is moderated to a chocolate color and cells in culture are somewhat reduced in pigmentation. In the absence of both RAB38 and RAB32 the loss of pigmentation in cultured cells is dramatic (Wasmeier et al. 2006), suggesting that more processes than TYRP1 transport may be affected. The impact of these defects on retinal pigment epithelium is greater than in the skin melanocytes (Brooks et al. 2007; Lopes et al. 2007).

In the rat, Hermansky–Pudlak syndrome-like hypopigmentation and platelet storage-pool deficiency result from a mutation within the *Ruby* (*Red eyed dilution, R*) locus, which is now known as the *Rab38* locus (Oiso et al. 2004).

Buff (Bf, Vps33a, Vacuolar protein sorting 33a)

One phenotypic allele.
Availability: JAX, MMRRC, and TIGM.

Suzuki et al. (2003b) demonstrated that the mouse *Buff* locus encodes VPS33A, a member of the class C vacuolar protein sorting (VPS) complex of four proteins. The VPS proteins are involved in Golgi-to-lysosome trafficking. Mutations in class C VPS proteins, which includes VPS33A, result in severe vacuolar protein sorting and morphology defects (Carim et al. 2000; Huizing et al. 2001). The complex is involved in tethering functions, interacting with attachment protein receptors, both early and late in the endocytic pathway, and may act at several steps along the way. Absence of buff protein appears to interfere with the structure of melanosomes between stages III and IV.

4.5.5 The BLOC proteins

The BLOCs (biogenesis of lysosome-related organelle complexes) are heteromeric protein complexes, required for routing the proteins needed by melanosomes, lysosomes, and platelet dense granules in eukaryotic organisms. The three BLOC proteins identified to date are obviously involved in serving the same organelles, because all known mutations at the murine loci that encode their component proteins produce symptoms that model human Hermansky–Pudlak syndrome, and most of the known human Hermansky–Pudlak syndrome loci encode BLOC components and are homologs of these mouse loci. As mentioned, Hermansky–Pudlak syndrome is characterized by oculocutaneous albinism resulting from melanosomal abnormalities; excessive bleeding due to defective platelet dense granules, and lysosomal abnormalities. However, there are differences between the disorders associated with defects in the different BLOCs, which reflect the different stages of protein transport for which they are required, as will be discussed.

The first clues to these differential functions were provided by phenotypes of the melanosomes and of the mutant mice (Nguyen & Wei 2002; Li et al. 2004; Raposo & Marks 2007). The BLOC1 mutants *Cappuccino* and *Pallid* show only early multivesicular body/stage I melanosomes. As a result, pallid and cappuccino mice are lightly pigmented (Fig. 4.25). The BLOC2 mutants *Ruby eye* and *Ruby eye-2* have intermediate levels of pigment (Fig. 4.25) and melanosomes mostly resembling stages I and II. Mice with the BLOC3 mutations light ear and pale ear are nearly wild type in intensity of hair pigmentation, but have pale skin. As mice have very little skin pigmentation, and most is in the ears, these mutants were initially recognized by their pale ears. Their melanosomes are relatively mature but tend to form clusters and fused giant melanosomes.

BLOC1

The BLOC1 complex has eight subunits (Figs 4.3 and 4.25); there are known mouse mutants at five of the encoding loci, as follows (Setty et al. 2007).

***Cappuccino* (*Cno*)**

MGI lists one spontaneous phenotypic mutation.
Availability: JAX.

Homozygous mutant animals exhibit much reduced coat and eye color due to a reduced number of melanosomes (Figs 4.3 and 4.25). Seventy-five per cent of homozygous mutant animals exhibit some form of posture or balance abnormality, with variable severity. Platelet dense bodies are markedly deficient leading to prolonged bleeding. *Cappuccino* is a mutation in a novel protein that is part of the BLOC1 complex. This allele comprises an 11-base-pair deletion causing a frame shift that alters the C-terminal third of the protein sequence. Expression and subcellular localization of the protein are not affected. The mutation does affect the ability of the protein to form a complex with other lysosome-related proteins (Ciciotte et al. 2003).

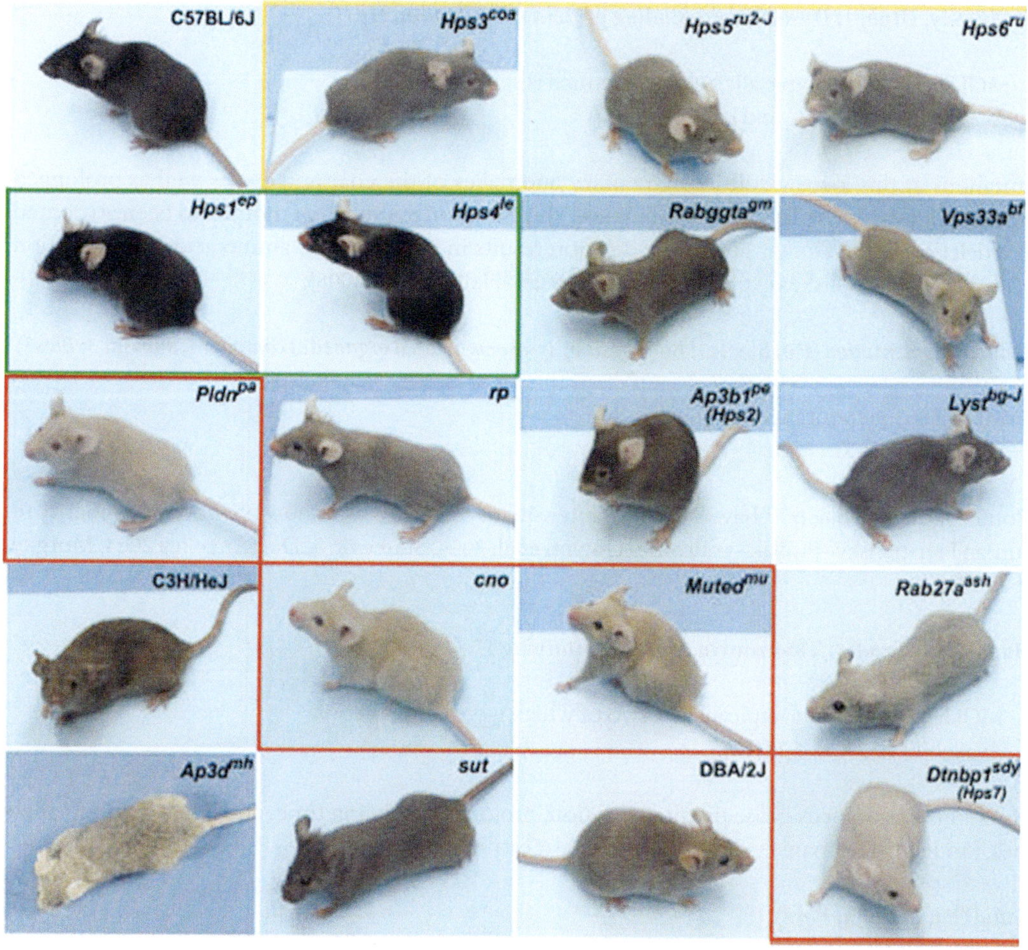

Figure 4.25 Coat colors of HPS mutant and control C57BL/6J, C3H/HeJ and DBA/2J mice. The members of the BLOC1 complex are outlined in red; BLOC2 members are outlined in yellow, and BLOC3 members are outlined in green. The closely related *Beige* (*Lyst*^{bg}) mutant, a model for Chediak–Higashi syndrome, is included for comparison. The eleven mutants above, *Cocoa* through *Beige*, are maintained as congenic stocks on the C57BL/6J background. The *Subtle gray* (*Slc7a11*^{sut}), *Cappuccino* (*Cno*^{cno}), and *Ashen* (*Rab27a*) mutants are maintained on the C3H/HeSnJ, C3H/HeJ, and C3H/HeSnJ inbred strains respectively. The *Mocha* (*Ap3d*^{mh}) mutant (photograph courtesy of Dr Margit Burmeister) is maintained on a heterozygous agouti stock comparable in coat color to the C3H/HeJ control. The *Sandy* (*Dtnbp1*^{sdy}) mutant is maintained on the DBA/2J strain. *Ruby eye-2* (*Hps5*^{ru2}), *Pale ear* (*Hps1*^{ep}), *Light ear* (*Hps4*^{le}), *Pallid* (*Pldn*^{pa}), *Mocha* (*Ap3d*^{mh}), and *Sandy* (*Dtnbp1*^{sdy}) are null mutations. It is highly likely that *Cocoa* (*Hps3*^{coa}), *Ruby eye* (*Hps6*^{ru}), *Cappuccino* (*Cno*^{cno}), *Muted* (*Txndc5*^{mu}), and *Ashen* (*Rab27a*^{ash}) are null mutations. Whether *Buff* (*Vps33a*^{bf}) is a null mutation is uncertain. The *Gunmetal* (*Rabggta*^{gm}) and *Pearl* (*Ap3b1*^{pe}) mutants are hypomorphs. However, *Pearl* is very nearly a null mutant since it produces no detectable protein product on Western-blot analyses and is only very slightly darker in color than the null allele. From Li et al. (2004), with the original legend and updated gene names and symbols.

***Sandy* (*Sdy, Dtnbp1, Dystrobrevin-binding protein 1, Dysbindin, Hps7*)**

MGI lists 20 phenotypic alleles, one of which is spontaneous.
Availability: MMRRC and JAX.

Mutations at this locus result in pigmentary anomalies of the coat and eye as well as prolonged bleeding times due to platelet abnormalities. The mutation in the sandy mouse has been attributed to a deletion in the *Dtnbp1* gene. The deletion results in the loss of 52 amino acids from position 119–172, which abolishes expression of the protein (Hattori et al. 2008).

***Reduced pigmentation* (*Rp, Bloc1s3; Biogenesis of lysosome-related organelles complex-1, subunit 3; Blos3*)**

MGI lists one spontaneous allele.
Availability: JAX.

Homozygotes for $Bloc1s3^{rp}$ have a reduced intensity of coat color and show phenotypic similarity to human Hermansky–Pudlak syndrome (Gwynn et al. 2004; Starcevic & Dell'Angelica 2004; Morgan et al. 2006).

***Muted* (*Mu, Txndc5, Thioredoxin domain containing 5*)**

MGI lists three phenotypic alleles, two of which were spontaneous.
Availability: JAX and TIGM.

Mutations at this locus cause pigment dilution, prolonged bleeding time, and inner-ear abnormalities, modeling Hermansky–Pudlak syndrome (Setty et al. 2007).

***Pallid* (*Pa, Pallidin, Pldn*)**

MGI lists one spontaneous phenotypic allele.
Availability: JAX and TIGM.

Homozygous mutation at this locus ($Pldn^{pa}$) results in a coat-color abnormality, abnormal platelet morphology, and impaired motor coordination, providing a model for Hermansky–Pudlak syndrome (McGarry et al. 2002; Salazar et al. 2006).

BLOC2

Three subunits are known for the BLOC2 complex (Figs 4.3 and 4.25), all of which have known mutations in both mice and human Hermansky–Pudlak syndrome subsets (HPS3, 5, and 6).

***Cocoa* (*Coa, Hps3, Hermansky–Pudlak syndrome 3 homolog*)**

MGI lists seven phenotypic alleles, six of which are spontaneous.
Availability: JAX.

Homozygotes for spontaneous null mutations exhibit hypopigmentation and prolonged bleeding associated with a platelet defect (Suzuki et al. 2001; Kaput et al. 2004).

Ruby-eye (*Ru*, *Hps6*, *Hermansky–Pudlak syndrome 6 homolog*)

MGI lists eight phenotypic alleles are reported, seven of which are spontaneous.
Availability: JAX.

Homozygous *Hps6*^{ru} mice have reduced pigmentation of eyes and hair, impaired secretion of lysosomal enzymes by renal proximal tubules and reduced clotting due to a defect in platelet dense granules (Zhang et al. 2003).

Ruby-eye 2 (*Ru-2*, *Hps5*, *Hermansky–Pudlak syndrome 5 homolog*)

MGI lists nine phenotypic alleles, six of which are spontaneous.
Availability: MMRRC, ORNL, TIGM, and JAX.

Homozygotes for *Hps5*^{ru2} (Fig. 4.3) have reduced pigmentation in the eyes and hair, impaired secretion of lysosomal enzymes by renal proximal tubules, and reduced clotting due to a defect in platelet dense granules. Homozygotes for one allele are less susceptible to diet-induced atherosclerosis (Zhang et al. 2003).

BLOC3

Proteins encoded at the following loci are the two known components of BLOC3 (Fig. 4.25) (Martina et al. 2003; Nazarian et al. 2003).

Pale ear (*Ep*, *Hps1*, *Hermansky–Pudlak syndrome 1 homolog*)

MGI lists two spontaneous alleles.
Availability: JAX, RBRC, and ORNL.

Mice that are homozygous for a spontaneous null mutation of *Hps1*, *Pale ear* (*Hps1*^{ep}), have slightly reduced pigmentation, evident mostly in the skin, increased bleeding time, impaired natural killer cell function, and reduced secretion of kidney lysosomal enzymes. Abnormal retinofugal neuronal projections characterize some alleles (Nguyen & Wei 2007).

Light ear (*Le*, *Hps4*, *Hermansky–Pudlak syndrome 4 homolog*)

MGI lists one spontaneous phenotypic allele.
Availability: JAX.

Mice that are homozygous for the spontaneous null mutation of *Hps4*, *Light ear* (*Hps4*^{le}), have slightly reduced pigmentation, evident mostly in the skin, prolonged bleeding associated with a platelet defect, reduced secretion of kidney lysosomal enzymes, and resistance to diet-induced atherosclerosis.

Chediak–Higashi syndrome

Chediak–Higashi syndrome is a human syndrome similar, but not identical, to Hermansky–Pudlak syndrome. Qualitative changes of leukocytes are found in Chediak–Higashi syndrome but not

Hermansky–Pudlak syndrome, and pigmented macrophages are found in Hermansky–Pudlak syndrome but not Chediak–Higashi syndrome (Perou et al. 1997; Introne et al. 1999). *Beige* is the mouse model for Chediak–Higashi syndrome.

Beige (Bg, Lyst, Lysosomal trafficking regulator)

MGI lists 14 phenotypic alleles, eight of which were spontaneous.
Availability: CMMR, JAX, MMRRC, ORNL, RBRC, and TIGM.

Homozygous *Beige* (*Lyst*^{bg}; Fig. 4.26) mice have a phenotype similar to human Chediak–Higashi syndrome patients, exhibiting lysosomal dysfunction with resultant protein storage, diluted coat color, abnormal melanogenesis, immune-cell dysfunction resulting in increased susceptibility to bacterial, viral, and parasitic infections, and decreased cytotoxic activity against tumor cells. The encoded protein was named after its function in lysosomal protein trafficking, but is evidently also needed for that of melanosomal proteins. As pointed out by Barbosa et al. (1997), many of the diverse clinical features of Chediak–Higashi syndrome can be caused by a defect in trafficking of proteins to secretory lysosomes. However, the exact function of the Chediak–Higashi syndrome gene product is uncertain.

Characteristically, *Beige* melanosomes apparently aggregate during their maturation to form larger clumps that are pigmenting asynchronously, apparently because the component melanosomes are at different stages of their development (Silvers 1979). As described in detail by Silvers, the beige phenotype (as it relates to hair pigmentation) is hypostatic to *Pink-eyed dilution* and epistatic to *Ruby-eye*, reinforcing the suggestion that the function of *Beige* has to do with routing and suggesting that its contribution to pigmentation is required after that of *Pink-eyed dilution* (*Oca2*) and before that of BLOC2, of which *Ruby-eye* (*Hps6*) encodes a component protein. Note that the 'clumping' of melanosomes in beige mice apparently occurs before the type of clumping that results from failure of melanosome transport in dilute (Griscelli syndrome) phenotypes. It is also very different at the ultrastructural level in that beige melanosomes appear as giant organelles anywhere in the cell

Figure 4.26 A *Beige* mouse is shown in the foreground, C57BL/6J-*Lyst*^{bg}/*Lyst*^{bg}. The mouse at the back is wild type.

whereas in the mutants that are defective in melanosomal transport (*Gunmetal, Dilute, Leaden*, and *Ashen*) the organelles are normal but clustered around the nucleus.

4.6 Melanosome transport

Gunmetal (Gm, Rabggta, Rab geranylgeranyl transferase, α subunit)

MGI lists one spontaneous recessive allele.
Availability: C57BL/6J-*Rabggta^{gm}*/J is available at JAX.

This geranylgeranyl transferase adds hydrophobic phenyl groups to the C-terminus of all RABs. It therefore contributes to the normal function of the RABs in routing proteins from the ER/Golgi to their respective lysosome-related organelle and in transporting the mature melanosome to the periphery of the cell. In the case of the gunmetal (*Rabggta^{gm}*) phenotype, the platelet defect has been most extensively studied.

Homozygotes exhibit diluted pigmentation, a platelet and megakaryocyte defect resulting in prolonged bleeding (Swank et al. 1993; Novak et al. 1995), macrothrombocytopenia and reduced platelet α- and δ-granule contents (storage-pool deficiency), impaired killing by cytotoxic T lymphocytes, high mortality, and poor breeding. Detter et al. (2000) noted that the phenotype resembles human gray platelet syndrome and platelet α- and δ-storage-pool deficiency.

By positional cloning, Detter et al. (2000) showed that *Gunmetal* results from a G-to-A substitution in a splice acceptor site within the α-subunit of RAB geranylgeranyl transferase (RABGGTA). Most RABGGTA mRNAs from *Gunmetal* tissues skipped exon 1 and lacked a start codon. Levels of the protein and of enzyme activity were reduced four-fold in *Gunmetal* platelets. Geranylgeranylation and membrane association of RAB27, a RAB GGTase substrate, were significantly decreased in *Gunmetal* mouse platelets. These findings indicated that geranylgeranylation of RAB GTPases is critical for hemostasis and for melanosome transport. The latter defect predominates in the pigmentary phenotype of the *Gunmetal* mutant mouse, which is essentially the same 'dilute' type phenotype that characterizes the three other primary mutants of the melanosome transport system. The 'dilute' phenotype is illustrated in Figs 1.11, 4.27, and 4.29 (see end of chapter), and in mice in Figs 2.4 and 4.4.

Dilute-type phenotypes in mice and other mammals (Figs 4.4, 4.27, and 4.29) involve the *Dilute* (*Myosin Va*), *Leaden* (*Melanophilin*) or *Ashen* (*Rab27a*) loci. *Dilute suppressor* (*Dsu*, which has been renamed *Melanoregulin*) affects the same protein complex, but its mode of action is not known. It functions to reverse the dilute phenotype in mice mutant at the dilute-type loci.

Homologous syndromes in humans are classified as Elejalde syndrome or Griscelli syndrome. Griscelli syndrome is a rare autosomal recessive disorder that results in 'silvery' pigmentary dilution phenotype of the skin and hair, the presence of large clumps of pigment in hair shafts, and an accumulation of melanosomes in melanocytes. Griscelli syndromes include: Griscelli syndrome type 1, which represents hypomelanosis with a primary neurologic deficit and without immunologic impairment or manifestations of hemophagocytic syndrome (Ménasché et al. 2002); Griscelli syndrome type 2, which is Griscelli syndrome with immune impairment, caused by mutation in the *Rab27a* gene; and Griscelli syndrome type 3, characterized by hypomelanosis with no immunologic or neurologic manifestations. It can be caused by mutation in the *Mlph* or *Myo5a* genes.

A 'dilute' phenotype in a mammal is striking, characteristic, and usually is easily recognizable in the animal (Figs 4.4, 4.27, and 4.29), which can be of any color, but of a paler and more silvery

Figure 4.27 A blue-cream calico cat. This cat is homozygous for the cat version of the 'dilute' phenotype. The white spotting is unrelated to the pigment colors, but does generate a modest 'calico' patterning (see Chapter 3). Because the cat is also a chimeric, it is possible to see the effect of 'dilute' on both pheomelanic and eumelanic pigmentation. In the mouse, the dilute phenotype may be caused by homozygosity at any of three loci, *Dilute* (*Myo5a*^{d}), *Ashen* (*Rab27a*^{ash}), and *Leaden* (*Mlph*^{ln}). Unfortunately the name 'dilute' in the cat was given to a mutation that corresponds with *Leaden* in the mouse. However, the phenotypes generated by mutation at any of these loci are similar, because they all interfere with the same process of transporting the melanosomes to the periphery of the melanocyte, where they are deposited into the growing hair. When the process breaks down for any reason, the melanosomes are clumped in the hair and scatter the rays of light that fall upon it, so that the color appears to our eyes to be diluted. See Fig. 4.4 for a dilute mouse. The effect is very similar in pheomelanic mice.

appearance than that of its non-dilute counterpart (note that this is not related to the 'silver' phenotype or genotype). On this basis, early in the past century, the phenotype name, 'dilute,' originated.

Somewhat later, it became evident, from examination of the hairs and evaluation of melanin content (Wolfe & Coleman 1966), that the dilute phenotype does not result from 'dilution' or depletion of the amount of pigmentation in the skin and hairs. Instead, the aberrant distribution of melanosomes in the melanocyte and in the hair shaft is thought to alter the optical qualities of the hairs. The pigment granules in the hairs are distributed in clumps, lumps, fine scattering, and abnormal empty spaces, very different from the normal orderly array. This disarray is thought to scatter light, and/or absorb it abnormally, resulting in the dilute appearance.

The abnormal hairs are easily identified by treating them in xylene for a couple of days, to remove air from the shaft, then transferring to absolute alcohol, then mounting and observing under a low power microscope. Also using a microscope it is easy to observe, in the melanocytes of animals that are homozygous for any of these mutant genes, that the melanosomes are clustered around the nucleus rather than distributed normally throughout the melanocyte. This was referred to as a nucleopetal distribution.

The term 'dilute' of course is often incorrectly used to describe other phenomena, but the distinctive phenotype characteristic of this syndrome is commonly recognized by pigmentary scientists and by people who raise livestock or pets. Fanciers of cats, dogs, mice, rats, guinea pigs, rabbits, and other species commonly give exotic names to such animals, depending on the base colors. The most common of these are 'blue' to describe dilute black, 'lavender' to describe dilute brown, and 'cream' to describe dilute yellow/red. A blue-cream tortoiseshell cat is chimeric for black and yellow color of pigment, as shown in Fig. 4.27, and as described in Chapter 3, and is also of the dilute phenotype, homozygous in this case for a mutation in *Melanophilin*.

To further add to the verbal confusion, the term 'dilute' has two distinct cross-species meanings. In the mouse, it is the name of the phenotype and the gene locus that encodes MYO5A. In the cat and dog it happens that the dilute phenotype is caused by mutation at the *Melanophilin* locus.

The 'dilute' phenotype is essentially the same relative to each of the three murine loci that cause it. These are the murine *Dilute* locus, *Myosin Va/Myo5a*, the murine *Leaden* locus, *Melanophilin/Mlph*, and the murine *Ashen* locus, *Rab27a*. A fourth locus, *Dilute suppressor* (*Dsu* or *Melanoregulin/Mreg*), when it is mutant, counteracts the effects of any of the dilute-type loci – *Dilute*, *Leaden*, or *Ashen* – returning to normal the phenotypes of the doubly mutant mice.

Differences in pleiotropic effects, particularly neurological effects, have been documented in mice and human mutants. These may result from the fact that the proteins are complex, especially myosin Va, with multiple domains, and have somewhat different functions in different cell types. Or there may be an involvement of linked loci.

Not surprisingly, the three dilute-type loci encode proteins that participate in the same multi-protein complex that is part of the melanosomal transport system. In the skin and hair follicles of normal mice, as the melanosomes mature, they are carried by the transport system of the cell from the area around the nucleus to the tips of the dendrites (see Fig. 4.28), where they are transferred into a neighboring keratinocytes of the skin or growing hair.

A lively debate in the early literature attempted to elucidate the relationship between the two characteristic anomalies of dilute animals: hair abnormalities and abnormal structure of the pigment cells (for more detail on this see Silvers 1979). Some favored the hypothesis that the melanocyte was mis-shapen, lacking the normal dendritic structure, and was thus prone to become dislodged from its moorings and incorporated into the growing hair. Others believed the defect lay in the distribution of the melanosomes within the normally formed melanocyte. These phenotypes are well described in Silvers (1979). Most of the early questions have been answered.

In the pigment cell, the dilute family of proteins functions particularly at the periphery of the cell where the melanosomes are transferred from the microtubule portion of the transport system to the actin network, which presumably chauffeurs them toward the unknown mechanism that will move them across the cell membrane into the adjacent keratinocyte. The transport system consists of a bi-directional microtubule-based process that initially carries the melanosomes to the periphery of the cell. It also carries them back to the nucleus if they are not removed from the microtubule and transferred to the actin portion of the cytoskeleton at the periphery of the cell. The microtubule portion of this operation is not defective in dilute-syndrome mice. It is the transfer to the actin portion that fails, with the result that melanosomes accumulate at the center of the cell (Wu et al. 1998) where microtubules predominate. The result, in mutant mice where the melanosomes are not properly removed to the actin system so they can be transferred out of the cell, is the nucleopetal cell (Fig. 4.28).

The protein complex that accomplishes this transfer is relatively well understand and thus serves as an excellent model of organelle transport (Nascimento et al. 2003). The murine melanosome

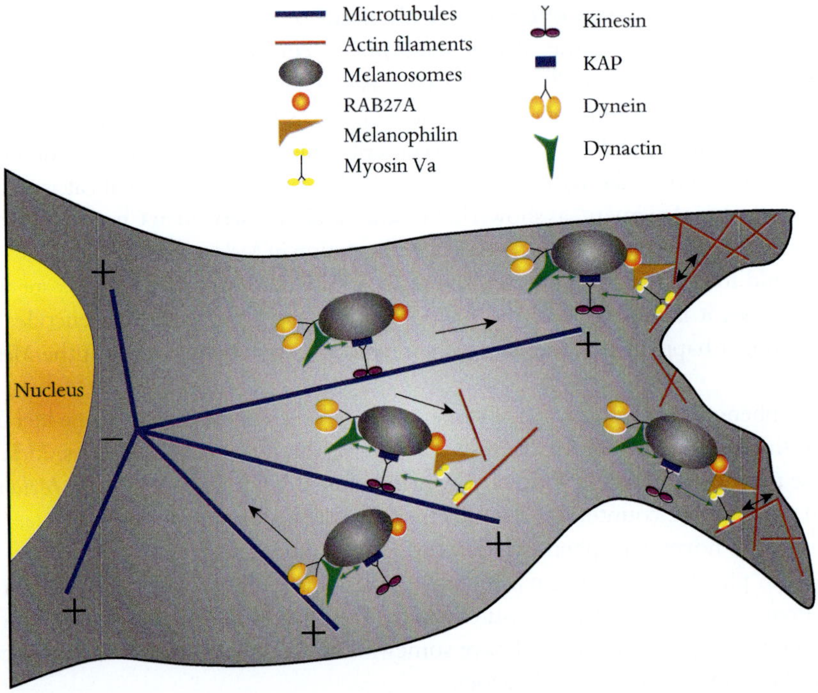

Figure 4.28 Molecular machinery involved in melanosome movement in melanocytes. The diagram shows that movement toward the plus end of microtubules at the cell periphery is achieved by kinesin. According to the cooperative delivery/capture model, the complex formed by RAB27A, melanophilin, and myosin Va allows the peripheral capture of melanosomes. In this region of the cell, melanosomes display short-range movements on actin filaments, indicated by the black double arrows. The tug-of-war model predicts that the microtubule- and actin-dependent motors are coordinated in achieving the intracellular dispersion of melanosomes. Green arrows denote interactions that are predicted to be necessary for the coordination of the different molecular motor activities. Movement toward the minus end of microtubules in the center of the cell is dependent on dynein. Single black arrows indicate direction of melanosome movement. Figure and description taken from Barral and Seabra (2004); reproduced with permission of Wiley-Blackwell. See that paper for a more detailed description of the melanosomal transport model.

transfer requires all three of the dilute-complex proteins. Melanophilin binds both myosin Va and RAB27A. RAB27A and melanophilin interaction is the primary targeting and stabilizing factor for melanophilin. The interaction of melanophilin with myosin Va allows the complex, with attached melanosome, to be captured into the peripheral dendrites (Wu et al. 2002; Hume et al. 2007; Li et al. 2007). The N-terminal region of melanophilin specifically binds RAB27A, while the C-terminal half directly binds the globular tail of myosin Va (Fukuda & Kuroda 2002). The binding of melanophilin to exon F of myosin Va activates its ATPase activity, energizing the movement of the protein complex with its melanosomal cargo (Li et al. 2005). The function of *dilute suppressor* is not known, and neither is the mode of transfer to keratinocytes (Berens et al. 2005; Hearing 2007).

Dilute (*D*, *Myo5a*, *Myosin Va*)

MGI lists 69 phenotypic alleles of which 52 are spontaneous.
Availability: JAX, KOMP, MMRRC, NMICE, ORNL, RBRC, and TIGM.

The *Dilute* locus encodes myosin Va (MYO5A) (Mercer et al. 1991; Pastural et al. 1997). Myosin Va is an actin-binding protein that belongs to the large myosin superfamily. The members of this superfamily share a common N-terminal motor domain (or 'head'), which binds actin and generates force through the hydrolysis of ATP. Expression of *Myo5a* is tissue-specific (Seperack et al. 1995) as a result of alternative-exon usage. The region of the gene encoding the tail contains three alternatively spliced exons designated B, D, and F that function differentially depending upon tissue type. Expressing tissues include neuromuscular junction (Röder et al. 2008), pigment cells, brain, and other tissues. Exon F is essential for the interaction of myosin Va with melanophilin, and consequently for its association with melanosomes. Myosin Va is involved in the process of melanosome transport as described above.

Ashen (*Ash*, *Rab27a*, *RAS oncogene family member RAB27A*)

MGI lists three phenotypic alleles, two of which are spontaneous.
Availability: CARD, CMMR, JAX, MMRRC, ORNL, and TIGM.

The *Ashen* locus encodes RAB27A (Wilson et al. 2000). RAB27A is a tissue-specific RAB that associates with lysosome-related organelles and secretory granules (Tolmachova et al. 2004). In melanocytes, RAB27A associates with the cytosolic leaflet of the melanosome membrane (Hume et al. 2001, 2007; Jancic et al. 2007) and appears to be the key melanosome-associated protein of the tethering complex that attaches the melanosome to the transport system (Wu et al. 2001; Hume et al. 2002).

Leaden (*Ln*, *Mlph*, *Melanophilin*)

Three phenotypic alleles, one of which is spontaneous.
Availability: JAX, RBRC, and TIGM.

The *Leaden* locus encodes melanophilin (Matesic et al. 2001; Provance et al. 2002) Melanophilin specifically interacts with the globular tail of myosin Va through its distinct coiled-coil regions (Wu et al. 2002; Fukuda et al. 2002; Strom et al. 2002; Nagashima et al. 2002). The C-terminal domain immunoprecipitates with actin and myosin Va (Hume et al. 2007). One can envision myosin Va as the engine that drives the complex along the actin transport system, while RAB27A grabs the melanosome and melanophilin connects the two.

Dilute suppressor (*Dsu*, *Melanoregulin*, *Mreg*)

One spontaneous phenotypic allele.
Availability: CARD, MMRRC, and TIGM.

The murine *Dilute suppressor* gene (Sweet 1983) functions in a myosin Va-independent fashion to suppress the dilute-type phenotype associated with mutation at *Ashen*, *Dilute*, or *Leaden* (Moore et al. 1994, 1988; Jenkins 1998; O'Sullivan et al. 2004).

Figure 4.29 Dilute yellow (cream) tabby cat; the dilute yellow phenotype is known as cream in the cat. Yellow in the cat is known as red to cat fanciers or orange in the genetics literature. The genotype of this cat might be written *d/d* by cat fanciers and in the older literature, but we can't do that here as it would suggest homology with the murine *Dilute* locus. Therefore, let us say *mlph/mlph r/Y*, or we could write it as *mlph/mlph* O^{-}/Y. The *Y* reflects the fact that this is a male cat and the yellow/red/orange gene is on the X chromosome. Therefore he has only one yellow gene locus. This cat is also a striped or 'lined' tabby, a phenotype that is not linked to either dilute or yellow. Striped tabby is dominant to other kinds of tabby, so can be designated as $T^{l}/–$. Probably this cat also carries the dominant *Agouti* allele (*A/–*) that is not able to fully express itself because pheomelanic pigmentation is epistatic to eumelanic pigmentation; that is, in the presence of a yellow genotype, the melanocytes do not make eumelanosomes. (There may be a few eumelanosomes in such an animal; yellow mice may have a few at the tips of the hairs, but as a generalization yellow animals do not appear eumelanic.) Therefore, and because the agouti pigmentation pattern consists of both eumelanin and pheomelanin, full expression of agouti is not possible in a pheomelanic mammal. Full expression of tabby is also not possible in the absence of eumelanin, for the same reason. The *Tabby* locus, with at least three alleles, appears to be unique to cats (Robinson 1971; Vella & Robinson 2003), although similar pattern loci that direct the location of the eumelanin and pheomelanin in agouti animals are also seen in other species (zebra, giraffe).

5

Pigment-Type Switching

Figure 5.1 A eumelanic black mouse, behind, and two pheomelanic mice. All of these mice are of the same inbred strain, C57BL/6J. The complexity of terms used for pigmentation requires that we define our terms precisely, and 'black' is a term often misused. Black mice are nonagouti, but not all nonagouti mice are black. The black phenotype is defined by the genotype at two loci: *Nonagouti* (*a*/*a*) and *Black* (*Tyrp1*^{B}/–). The *Black* genotype is determined by the black versus brown (*Tyrp1*) locus, with *Black* (*Tyrp1*^{B}) dominant to *Brown* (*Tyrp1*^{b}). Any eumelanin that is found on a mouse of the genotype *Tyrp1*^{B}/– will be black, rather than brown, as discussed in Chapter 4. Accurately stated, black mice are nonagouti black and brown mice are nonagouti brown. The *Agouti* locus determines whether or not pheomelanin will be produced. Pheomelanic phenotypes are epistatic to eumelanic ones. The agouti phenotype describes an animal that is patterned in eumelanin and pheomelanin, and in the mouse describes animals in which most of the hairs are individually banded in eumelanin and pheomelanin (see Figs 5.8 and 5.9). All of these mice in the photograph are *Nonagouti* (*a*/*a*). *Lethal yellow* (*A*^{y}/–) (the genotype of the mouse at the front) is dominant to *Agouti* (*A*/–) and epistatic to *Black* or *Brown*. The mouse to the right is *Recessive yellow* (*e*/*e* or *Mc1r*^{e}/*Mc1r*^{e}). Again, the yellowness caused by this mutation is epistatic to black or brown. Recessive yellow mice tend to have a bit of eumelanic pigmentation at the hair tips, especially dorsally; agouti-yellow phenotypes tend to be very sensitive to background genome and may have more or less eumelanic pigmentation at the hair tips and a considerable variability of phenotype depending upon the strain, as shown in Figs 5.6 and 5.7. *Recessive yellow* mice are not phenotypically highly variable. All these patterns of expression in inbred mice are consistent and predictable.

The Colors of Mice: A Model Genetic Network, 1st edition. By M. Lynn Lamoreux, Véronique Delmas, Lionel Larue, and Dorothy C. Bennett. Published 2010 by Blackwell Publishing Ltd.

5.1 Introduction

We have been approaching the description of pigmentation based in the perspective of a developmental timeline, with Chapter 3 describing the embryological development of the melanocyte (and defects that result in white spotting or fading with age) and Chapter 4 covering the differentiation and distribution of the melanosome. In this chapter we will describe the controls that regulate the two major chemically distinct types of melanin pigment that are produced by mammals. These are yellow pheomelanin and nonyellow eumelanin (Figs 5.1–5.7). The complex eumelanin/pheomelanin switching mechanism determines which type of pigment will be produced in differentiating melanosomes.

The eumelanin/pheomelanin switch is a beautiful model of genetic control over the biology of the organelle, cell, and tissue. Again we find that this elegant pigmentary model has led to discoveries well beyond the significance of pigmentation, in such widespread areas as basic mechanisms of cell biology, brain chemistry, obesity, trans-generational epigenetic inheritance, and mechanisms of information transduction from the environment into the cell (Voisey & Van Daal 2002). In this chapter we focus on the major gene loci that regulate the eumelanin/pheomelanin switch mechanism of melanocytes.

The normal controls over melanogenesis differ somewhat in humans compared with mice (Boissy et al. 1998). Therefore, human melanocytes should not be used as controls for mutant

Figure 5.2 Summer is a chestnut (sorrel) Oldenburg breed of horse. Chestnut and sorrel are two of the terms breeders use for yellow (red) horses. Her genotype is probably *Mc1r*^{e}/*Mc1r*^{e}. Photograph courtesy of TripleJ Sporthorses.

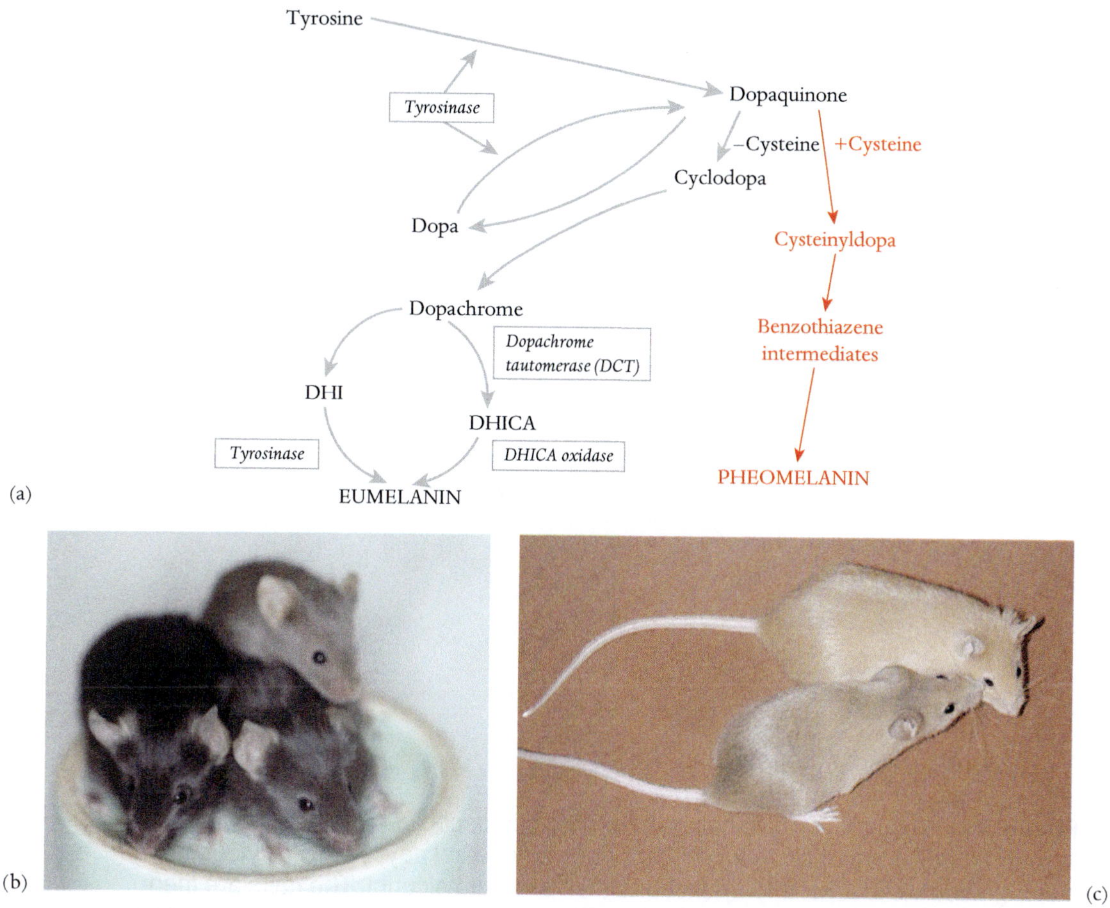

Figure 5.3 (a) The melanogenic pathway (simplified from Ito & Wakamatsu 2008; reproduced with permission of Wiley-Blackwell). Enzymes are boxed. The rate-limiting enzyme is tyrosinase; mutation at the *Tyr* (*Albino*) locus reduces the amount of melanin that the pigment cell can make. (b) In nonagouti mice that are otherwise wild type at pigment loci, mutation of the gene that encodes dopachrome tautomerase (DCT) result in slaty mice (b, lower right). Mutation of the gene encoding 5,6-dihydroxyindole-2-carboxylic acid (DHICA) oxidase activity (*Tyrp1*) results in brown mice (b, rear; see Chapter 4). (c) Pheomelanogenesis is primarily regulated by a cell-surface receptor (MC1R) and its ligands, melanocyte-stimulating hormone (MSH) and agouti signal protein (ASP), which control whether or not the cell will make yellow pigment. Pheomelanogenesis is epistatic to eumelanogenesis. These two pheomelanic mice are *Recessive yellow* (front) in which the MC1R is mutant, and *Lethal yellow* (back) in which the *Agouti* locus that encodes ASP is mutant.

murine melanocytes, or the reverse, when evaluating cells in culture. Below we discuss the murine condition, using *in vitro* results only rarely, when they are useful to interpret phenotypes. The phenotypes are described in section 5.2. The melanogenic pathway and phenotypic results in mice are shown (simplified) in Figure 5.3, and the physical nature of the pigment is discussed in section 5.3. At the level of the genes and their products, multiple proteins are switched on or off

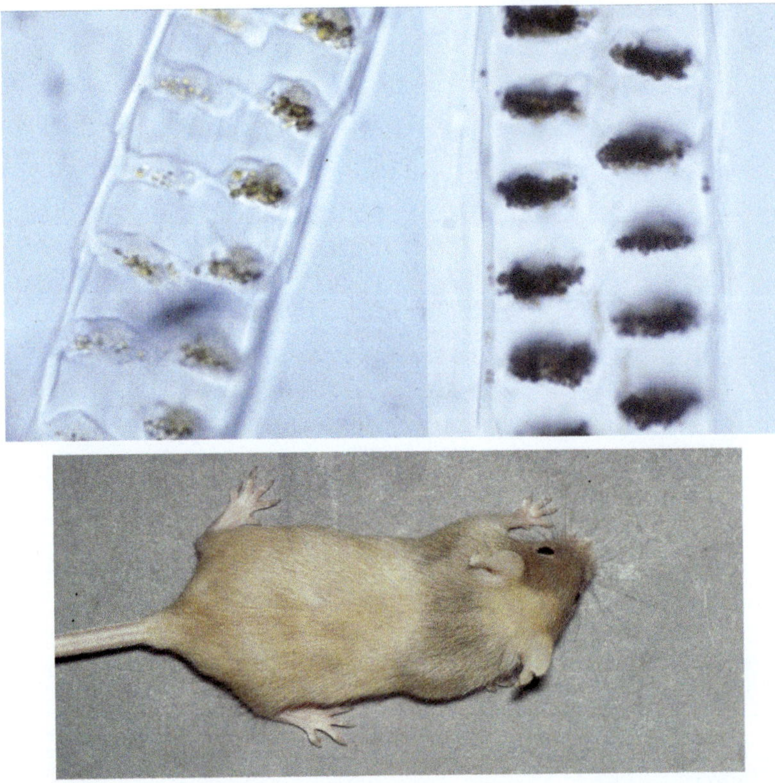

Figure 5.4 The hairs of mice can be observed microscopically after removing the internal air pockets by treatment successively, for several days each, in xylene, then a mix of absolute alcohol and xylene, and then absolute alcohol. The hairs are then quickly covered in an alcohol-soluble mounting medium and a coverslip applied. Notice, at this level of resolution, that the melanosomes appear for the most part to be either pheomelanic or eumelanic, rather than an in-between shade. The hairs are taken from a mouse congenic with the one shown below. This mouse is DK/Lm-A^y/a $Tyrp1^b/Tyrp1^b$. The mouse is in the middle of its first molt, showing the highly pheomelanic first coat and the extent to which the background genome influences the amount of eumelanin in the coat after the first molt. When the mouse completes its molt it will be uniformly of the darker color (compared with the adult C57BL6J-A^y/a $Tyrp1^+/Tyrp1^+$ mouse shown at the front in Fig. 5.1). The JU/CtLm inbred strain is similar in this respect and is closely related. DK/Lm is extinct, but JU/CtLm is available at MMRRC.

differentially to create the organelle and regulate pigment deposition. These are discussed in section 5.4, although we are just in the early stages of understanding this parameter of the pigment switch. At the cellular level, the switch is regulated by a signaling pathway shown in a basic way in Figure 5.15 and discussed in section 5.5. Finally, the gene loci are reviewed in section 5.6.

The perceived colors of eumelanic (nonyellow) and especially of pheomelanic (yellow) animals vary widely in mice (Figs 5.6 and 5.7), and even more across species lines, which is one reason why we provide pictures of other species, to present a broad perspective of the phenomenon. Even within species; imagine an Irish Setter breed of dog that appears red, and a Golden Retriever that

Figure 5.5 Sitatunga, an African antelope (*Tragelaphus spekei*). In this species, the males develop eumelanic pigmentation at maturity. A similar but somewhat less dramatic change occurs in cattle that have a wild-type phenotype.

appears yellow. Both are pheomelanic because of mutation at the same *Recessive yellow* (*E*, *Extension*, *Mc1r*) locus (Newton et al. 2000). Clearly, unknown modifying genes must be invoked to explain the dramatic difference in shade of yellowness among these dogs. Although the structure of the hair (or feathers if we were to extend the comparison to the Rhode Island Red breed of chicken) is also important to our perception of the colors.

The relationship between hair or feather structure and perceived colors of melanin pigmentation have been little studied; many genes, known and unknown, influence the structure of hairs and feathers and contribute to the color of the light that is reflected from them (Doucet et al. 2004), but we do not treat these as pigmentary genes, a distinction we reserve for factors that regulate pigment and pigment cells.

Because of the broad and varied appearances of eumelanin and pheomelanin, the inbred mouse has been invaluable in identifying major phenotypic classes of pigment colors that we describe, and assigning their biological causes. With inbred mice, we can assume that segregation of modifying genes has been eliminated in the process of inbreeding. On the contrary, of course, inbred mice are not so useful for determining the numbers and characteristics of modifying genes. For that purpose we can make comparisons with mice of different strain composition and with other mammals, which often have a higher range of variability associated with the classical pigmentation genes. With the newer techniques in genetics (Chapter 6) and in genomics, information gained from other species can be combined with information from the inbred mice to clarify further the processes of pigmentation. We have done this to some extent throughout, and in Chapter 7 will provide a summary of resources that relate to some other species, mostly domestic mammals that have been selected by breeders for interesting color variation.

5.1.1 Terminology again

Unfortunately, the comparison of different species, especially domestic ones, brings with it a tangle of terminology. Livestock and pet breeders, and even scientists, often use different words for the same phenotype, and sometimes apply the same word to different genotypes. Yellow, for example, is the common term for pheomelanic pigmentation in the laboratory mouse, but yellow cats have been termed orange in the technical literature and red (and cream) by fanciers. 'Yellow' cattle, dogs, horses, rabbits, and guinea pigs are referred to as red or yellow or sorrel or chestnut, or by other terms, depending upon breed and species.

Our purpose is not to bring logical order to the nomenclature of animal breeders, although we would prefer that scientific notation be precise. Our purpose primarily is to define and elucidate the series of steps that lead from the action of a gene (which is usually to make a protein) through the cellular results of this action, to the phenotype; and conversely to use phenotypic observations to clarify the functions of genes. For this we require a phenotypic terminology that is tightly correlated with what we know about the relevant gene function. For example, the term brown is sometimes applied to agouti animals that contain no brown pigment, while red is sometimes applied to animals that do make brown pigment (and no red/yellow pigment). Similarly, mislabeling the function of a gene can create confusion. For example, see the legend to Figure 5.1.

We will use the term yellow to refer to pheomelanic phenotypes; black for phenotypes that involve black eumelanin; and brown for phenotypes that are based in brown eumelanin pigment. These form the foundation for all the color phenotypes except white. White results from absence of pigment in the hairs or skin, either because the pigment cells are absent (Chapter 3) or because the pigment itself is absent (Chapter 4). The pinkish or bluish colors that result from blood in the tissues (see Quevedo & Holstein 2006) are not related to melanin pigments.

5.1.2 The background genome

Background genome strongly influences the phenotype of mice that have a normal wild-type melanocortin 1 receptor (MC1R) complex. Wild-type $Mc1r^{+}/-$ mice are responsive to agouti signaling protein (ASP). They are also responsive to differences at the *Mahogany* and *Mahoganoid* loci (Fig. 5.6), and to differences in the background genome (Fig. 5.7). Each of the two yellow mice in Figure 5.7 and the yellow mouse at the lower right in Figure 5.6 carry the A^{y} allele from the same source on differing genetic backgrounds. Thus the A^{y} allele is not variable among these mice of different phenotypes. All of them are wild type at the *Mc1r*, *Mahogany*, and *Mahoganoid* loci. Thus the phenotypic differences among these three yellow mice are imposed by their differing genetic backgrounds. C3H/HeJ-A^{+}/a mice (at the rear of Fig. 5.7) are considerably darker than C57BL/6J-A^{y}/a mice. JU/CtLm-A^{y}/a become very dark at the first molt, but are yellow until that time.

When JU mice and C57BL/6J mice were crossed, the color of the yellow offspring, after first molt, was highly variable in the F_2 generation. Observations seemed to show that darkening of the fur (and obesity) is modulated by gender and by social relationships, in addition to, or rather as a component of, the impact of background genome, judging by the differences between mice that are caged individually and those that are caged in various social combinations, and their interactions such as aggression (M.L. Lamoreux, unpublished observations; Lamoreux & Galbraith 1986).

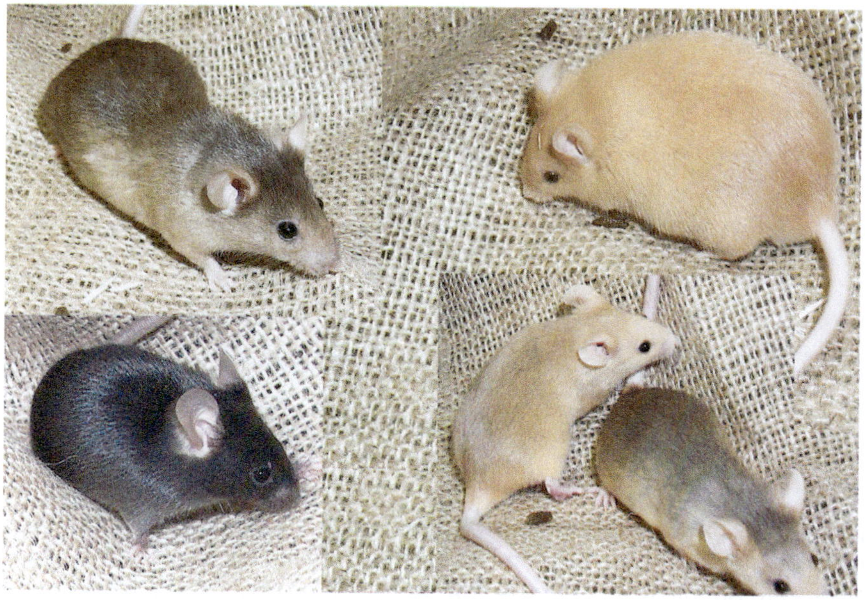

Figure 5.6 Variability in the yellow phenotype that is caused by modifying genes. The mouse on the lower left is a C57BL/6J *Nonagouti Black* eumelanic control. All the other mice are genetically yellow because of mutation at the *Agouti* locus (A^y/a). On the lower right are two mice, one young and the darker one mature, of the genotype JU/CtLm-A^y/a (see also Fig. 5.4). The mouse in the upper right is C57BL/6J-A^y/a, the most common 'yellow' laboratory mouse, and in the upper left corner a similar mouse that is also mutant at the *Mahogany* (*Attractin, Atrn*) locus, genotype C57BL/6J-A^y/a, $Atrn^{mg}/Atrn^{mg}$. The pigment phenotype of *Mahoganoid* (*Mahogunin ring finger 1*, $Mgrn1^{md}/Mgrn1^{md}$) mice that are yellow because of mutation at the *Agouti* locus (A^y/a) is similar to that of *Mahogany* mice. Specific mutant alleles are discussed in section 5.6.

Supporting these observations, similar results were noted by Suto et al. (2000) in published work. F_2 mice generated by an original cross between KK-A^y/a males and C57BL/6J-a/a females were used to identify several influences of background genome over obesity (Suto et al. 2000) and over the degree of yellowness of A^y/a mice. KK-A^y/a mice are nearly as dark as JU-A^y/a mice. In the F_2 crosses Suto et al. (1998) found a continual range of shades that was reportedly caused solely by an allele carried on chromosome 15 of the KK mice. The allele was described as responsible for pheomelanic content of the hairs (measured as aminohydroxyphenylalanine (or AHP) and visually). A modulator of eumelanin content (measured as pyrrole-2,3,5-tricarboxylic acid, or PTCA) was not found. Clearly, yellow mice have much yet to tell us about genetic regulation of complex systems; identifying these unknown modifying genes, using the available inbred strains, should be well worth the effort.

In addition to the pigmentation differences we will discuss here, it is useful to know that C57BL/6 mice are among the most tumor- (and carcinogen-) resistant strains of mice available, while C3H mice are much more tumor-susceptible (Strickland et al. 2003). This is particularly true for skin cancer, including melanoma, where under identical conditions C3H mice were much more sensitive than C57BL/6 to ultraviolet radiation and chemical-induced melanoma carcinogenesis (Romerdahl et al. 1989; Kripke 1977).

Figure 5.7 Alleles have been backcrossed from C57BL/6J on to a C3H background, where yellow phenotypes are differently expressed. From the left, counterclockwise, the genotypes are *Agouti* (*A*/–), *Lethal yellow* (A^y/–), *Albino* (Tyr^c/Tyr^c), *Nonagouti* (*a*/*a*), and at the rear is another *Lethal yellow* (A^y/–) mouse. All mice are $Tyrp1^B$/–. The difference between the two *Lethal yellow* mice is caused by strain background: the one at the front is C57BL/6J and the one at the rear is C3HFf. Photograph courtesy of Dr. Faith Strickland.

5.2 Yellow phenotypes

In Chapters 3 and 4 we discussed loci that modify the perceived colors of mice and other mammals. In this chapter we discuss the three basic colors of melanin pigment, primarily as they exist in the hairs of mammals. The varied and lovely phenotypes generated by the many interacting functions of the eumelanin/pheomelanin switch mechanism have provided clues to their genetic regulation and are the result of different combinations and patterns of yellow pheomelanin and black or brown eumelanin. The switch mechanism determines whether or not the pigment will be yellow. The unrelated black/brown (*Tyrp1*) locus determines which of the two basic colors of eumelanin will be produced (Lamoreux 1973). Pheomelanic pigmentation is epistatic to eumelanic. The MC1R complex determines the type of pigment that is produced inside the pigment cell, in normal mice, under instructions from outside the cell (see below).

ASP, encoded at the *Agouti* (*A*) locus, regulates patterns of pigment deposition in individual hairs (Figs 5.8 and 5.9) and the 'black-and-tan' (or brown-and-tan) patterns of mice and dogs (Fig. 5.10).

In laboratory agouti mice, in general, most of the dorsal hairs are eumelanic at their base, with (usually) one subterminal band of pheomelanin, and then eumelanin at the tips. The thicker overhairs are eumelanic. Usually the hairs behind the ears and around the nipples and perianal region are mostly pheomelanic, and the belly hairs are tipped with pheomelanin. The *Agouti* locus also separately controls the dorsoventral distribution of eumelanin and pheomelanin, as seen in 'black-and-tan' (black and yellow or brown and yellow) mice and dogs (Fig. 5.10).

The long allelic series of *Agouti*-locus mutants (some of which are discussed below in section 5.6) result in differences in deposition of eumelanin (versus pheomelanin) in individual hairs and/or

Figure 5.8 The classic agouti pattern is described as consisting of hairs that are banded in eumelanin and pheomelanin. The eumelanin may be black or brown, depending upon the genotype at the *Tyrp1* (black versus brown) locus. The *Brown, Agouti* mouse is often called (by fanciers) cinnamon. The pattern of banding is very precise at a specific location on a specific inbred animal; however, among species and locations, hairs may be banded, not banded, or multiply banded or tipped with yellow pigment. These patterns are controlled at the *Agouti* locus, but the relative amounts of eumelanin and pheomelanin are also influenced by other gene loci, called modifying genes, which are either other proteins of the MC1R complex or unknown modifiers.

different regions of the body. The patterns of eumelanin and pheomelanin within individual hairs as well as differences over the body, where the dorsum is eumelanic and the ventrum pheomelanic, have been attributed to control of the eumelanin/pheomelanin switching mechanism by region-specific promoters of the *A* locus (Vrieling et al. 1994; Millar et al. 1995) or chimeric patterns such as those of *Viable yellow* ($A^{vy}/-$) mice (see Fig. 5.19) or 'brindle' cattle (Girardot et al. 2006) (see Fig. 5.11).

In mice and other mammals the agouti phenotype does not imply an exact banding pattern. However, within a given species; and depending upon the genotype and the background genome, the location on the body, and the type of hair; banding patterns are very tightly controlled (Galbraith 1964). In leopards, the banding is so precise that eumelanin production stops (and starts) in exactly the right location of each of the individual hairs so that the spot is precisely defined at all levels of the overlapping hairs. Clearly the *Agouti* locus must be exquisitely sensitive to environmental cues, to which it responds by production (or not) of ASP.

Figure 5.9 The mouse in the foreground is agouti in phenotype, with most of the dorsal hairs banded as shown in Fig. 5.8. Most of the ventral hairs of most agouti animals are either banded or tipped with pheomelanin. The mouse behind is *Nonagouti Black* (*a*/*a*, $Tyrp1^B/Tyrp1^B$) in genotype and phenotype. Both mice are inbred (C57BL/6J).

Figure 5.10 Both these dogs are heterozygous at the canine *Merle* locus (which breeders term blue merle, discussed in Chapter 4, but the basic color is actually black). The dog in front also has white spotting (discussed in Chapter 3) around the collar and down the midline of the face, and additionally he is black-and-tan. *Black-and-tan* (a^t) is an allele at the *Agouti* locus. Thus, in addition to determining *when* pheomelanin or eumelanin will be produced during development of the phenotype, the *Agouti* locus determines *where* on the body it will be produced.

Figure 5.11 Commonly called 'brindle' patterning in the cow, the mutation is not related to CBD103 (as is the 'brindle' pattern of the dog), but is similar to *Viable yellow* in the mouse. The defect is caused by insertion of a transposable element into the *Agouti* locus (Girardot et al. 2006). The brindle cow at the rear may also be carrying the mutation found in Charolais cattle (see Chapter 4) that has been attributed to mutation at the *Silver* locus, as probably is the pale cow on the right. The dark animal at the rear is wild type in color, with eumelanic pigmentation predominating over the head and shoulders. Photograph courtesy of Britten Trust.

Thus, the product of the *Agouti* locus is not cell autonomous. For this reason, the *Agouti* locus does not regulate all types of eumelanin/pheomelanin patterning. It does not regulate tortoiseshell or calico patterns of pigmentation that we discussed in Chapter 3. These patterns (see Fig. 5.12) are the result of chimerism for pigmentary traits that are autonomous to the melanocyte.

The MC1R is embedded in the cell membrane; thus its mutant expression is autonomous to the pigment cell (Lamoreux & Mayer 1975; Poole & Silvers 1976). Chimeric mice that are composed of *Agouti* embryos merged with *Nonagouti* embryos display a different pattern of pigmentation from mice, such as those mentioned in Chapter 3, that are chimeric for cell-autonomous traits. The agouti chimeric pattern seen in these artificial chimeras, and in the *Viable yellow* ($A^{vy}/-$) mutant mouse (Fig. 5.19), is based in chimerism of the mesodermal tissues, where the ASP is produced (Mintz 1971a, 1971b).

A third chimeric pattern of eumelanin/pheomelanin distribution is found in the 'brindle' pattern of dogs which apparently reflects a chimerism of keratinocytes, implicating a third cell type in control over the eumelanin/pheomelanin switch. In dogs, the brindle pattern (see Fig. 5.13) is not defined at the *Agouti* locus, but at the *K* (*Dominant black*, *CBD103*, *Canine β-defensin 103*) locus which encodes another ligand of MC1R that promotes eumelanogenesis (Candille et al. 2007). The brindle phenotype of dogs is a member of the 'dominant black' allelic series. This series of phenotypic alleles is black/brindle/normal, with black dominant to brindle and epistatic to yellow.

Figure 5.12 Mini Mintzmeat. A tortoiseshell cat of the genotype *Tyrp1^b/Tyrp1^b* (*Brown*) autosomally, and X^O/X^{O+}. The cat has nonagouti brown stripes and yellow stripes. The chimeric pattern is the result of Lyonization and the fact that the cell-autonomous yellow (*Orange*) mutation is X-linked, and in this female cat is heterozygous (a homozygous yellow (orange, red) cat is shown in Fig. 5.13). The pattern of distribution of eumelanin and pheomelanin in tortoiseshell cats is a result of chimerism and is not determined at the *Agouti* locus. See also Figure 1.1.

Other loci (including *Mahogany* (*Atrn, Attractin*) and *Mahoganoid* (*Mgrn1, Mahogunin, Mahogunin ring finger 1*)) have been referred to as 'modifiers' of the functions of the *Agouti* and *Extension* loci in regulating the eumelanin/pheomelanin switch mechanism. This rather vague terminology is growing more specific as we learn more about the sensitive controls over melanin type. Other 'modifers' exist.

The striped distribution of agouti hairs in tabby (striped or spotted) cats is not a chimeric pattern, but is more akin to the precisely defined patterns of the giraffe, zebra, and various wild cats. Tabby striping in the cat is another example of eumelanin/pheomelanin patterning that is not controlled

Figure 5.13 A yellow cat (X^O/X^O) and a yellow and black brindle dog ($Tyrp1^B/-\ K^{br}/-$) with white spotting.

by the *Agouti* locus. Or rather, the *Tabby* locus apparently directs the *Agouti*-locus banding of the hairs to form the pattern of stripes or spots. Tabby cats are agouti, because their hairs are banded with yellow and nonyellow pigment. Based on phenotype, one can envision the *Tabby*-locus function to somehow direct the *Agouti* locus when and where to produce ASP. The feline *Tabby* locus has recently been identified (Lyons et al. 2006), and probably will have some new information to add to this complex story. The mouse equivalent of the cat *Tabby* locus is not known, nor do mice have tabby stripes, although mice do have a mutation of that name.

5.3 Melanin pigment

Pheomelanin can be compared chemically with eumelanin by a specific process of degradation followed by measuring the ratio of pyrrole-2,3,5-tricarboxylic acid (PTCA) to aminohydroxyphenylalanine (AHP), with high levels of the latter defining pheomelanin at the chemical level (Lamoreux et al. 2001; Ito & Wakamatsu 2006, 2008). At the chemical level, mammalian melanosomes contain the necessary enzymes that catalyze the deposition of either eumelanin or pheomelanin in melanosomes. However, to say that mammalian melanocytes produce either eumelanosomes or pheomelanosomes is simplistic. 'Mixed melanin' has been described at the chemical level (Ito & Wakamatsu 2008) (Fig. 5.14), and seems to have been adequately demonstrated.

However, the phenotypes of inbred mutant mice demonstrate that melanogenesis must be under very precise control by the pigment-type switching mechanism. If the pigments were indeed mixed, by production of eumelanin and pheomelanin at the same time, as a continual variable, one would be hard pressed to explain the high level of precision of their deposition. For example,

In addition to a reduction of TYR activity, the switch involves modification of a whole panel of other gene loci. Precisely how this condition is brought about is not clear, although one actor is known to be cAMP, and the pathway probably involves microphthalmia-related transcription factor (MITF). Another intriging possibility is that MGRN1 may be involved in the routing (trafficking) of the proteins required to make melanosomes (Kim et al. 2007).

It has often been proposed that the availability of thiol compounds is the key to the production of pheomelanin. Pheomelanic pigmentation is indeed reduced in pheomelanic regions of mice mutant at the *Subtle gray* (*Sut*, *Slc7a11*) locus (see Fig. 5.23, below), which encodes a protein that functions in transport of sulfur-containing amino acids into the cell. Le Pape et al. (2009) report that application of ASP to cultured pigment cells modulates glutathione synthesis. Several other loci that appear to influence pheomelanic pigmentation more than eumelanic pigmentation in mutant mice have not been studied in this regard. These include *Grey-lethal* and *Grizzled* (Silvers 1979; see also the Appendix in Chapter 1).

Similar to the *Subtle gray* mice, mice that are mutant at the *Albino* locus also exhibit a greater reduction in pheomelanic than eumelanic pigmentation. For example *Chinchilla* ($Tyr^{c\text{-}ch}/Tyr^{c\text{-}ch}$) black mice are barely different from normal black mice, but pigmentation is much reduced by the *Chinchilla* mutation in mice that are yellow, or in the yellow bands of agouti hairs of mice and other animals (including 'white' tigers and chinchillas, from which the mutant gets its name). The exception to the rule that pheomelanin is reduced more dramatically than eumelanin in mice mutant at the *Tyr* locus is seen in mice homozygous for the *Platinum* ($Tyr^{c\text{-}p}$) allele. As TYR in *Platinum* mice apparently does not reach the melanosome (Orlow et al. 1993; Beermann et al. 1995, 2004), this suggests that the observed difference between the pheomelanic and eumelanic response to the same albino condition in inbred mice occurs at the level of the melanosome, after the panel of proteins that are involved with eumelanogenesis has been activated, before they are successfully routed to the developing organelle.

In quite the reverse interaction, *Pink-eyed dilution* ($Oca2^{p}/Oca2^{p}$), which is also switched off during pheomelanogenesis, strongly reduces eumelanic but not pheomelanic pigmentation (Chiu et al. 1993). A similar relationship has been observed in humans that are pheomelanic (or not) because of their genotype at MC1R (King et al. 2003b). It is conceivable that OCA2 may function in availability of substrate or in pH control (Chen et al. 2002, 2004).

In sum, all of the above factors are changed during pigment-type switching, but none of them individually causes the switch. Whether viewed from the perspective of the phenotypic relationships discussed above, or from the perspective of gene-expression profiles of cultured melanocytes as reported by Le Pape et al. (2009), the eumelanin/pheomelanin switch appears to be a choice between two extensive programs of gene expression that are all 'switched' together. The primary cause of all this activity is the interaction between the MC1R at the cell surface, and its ligands, ASP, MSH, and, at least in the dog, canine β-defensin 103 (CBD103).

5.5 Signaling the switch mechanism at the cellular level

We have emphasized that the pigment cell is highly responsive to the tssue environment within which it functions, and we have shown that the switch is mediated by MC1R, encoded at the *Recessive yellow* (*Mc1r*) locus, a protein that resides in the cell membrane and transduces the environmental messages into intracellular responses. Arriving messages are generated in the dermis and epidermis (April & Barsh 2006) as well as in the endocrine system and probably the nervous system. These include ASP encoded at the *Agouti* locus and expressed in the mesoderm (Aberdam et al.

1998), melanocyte-stimulating hormone (α-MSH) encoded at the *Pro-opiomelanocortin* (*Pomc*) locus, which can be pituitary or epidermal in origin, and CBD103, which is expressed in keratinocytes. Attractin (ATRN) encoded at the *Mahogany* locus may be associated with the receptor as a modulator of ligand binding and mahogunin ring finger 1 (MGRN1) encoded at the *Mahoganoid* locus may be required for transmission of the message.

The best-known result of this signaling system in the case of eumelanogenesis is increased levels of cAMP within the cytoplasm and activation of the cAMP signaling pathway, or the absence or reduction of this signaling in the case of pheomelanogenesis. However, it has also been shown that ASP signals through two different pathways when pheomelanic cells are exposed to it in culture (Hida et al. 2009), and that CBD103 exerts its influence via a cAMP-independent pathway (Candille et al. 2007).

MITF has been implicated as a transcription factor that influences production of many proteins involved in melananogenesis (see Chapter 3), but the details of the signaling pathway from the receptor to the actual switch of pigment type are not known. Little work has been done to study pheomelanogenesis in mice that are mutant at loci known to be inactivated during pheomelanogenesis, or to directly confirm a relationship between MITF and the switch mechanism. It is difficult to study pheomelanogenesis in cell culture, although apparently normal pheomelanogenesis has been achieved in hair follicles in tissue culture (Tamate & Takeuchi 1984) and cell-culture conditions continue to be refined. This difficulty may be the result of the complexity and multicellular origin of the external signals required for normal regulation of the switch mechanism, combined with the fact that MC1R maintains a constitutive level of activity in normal cells (Bennett 1989; Bertolotto et al. 1998; Huver et al. 2003; Sánchez-Más et al. 2004; Levy et al. 2006).

MC1R is a G-protein-coupled cell surface receptor that can be bound by α-MSH or CBD103 as 'positive' agonists. Activation of MC1R by these agonists stimulates adenylate cyclase and increases the intracellular levels of cAMP (Suzuki et al. 1996) (see Fig. 5.15) which signals production of eumelanin. Binding of ASP prevents binding of MSH, and transmits alternative unidentified signals independent of cAMP and directs the production of pheomelanin (Abdel-Malek et al. 2001; García-Borrón et al. 2005; April & Barsh 2006; Le Pape et al. 2008; Hida et al. 2009). An agonist binds a receptor and changes its activity; an antagonist binds a receptor as a competitive inhibitor, but does not independently change its activity. ASP binds the receptor competitively, but it also independently reduces cAMP levels and produces an alternative signal; therefore it is correctly termed an inverse agonist and is the primary signal for the eumelanin/pheomelanin switch (Abdel-Malek & Supp 2008). However, it seems that MSH, cAMP, and CBD103 are not solely responsible for pigment-type switching, or at least not solely responsible for regulating the sensitivity of the receptor complex to the external stimuli it receives. As we have already stated a number of modifiers of the switch response exist.

The eumelanin/pheomelanin switch mechanism is a remarkable example of cellular responsiveness to its environment. The more we learn about it, the more it moves us forward to yet more interesting questions. To what is the *Agouti* locus regulatory region responding? How does it know which pigment to put where (Weiner et al. 2007; Enshell-Seijffers et al. 2008)? How does the message from the receptor and its ligands become so intricately and accurately transduced into highly variable pigmentary phenotypes? And perhaps most interesting, how does such a complex switch, involving dozens of proteins inside and outside the cell, evolve?

One wonders whether the rapid changes between production of eumelanin and of pheomelanin may represent a controlled reversion to and from a more primitive condition. And if so, which is the more primitive condition? The answers to these and similar questions will no doubt tell us a great deal about the processes of life, well beyond the functions of the pigment cell.

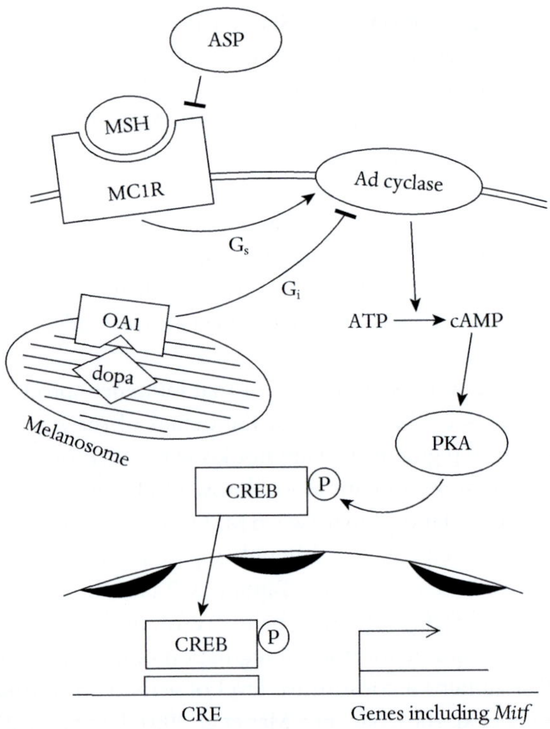

Figure 5.15 Basic components of cAMP signaling in the melanocyte lineage. Symbols and abbreviations are as in Fig. 3.9. See text for more details and sources. Ad cyclase, adenylate cyclase; ASP, agouti signal protein; CRE, cAMP-response element; CREB, CRE-binding protein; OA1, ocular albinism 1 protein (a receptor for L-dopa); P in circle, phosphate group.

5.6 Yellow genes

Recessive yellow (E, Extension, Mc1r, Melanocortin 1 receptor)

MGI lists seven phenotypic alleles, of which six were spontaneous.

Availability: JAX, MMRRC, ORNL, and TAC (see Appendix in Chapter 1 for details about repositories).

MC1R is an integral membrane protein of 317 amino acids in humans and 315 in mice, with an extracellular N-terminus, seven transmembrane domains, and an intracellular C-terminal extension (García-Borrón et al. 2005) (Fig. 5.16). The *Recessive yellow* locus encodes MC1R, which functions as the key switching element of the MC1R complex of the pigment cell. The normal, wild-type MC1R responds to multiple cues, including melanotropic hormones encoded at the *Pro-opiomelanocortin* locus, CBD103, and ASP encoded at the *Agouti* locus. The result, in normal mice and most other mammals, is the agouti pattern of distribution of yellow and nonyellow pigmentation that is described above (Figs 5.8, 5.9, and 5.17). With certain mutations at the *Mc1r*

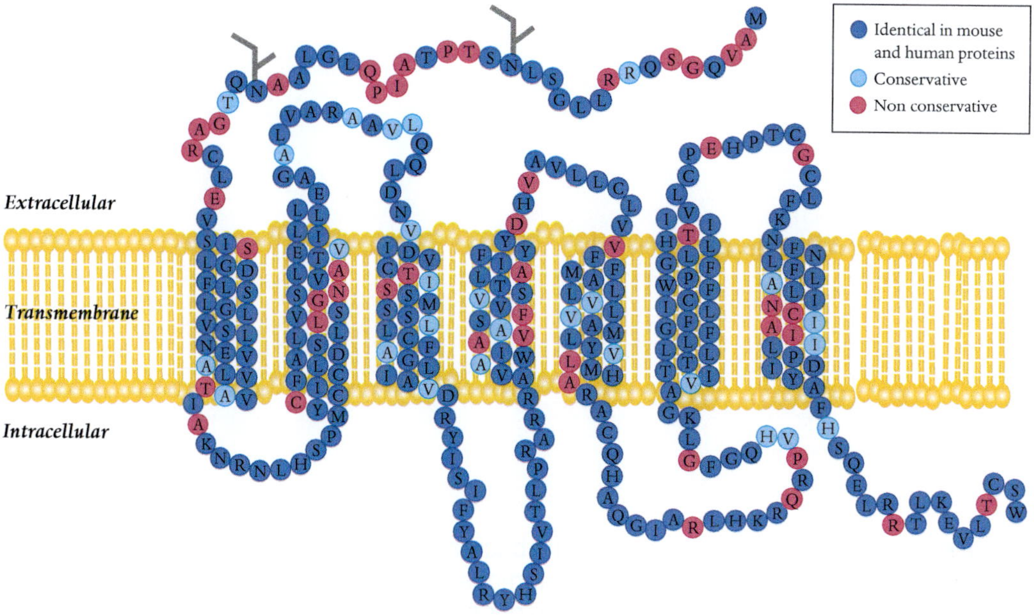

Figure 5.16 Generalized diagram of MC1R residing in the cell membrane, comparing the human and mouse proteins. In the living cell, of course, this protein is closely associated with or interacts with other proteins discussed in this chapter to make up the melanocortin receptor complex (Pérez-Oliva et al. 2009). Diagram courtesy of García-Borrón, Sánchez-Laorden and Jiménez-Cervantes 2005; reproduced with permission of Wiley-Blackwell.

Figure 5.17 Mice showing the normal agouti phenotype on two different inbred backgrounds. Function of the *Agouti* locus is very sensitive to background genome. For example, all the alleles tend to be darker (more eumelanic) on the C3H background than they are on the C57BL/6J background.

Figure 5.18 Yellow mice. The mouse at the back is an *Agouti*-locus mutant, C57BL/6J-A^y/a, $Mc1r^E/Mc1r^E$. The *Recessive yellow* mouse in the foreground is C57BL/6J-*a*/*a*, $Mc1r^e/Mc1r^e$. Notice that the characteristic minor eumelanic tipping of the hairs of the recessive yellow mouse is greater than that seen on the *Lethal yellow* mouse when the background is C57BL/6J. *Agouti*-locus mutant phenotypes are more highly variable in this respect, depending on strain background.

locus, the MC1R cannot implement the switch. Pigmentation will then be either pheomelanic, as in *Recessive yellow* ($Mc1r^e$) (Fig. 5.18) in mice or similar alleles in humans (Krude et al. 1998; Beaumont et al. 2008) and other animals, or eumelanic as in *Sombre* ($Mc1r^{E\text{-}So}$) or similar alleles. The eumelanin will be black or brown depending upon the genotype at the unrelated *Tyrp1* locus, or will show modifications of the three major pigment colors if other relevant mutations are present. The wild-type MC1R maintains a constitutive level of activity that results in a eumelanic mouse, even in the absence of signaling from melanotropic hormones (Bennett 1989; Bertolotto et al. 1998; Levy et al. 2006).

MC1R can be artificially activated by injection of α-MSH during a hair growth cycle of a mouse that is making pheomelanin under control of the agouti locus (Geschwind 1966), or by addition of MSH to cultured melanocytes. α-MSH is encoded at the *Pro-opiomelanocortin* (*Pomc*) locus, discussed below. The receptor switch in mice with normal MC1R can be artificially switched to eumelanogenesis by injection of α-MSH. If the cell is producing pheomelanosomes, it will switch to eumelanosomes, without interrupting the processes involved in generating the melanosomes and delivering them to keratinocytes. The changing patterns of melanogenesis suggest that circulating MSH is not the normal signal for the MC1R (Millar et al. 1995) and that the situation is considerably more complex and involves at least three additional loci, and regulation by both dermal and keratinocyte (Hirobe 2005) elements of the skin, as discussed above.

Binding of ASP (encoded at the *Agouti* locus) to MC1R prevents the upregulation of cAMP that is associated with activation of eumelanogenesis and causes the cell to switch its programming toward pheomelanogenesis. Abnormal pheomelanogenesis may be caused by mutation at the *Extension* locus, as in *Recessive yellow* mouse, or at the *Agouti* locus. Mice of both types are shown in Figure 5.18.

Constitutively active MC1R (as in *Sombre*/*Mc1r*$^{E\text{-}so}$ or *Tobacco*/*Mc1r*$^{E\text{-}tob}$ mice) or a loss-of-function mutation in *Agouti* (*Nonagouti*, *a*, or *Extreme nonagouti*, a^e, for example) results in nonagouti eumelanic mice (Robbins et al. 1993; Jackson 1994).

In human skin, MC1R is importantly involved in the response to ultraviolet irradiation and protection from skin cancer, as well as determining pheomelanic versus eumelanic pigmentation (Bonilla et al. 2005). A number of recent reviews deal with the relationship of mammalian MC1R with melanocyte function, skin pigmentation, and skin-cancer risk (Robinson & Healy 2002; Sturm 2002; Bataille 2003; Rees 2003; Healy 2004; García-Borrón et al. 2005; Rouzaud et al. 2005).

Agouti (*A*, *Nonagouti*, *a*)

MGI lists 101 phenotypic alleles, 58 of which were spontaneous.
Availability: HAR, JAX, MMRRC, ORNL, and RBRC.

A note about nomenclature: the official name of the *Agouti* locus has been changed to '*Nonagouti*.' However, from the initiation of the science of mammalian pigmentary genetics over 100 years ago, until very recently, this locus has been named *Agouti*, and most of the literature on the subject of pigmentary phenotypes to this day refers to the *Agouti* locus, especially across species lines. Most of the phenotypic literature will be accessible using the term *Agouti* in preference to *Nonagouti*, which is the name of one of the recessive alleles at the locus.

The *Agouti* locus (Fig. 5.17) encodes ASP, which functions as an inverse MC1R agonist (see sections 5.4 and 5.5), not only preventing other ligands from binding the MC1R complex, but also blocking its constitutive activity, generating a different signal of its own (Hida et al. 2009) and so preventing formation of eumelanosomes and activating pheomelanin synthesis. The presence or absence of ASP is a primary external control of the switch mechanism in mice (Siracusa 1994).

Mice that are yellow because of mutation at the *Agouti* locus are also obese, approximately in proportion to the ratio of yellow (compared with nonyellow or agouti) hairs in their fur. Therefore it is not surprising that elucidation of the mechanism of action of the pigment defect has also informed our understanding of obesity. Hyperphagic obesity of *Agouti*-locus yellow mutant mice relates to the fact that the melanocortin 3 and 4 receptors (MC3R and MC4R) are capable of binding ASP. These receptors are found in areas of the brain, including the hypothalamus, that regulate energy metabolism. When ASP is normally expressed, only MC1R (on melanocytes) has access to it. When ASP is expressed ubiquitously (see next section), MC4R and MC3R are able to bind it (Ollmann et al. 1997: Shutter et al. 1997; Yang et al. 1999; Barsh et al. 2000).

Yellowness in mice influences size of white spots (Lamoreux 1983) and colors of mice mutant at *Mitf* (Lamoreux 1981).

In normal mice, the pheomelanic alternative is determined by the genotype of dermal tissues that produce ASP at appropriate times and locations (Markert & Silvers 1956; Mayer & Fishbane 1972; Millar et al. 1995; Hirobe et al. 2004b). Postnatally, ASP is produced by the dermal hair papilla, a core of dermis inside the hair bulb and adjacent to the follicular melanocytes (Millar et al. 1995). In normal mice, the location and timing of ASP expression is controlled by region-specific and hair cycle-specific promoters of the *Agouti* gene (Vrieling et al. 1994; Millar et al. 1995), presumably in coordination with or in response to signaling from modifying loci. One of these may be *Tbx15* (Candille et al. 2004). *Agouti* transcripts of mice contain four exons: exons 2–4 contain the region that encodes the protein. There are several alternatively spliced transcripts containing different exons 1 (not translated), which are expressed from different promoters. Four separate exons 1 have been identified: 1A and 1A′ are found only in mRNA from ventral skin, whereas exons 1B and 1C are specific to the hair growth cycle (Hustad et al. 1995).

Figure 5.19 YS/WfLm-A^{vy}/*a Ednrb*^{s}/*Ednrb*^{s}. A chimeric (A^{vy}/*a*) mouse with *Piebald* white spotting. Some mice of this genotype are entirely agouti in appearance, while others are entirely yellow, and anything between is possible. Notice that the spotting does not affect pattern or size of the stripes, which are dermal in origin. The calico phenotype, in which distribution of pigment is modified by the presence of some genotypes of white spotting, affects only cell-autonomous chimerism (Chapter 2).

Lethal yellow **(A^{y})** The *Lethal yellow* mutation of the mouse (Figs 5.1 and 5.18) was one of the earliest described mammalian genetic anomalies (Cuénot 1902) and continues to be one of the most instructive. Lethal yellow mice are pheomelanic; they develop adult-onset hyperphagic obesity and obesity-related problems such as diabetes. On some genetic backgrounds they have a higher incidence of tumors than nonagouti or agouti mice; and homozygous embryos die at implantation. The A^{y} allele is dominant to the wild-type (*Agouti*) allele, *A*. The lethality of the A^{y}/A^{y} genotype in mice has been explained by the constitutive and nearly ubiquitous expression of the ASP, the protein encoded at the *Agouti* locus. In this mutant allele, the structural gene for ASP is aberrantly linked to the promoter of the *Raly* gene, which is ubiquitously expressed and encodes an RNA-binding protein important to embryogenesis but not related to pigmentation (Bultman et al. 1992; Miller et al. 1993; Michaud et al. 1993, 1994; Duhl et al. 1994). The rearrangement results in loss of RALY expression, thus explaining the lethality of homozygosity for the *'lethal yellow'* allele.

In addition to *Lethal yellow*, several other *Agouti*-locus alleles have been identified that are dominant to the wild type. Some are detailed below. None is lethal, but they all result in mice with some increased amount of pheomelanic pigmentation compared with wild-type agouti mice. Some are chimeric, with variable patches of yellow fur and wild-type agouti fur (Fig. 5.19) that depend upon expression of the gene in the mesodermal tissues. Similarly, black/agouti chimerism is generated by other alleles that are lower in the dominance series than *Agouti* (a^{m} for example, Fig. 5.20). The cause of the chimerism is not clear in all cases (but see *Hypervariable yellow* below) but chimeric phenotypes and their implications are discussed in Chapter 3 and above.

Viable yellow **(A^{vy})** This allele (Fig. 5.19) has been used extensively to study the pleiotropic effects associated with the agouti-yellow phenotype on the C3H/HeJ and the VY/Wf and YS/Wf backgrounds. Homozygotes (A^{vy}/A^{vy}) and heterozygotes (A^{vy}/A and A^{vy}/a) show considerable variation in appearance, ranging from clear yellow, through mottling with dark patches,

Figure 5.20 C57BL/6J-a^m/a^m (on the left) and JU/CtLm-a^m/a^m (on the right). These mice are available at MMRRC.

Figure 5.21 C57BL/6J. These are *Mahogany* mice: the phenotypes of *Mahogany* and *Mahoganoid* mice are similar. The one at the bottom is nonagouti black, and the one at the top is yellow. These mice are available at MMRRC.

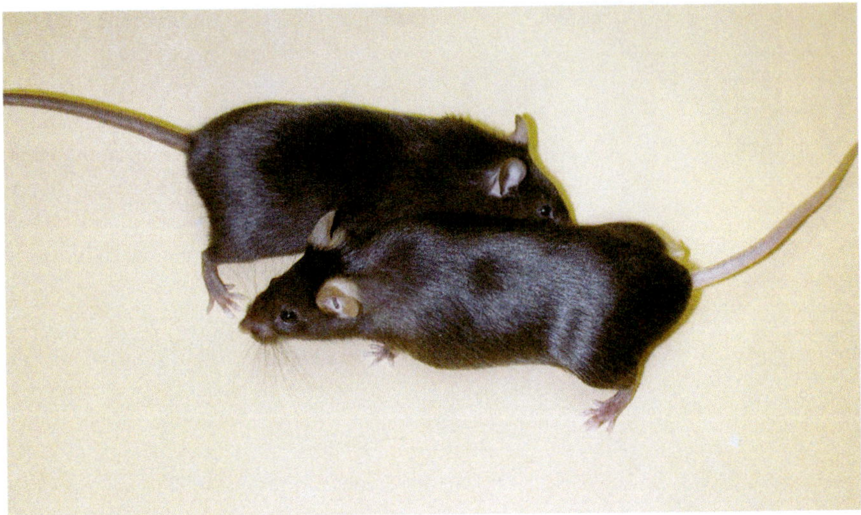

Figure 5.22 *Extreme nonagouti*, C57BL/6J-a^e/a^e. Notice that the mutant mouse is more eumelanic than the control C57BL/6J-*a*/*a* mouse in the foreground, particularly behind the ears and also in the tail and peri-anal region, and around the nipples.

to a completely agouti-like coat. The variation is strongly influenced by the *Agouti*-locus genotype and strain genome of the dam, by genetic background, and also by some environmental conditions including diet (Wolff 2003). Homozygotes and heterozygotes tend to become obese, and the degree of obesity is correlated with the amount of yellow in the coat. Some types of tumor are preferentially expressed in the yellow mice compared with the nonyellow (agouti) mice. The work, undertaken notably by Heston and Vlahakis (1962) and by George Wolff, and their co-workers, provided extensive data relative to the correlation of the yellow phenotype with obesity, tumor incidence, diabetes, higher growth rates, several biochemical parameters, and also maternal influence, in mice that were of the same genotype but differing phenotypes. This ongoing lifetime of work was summarized in a recent review (Wolff 2003).

***Hypervariable yellow* (A^{hvy})** This mouse has an even broader range of phenotypes than the *Viable yellow*, from pheomelanic through eumelanic nonagouti. Carriers' coat colors range from mostly yellow to almost completely nonagouti (black, on the C57BL/6J background) and are usually patchy or striped. Mice with coats on the yellow end of the spectrum tend to develop obesity while those with more eumelanic coats do not. Homozygotes are viable and fertile, and are more likely to have yellow or mostly yellow coats than are heterozygotes. A^{hvy} is recessive to A^y, and dominant to a and a^e (*Extreme nonagouti*, Fig. 5.22). *Hypervariable yellow* was backcrossed on to C57BL/6J and used (Argeson et al. 1996) to demonstrate the cause of the chimerism as described below. The allele is available on the C3H background at JAX. The C57BL/6J-A^{hvy}/a mouse was lost.

In the A^{hvy} mouse, the yellow/nonyellow chimerism is attributed to an intracisternal A particle (IAP; essentially a transposable element) inserted into the agouti exon 1C in an antisense orientation. Exon 1C normally controls yellow/nonyellow banding in the hairs by regulating expression of ASP in the dermal papilla cells (Argeson et al. 1996). However, this 5.4 kb insertion results in transcription of the full-length coding sequence from the long-terminal repeat (LTR; promoter) of this IAP. Thus, ectopic expression is found in a manner that parallels the broad variation in coat color routinely seen within a population of A^{hvy} mice. Specifically, A^{hvy} mice with a predominantly yellow coat color have a greater ectopic expression of ASP than do those which display a predominantly nonyellow (nonagouti black in this work) coat color. This variation in expressivity has been attributed to variation in methylation of the inserted IAP (Argeson et al. 1996). The methylation state is clonally derived, is to some extent heritable, and is influenced both by the pigmentation ratio of the mother and by environmental factors that include diet (Wolff 2003).

The correlation of obesity with yellowness is found in all the dominant-yellow chimeric genotypes of mice mutant at the *Agouti* locus, including A^{vy}/a mice, which are phenotypically variable though co-isogenic (Lamoreux 1983; Wolff 2003). For this reason, the A^{vy} and A^{hvy} mutant inbred mice are importantly used in studies of transgenerational epigenetic phenomena (Morgan et al. 1999; Wolff 2003; Blewitt et al. 2006; Dolinoy et al. 2006; Dolinoy 2008).

***Agouti mottled* (a^{m} *and* $a^{m\text{-}J}$)** *Agouti mottled* mice are also interesting because of their chimerism. Their phenotypes on a C57BL/6J background range between agouti and nonagouti. In the $a^{m\text{-}J}$ mouse, the cause of chimerism is the replacement of one of the normal exons of the control region of the transcript by a nucleotide sequence that is normally located in the genomic DNA upstream of the coding regions. These mice are available from JAX on the C57BL/6J background. The phenotype on the JU/CtLm background differs quite dramatically. Lamoreux (unpublished results) obtained the a^{m} allele from George Wolff and backcrossed from the same initial source on to both C57BL/6J and JU/CtLm (Fig. 5.20). On the B6 background, the phenotype is mottled as expected. On the JU background the mouse appears like a normal agouti animal with no mottling. Both these stocks are available from MMRRC.

***Agouti* (*A*)** (Figs 5.9 and 5.17) This is the phenotype most often referred to as wild type in *Mus musculus*, as described in section 5.2 (although see also the following allele). Briefly, the phenotype is the result of banding of eumelanic and pheomelanic pigmentation within the hairs, especially dorsally. This patterning is imposed by the *Agouti* locus promoter which directs ASP production during the hair cycle (Millar et al. 1995). The number of bands on hairs of agouti mammals, the width of the bands, and their locations are tightly controlled genetically and at least in cats are selectable. This probably reflects interactions of the *Agouti* promoter with unknown 'modifying genes.'

Inbred *Agouti* mice are available on C57BL/6J and C3H inbred backgrounds. It is useful to note that C3H *Agouti* mice have often been used as a wild-type 'control' for C57BL/6J mutant experiments. This practice is undesirable, because the phenotype of *Agouti*-locus mutants is significantly different on the two backgrounds in terms of color (C3H mice are darker, more eumelanic; Fig. 5.7) and of other traits mentioned above in the section on background genome.

***White-bellied agouti* (A^{w})** *White-bellied agouti* is a common wild mouse phenotype that lacks eumelanic banding of its ventral hairs, which also are reduced in pheomelanic pigmentation and so are white or cream. The mice are dorsally agouti in phenotype. There are many similar alleles. In the original A^{w} allele, the locus contains a large insertion in the first intron of an LTR of a transposable element. This gives additional expression of abnormally spliced ASP transcripts.

***'Black-and-tan'* (a^{t})** As is true of all *Agouti*-locus phenotypes, these mice are black and yellow (or brown and yellow depending upon the genotype at the *Tyrp1* locus). In the case of 'black-and-tan' animals, however, the coloring is much more dramatic because the hairs are not of the agouti type. The distribution of the tan (yellow) and nonyellow pigmentation depends upon the location on the body (yellow is on the belly and may also be expressed in other locations, in other species, as dogs). In mice, the belly ranges from vibrantly pheomelanic to nearly white, depending upon the background genome, while the entire dorsum is nonagouti and eumelanic. This phenotype beautifully illustrates control of the *Agouti* locus over spatial distribution of the eumelanin/pheomelanin alternative (Millar et al. 1995). Mouse fanciers have bred mice that are homozygous for *Black-and-tan* and for *Pink-eyed dilution*, taking advantage of the fact that *Pink-eyed dilution* reduces the eumelanic pigmentation but not the pheomelanic pigmentation. The result is a pretty gray mouse with a bright yellow belly (p. 105 of Jones 1979). *Black and tan* is dominant to *Nonagouti*.

***Nonagouti* (*a*)** This is most commonly represented by the standard eumelanic C57BL/6J mouse (Figs 5.9, 5.22, and many other figures in this book), always available on the shelf at The Jackson Laboratory, which is shown as the control in most of the pictures in this book. The inbred C57BL/6J mouse is *Nonagouti* (*a*/*a*) and is wild type at the *Tyrp1* and other known color loci; thus, nonagouti black in phenotype. It has a few yellow hairs behind the ears and around the perianum and nipples. Correctly stated, homozygosity for the *a* allele results in a nonagouti eumelanic mouse: one that lacks the subterminal pheomelanic banding characteristic of agouti mice. *Nonagouti* does not define the color of the eumelanin. Similarly, C57BL (or B6 or C57) are not synonyms for black color but rather define the strain background of the mouse.

The *Nonagouti* allele (*a*) contains an 11-kb insertion in the first intron of the *Agouti* transcript. This 11-kb insertion consists of a 5.5-kb VL30 element with an additional 5.5 kb of repeat-flanked sequence inserted between base pairs 2474 and 2473 of this VL30 element, in opposite transcriptional orientation from *Agouti*. This apparently disrupts splicing, since the result is a smaller and much less abundant transcript, hence the greatly reduced ASP activity.

Nonagouti mice are available on several inbred backgrounds, including C57BL/6J, C3H/FeLa, JU/CtLm (the *a* allele in the latter case originally from HAR, rather than from JAX). Many other natural and created alleles of the *Agouti* locus are available. They consistently maintain a dominance order that correlates with phenotype, from A^y, the yellow top dominant, to a^e, *Extreme nonagouti*, and several other alleles (Fig. 5.22), homozygotes for which, as with a^e, are profoundly nonagouti with no evidence of yellow hairs. a^e appears to be a null allele for ASP; it is a mutation in the ATG methionine codon that signals the start of translation, resulting in no translation (Hustad et al. 1995).

Pro-opiomelanocortin (*POMC*)

Four phenotypic alleles are listed at MGI, none of which are spontaneous.
Availability: JAX.

The *POMC* locus encodes MSH, adrenocorticotropin (ACTH; corticotropin), and other melanocortins that function in an integrated fashion to regulate the functions of corticosteroids (Karpac et al. 2008). In terms of the switch mechanism, POMC is a large molecule that is chopped up to release α-MSH, which is a ligand of MC1R and is capable of competing with ASP to switch the signal from pheomelanin to eumelanin. This function of α-MSH was discovered accidentally in an *Agouti* mouse that had a tumor of the pituitary and an unexpectedly black phenotype. The ability of MSH to activate the switch from pheomelanic to eumelanic was subsequently confirmed by injection of MSH into *Lethal yellow* mice during hair growth (Geschwind 1966). However, the presence or absence of a circulating hormone is not sufficient to explain all the subtleties of phenotype generated by the switch mechanism. Much later it was demonstrated, rather surprisingly, that nonagouti POMC-knockout mice, which are unable to produce MSH, are visually as black as *Nonagouti* POMC-expressing mice (Slominski et al. 2005) although apparently not as totally black as *Extreme nonagouti* mice or mice that are black because of an activating mutation of the MC1R, such as *Sombre* (*Mc1r*$^{E\text{-}So}$). Evidently the receptor has some inherent activity (García-Borrón et al. 2005; Slominski et al. 2005) or is being influenced by signals other than ASP (April & Barsh 2006; Candille et al. 2007).

ACTH also is capable of promoting eumelanin synthesis, especially in humans. Keratinocytes of the skin and hair follicles, adjacent to the melanocytes, can synthesize MSH and POMC (Yamaguchi & Hearing 2009) and (at least in dogs and wolves) CBD103 also promotes eumelanogenesis.

Mahogany (*Mg, Atrn, Attractin*) and *Mahoganoid* (*Md, Mgrn1, Mahogunin ring finger 1*)

Mahogany: MGI lists seven phenotypic alleles, five of which were spontaneous.
Mahoganoid: MGI lists seven phenotypic alleles, five of which were spontaneous.
Availability for both: CMMR, JAX, MMRRC, ORNL, and TIGM.

The *Mahogany* locus encodes attractin (ATRN) (Gunn et al. 1999), a large transmembrane protein with several extracellular interaction domains including a CUB domain, EGF domains, a C-type lectin domain, and a γ/β common cytokine receptor-binding motif (Gunn et al. 1999, 2001; Nagle et al. 1999).

He et al. (2001) and Phan et al. (2002) cloned the mouse *Mgrn1* gene, mutant in *Mahoganoid* mice. Both groups identified the human homolog as KIAA0544 (Nagase et al. 1998), a ubiquitously expressed protein with a C3HC4 RING that has E3 ubiquitin ligase activity *in vitro*.

The pigment phenotype of homozygous *Atrn*^{mg} (*Mahogany*) mice is indistinguishable from that of *Mgrn1*^{md} (*Mahoganoid*) on the same inbred background. An example is shown in Figure 5.21. Both alleles cause darkened coat color due to increased production of eumelanin (compared with pheomelanin). $A^{y}/-$, *Atrn*^{mg}/*Atrn*^{mg} mice are darkly eumelanic dorsally (Fig. 5.21) but yellow on the belly, whereas $A^{y}/-$, *Atrn*$^{mg\text{-}3J}$/*Atrn*$^{mg\text{-}3J}$ mice are entirely black with dark ears and tail. Mice homozygous for the *Mahoganoid* mutant allele are characterized by reduction of yellow pigment. Complete loss of function (homozygous *Mgrn1*^{nc}) causes absence of yellow pigment.

In addition to reduction in yellow pigmentation the phenotypes include suppression of the obesity associated with mice that are yellow because of mutation at the *Agouti* locus. Mice lacking mahogunin ring finger 1 (MGRN1) or attractin develop age-dependent spongiform neurodegeneration through an unknown mechanism (Dinulescu et al. 1998; Walker et al. 2007). However, this may be a secondary result of mitochondrial dysfunction and oxidative stress that are evident before onset of vacuolation in the brains of mice homozygous for mutants at *Mgrn1* or *Atrn*. Many mitochondrial proteins are reduced in *Mgrn1* mutants (Sun et al. 2007), although the situation seems different in cultured melanocytes, where proteins downregulated by ASP, with the presumed help of normal MGRN1, prominently included 61 mitochondrial proteins (Le Pape et al. 2009).

Dinulescu et al. (1998) reported that mice homozygous for *Atrn*^{mg} on a C57BL/6J congenic background have increased night-time locomotor activity, a 0.5°C increase in body temperature, and increased basal metabolic rate, and are hyperphagic. Gunn et al. (1999) reported that mice homozygous for *Atrn*$^{mg\text{-}3J}$ on the C3H/HeJ background also have increased night-time locomotor activity but that they have normal food intake and decreased body weight associated with decreased adiposity. Mice homozygous for *Atrn*$^{mg\text{-}3J}$ have decreased fat storage and are resistant to weight gain when fed a high-fat diet. *Atrn* mutations do not alter the obesity caused by a null mutation of MCR4 or transgenic expression of agouti-related protein, nor do they inhibit obesity caused by mutations *tub*, Cpe^{fat}, $Lepr^{ob}$, or $Lepr^{db}$ (Jackson 1999; Nagle et al. 1999; He et al. 2001; Dinulescu et al. 1998). The severity of the vacuolation and also of behavioral defects increases with age and varies between strains (He et al. 2001).

Each mutant allele at *Mahogany* or *Mahoganoid* is recessive, and, as stated, when homozygous each reduces the amount of pheomelanin that is predicted by the *Agouti*-locus genotype, assuming that the mouse is wild type at the *Mc1r* locus and able to respond to its ligands (Bagher et al. 2006). This supports the idea that ATRN and MGRN1 proteins act in the same signaling pathway as ASP. ATRN is proposed to be a co-receptor for ASP (see above), while MGRN1 is an E3 ubiquitin ligase (Barsh 2006), an enzyme that adds the small peptide ubiquitin to another protein. This addition often signals degradation of that protein by proteasomes or lysosomes. MGRN1 was recently proposed to act downstream of the MC1R to interrupt a positive feedback in cAMP/eumelanogenesis signaling, possibly by inducing degradation of a key signaling component (Hida et al. 2009). The precise mechanism of function of these two proteins is still in doubt, but it seems clear from the murine phenotypes that they function together with the MC1R to refine the signal received from its ASP ligand.

In vitro, ATRN specifically binds ASP encoded at the *Agouti* locus, where it may function to facilitate the interaction of ASP with the MC1R complex (He et al. 2001). A similar protein has been found associated with the melanocortin 4 receptor (Haqq et al. 2003). MC4R also recognizes ASP, suggesting that further study of the eumelanin/pheomelanin switch mechanism will continue to inform at least two other functional sets of proteins: MC4R and MC3R and their associated proteins, which regulate important tasks in the mammalian brain (Czyzyk et al. 2008).

Dominant black (K, CBD103, Canine β-defensin 103)

Not listed at MGI or IMSR.

The work with the *Canine β-defensin 103* locus, which encodes CBD103, has been done primarily in dogs, and is discussed above as it relates to the eumelanin/pheomelanin signaling pathway. However, in transgenic mice CBD103 has been shown to bind with high affinity to the MC1R and has a simple and strong effect on pigment type, generating eumelanic skin and hair if either the normal or mutant canine sequence is expressed from an epidermal keratin promoter (fig. 4 of Candille et al. 2007). In human skin, the CBD103 homolog has cutaneous defence functions well beyond coloration (April & Barsh 2006, 2007; Kadekaro & Abdel-Malek 2007). These results expand the functional role of β-defensins, a protein family previously implicated in innate immunity, and identify an additional class of ligands for eumelanic signaling through melanocortin receptors.

Subtle gray (Sut, Slc7a11)

The new name of this locus does not fit in the heading above. It is *Slc7a11, Solute carrier family 7 (cationic amino acid 9 transporter, y+ system), member 11.*
MGI lists two phenotypic alleles, one of which is spontaneous.
Availability: JAX on the C3H and C57BL/6J backgrounds.

Homozygous mutant $Slc7a11^{sut}$ mice show a reduction in pheomelanic but not eumelanic pigment. In this respect the phenotype is similar to phenotypes of mice with mild mutations at the

Figure 5.23 A pheomelanic *Subtle gray* mouse on the left and a comparable yellow mouse that is not mutant at the *Sut* locus.

Tyr (*Albino*) locus. The A^y/a, *sut/sut* (*Lethal yellow, Subtle gray*) mouse shown in Figure 5.23 appears very similar to mice of the same inbred strain that are A^y/a $Tyr^{c\text{-}ch}/Tyr^{c\text{-}ch}$ (*Lethal yellow, Chinchilla*). Yellow mice are much reduced in pigmentation except for the normally eumelanic tips of their hairs. No other obvious similarities exist between the two loci. The protein encoded at *Sut* is a cell membrane transporter of cystine. Cystine is necessary to the melanocyte for normal production of pheomelanin (Chintala et al. 2005; Sato et al. 2005). *Sut* is also expressed in the brain (Burdo et al. 2006; Liu et al. 2007) and in neutrophils (Sakakura et al. 2007).

Other loci that influence pheomelanogenesis

Other loci that have been reported to reduce pheomelanic pigmentation include *Gray lethal* and *Grizzled* (Silvers 1979); see also the Appendix in Chapter 1. Further study of such loci will no doubt enhance our understanding of the switch mechanism.

5.6.1 Summary

The modulation of phenotypes in mice containing various combinations of mutant gene loci permitted pigment geneticists to devise a functional pathway that accurately summarized the eumelanin/pheomelanin switch mechanism and has been confirmed, with occasionally surprising nuances, through application of newer molecular and cellular technologies. Because of the multiple genetic and environmental influences on the switch mechanism, and its exquisite sensitivity to these influences, the phenotypes of inbred mice carrying various combinations of mutant genes continue to provide useful input into these processes. The availability of congenic inbred strains carrying the different pigment mutants makes possible our observation of the functional relationships among these different gene loci that influence the same process. In the case of the eumelanin/pheomelanin switch mechanism, this includes many of the loci discussed throughout this book.

Figure 5.24 Silvered langur with infant. Photograph courtesy of Brenda Bradley (Bradley & Mundy 2008).

Part III
Technology and Resources

Figure 6.1 Left: a C57BL/6J control mouse; right: a bcat* genetically engineered mouse (Delmas et al. 2007).

6

Novel Mouse Pigmentary Mutants Generated by Genetic Manipulation

6.1 Introduction

The production and development of genetic mouse models began a hundred years ago with the identification of spontaneous mutations within mouse colonies. Among the mutations isolated in the various mouse colonies, coat-color mutations have been studied particularly extensively because they are easily detectable and attractive. Many coat-color mutants in mouse colonies isolated spontaneously or after chemical or radiation treatment have been described by Silvers (1979). Work with these classical coat-color mutant mice and newer mutants that have been selected or created has led to the identification of numerous loci and genes controlling the pigmentation system (1360 genotypes with pigmentation phenotype; see www.informatics.jax.org/javawi2/servlet/WIFetch?page=mpAnnotSummary&id=MP:0001186). The functions of these genetic determinants of the pigmentation system have been studied extensively in the mutant mice and in melanocytes, the pigment cells of the skin. The current list of affected loci is given in the Appendix to Chapter 1 and also at www.espcr.org/micemut/ (Montoliu et al. 2009).

The visible and viable phenotypes of mouse coat-color mutants have permitted the collection of a unique repertoire of factors that are important to various aspects of pigment cell biology. However, there are clear limitations to the isolation and characterization of the genes involved in genetic disorders by techniques termed 'reverse genetics' or 'forward genetics' (from phenotype to gene identification). First, the characterization of the genes involved in a phenotype is not necessarily straightforward: half of coat-color mutants have still not been fully characterized at the molecular level. Second, spontaneous mutations are relatively infrequent events. The occurrence of mutations of biomedical interest is extremely rare; and even then, this approach only allows the identification of single-gene disorders.

The development of transgenic technology in the early 1980s has facilitated the generation of new mouse mutants that affect specific gene loci. Furthermore, genetically engineered mice can be used to study the effects of genetic alterations that are not likely to occur naturally.

The Colors of Mice: A Model Genetic Network, 1st edition. By M. Lynn Lamoreux, Véronique Delmas, Lionel Larue, and Dorothy C. Bennett. Published 2010 by Blackwell Publishing Ltd.

Biological and biomedical research today have access to the complete gene sequence of the mouse. The next challenge is to understand the relationship between the gene sequences and the functions of the genes we have identified. This effort, originally referred to as physiological genetics, and more recently as phenomics, is foundational to the genotype–phenotype relationships discussed throughout this book. With continuously expanding databases of sequenced genes, the laboratory mouse has become an increasingly important tool for generating data about the functions of genes *in vivo* and in tissues and cells of mutant mice.

The mouse has emerged as the animal of choice for *in vivo* experimentation for several reasons:

- the transgenic technology is relatively easy;
- colonies are straightforward to establish and maintain; and
- there is extensive genetic homology between mice and humans.

For the past 25 years, researchers have used transgenic mice to help understand the basic mechanisms associated with inherited human and animal disorders. From these experiments, new coat-color mutants have emerged. Some were expected from our understanding of the pigmentary system, but others were more surprising. The use of genetically modified mice is beginning to yield answers to long-standing questions in the field concerning the development and homeostasis of melanocytes and the relationships between the pigmentary system and other systems of the body. The aim of this chapter is to provide a framework of information about transgenic mouse methodologies that can be applied to the pigmentary system, including pronucleus injection, homologous recombination using embryonic stem (ES) cells, and conditional mutagenesis, and to describe the new coat-color mutants they have generated.

6.2 Mouse transgenesis: generation of genetically engineered mice

The production of the first transgenic mice was preceded by several technological and scientific advances, including increasing knowledge of the mouse reproductive system, the ability to culture early mouse embryos *in vitro*, the development of embryo micromanipulation, and the substantial progress in molecular biology technology. All these experimental tools and techniques have been necessary for engineering genetic modifications in the mouse. The first transgenic mice were obtained almost simultaneously in the laboratories of Brinster and Palmiter, and of Mintz prior to the Costantini and Ruddle laboratories (Palmiter & Brinster 1986). In 1982, Mintz and colleagues reported transgenic mice that were able to transmit injected human β-globin gene sequences to their progeny (Steward et al. 1982). The same year, the first transgenic mouse with a clear phenotype was generated; it ectopically expressed growth hormone which led to dramatic growth of the animal (Palmiter et al. 1982). These mice were created by injecting DNA into the pronucleus of one-cell mouse embryos. The technique has proved immensely useful for creating mouse mutants for research purposes. The next advance in genetically engineered mice technology, homologous recombination technology, made it possible to target genetic changes to specific gene loci and earned the 2007 Nobel Prize in Physiology or Medicine for Mario Capecchi, Martin Evans, and Oliver Smithies. This gene-targeting technique has revolutionized the study of mammalian biology and allowed the creation of animal models for hundreds of human diseases. In particular, for example, mouse models of human cancers have been obtained. Below we describe pronuclear injection, homologous recombination using ES cells, and related technologies that are used in transgenesis.

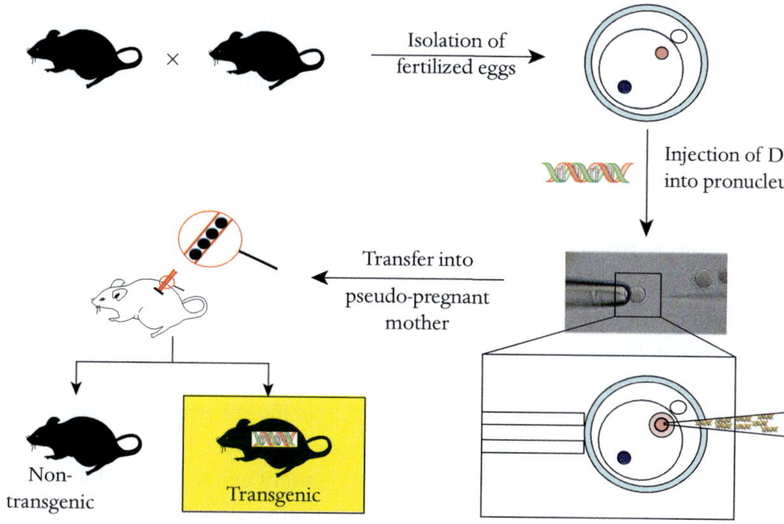

Figure 6.2 Schematic representation for producing transgenic mice. Single-cell eggs are obtained from day 1 pregnant females. The transgene of interest is injected into pronuclei of the eggs, visualized by the swelling of the pronuclei. The transgene integrates at random into the genome. The developing embryos are transferred to the oviduct of a pseudo-pregnant mother. After normal development of the embryos, newborn pups are screened for a potential integration of the transgene. The photo inset shows an egg held at the end of a holding pipette by suction; the injection needle containing the transgene solution is located to the right of the egg.

6.2.1 Pronucleus injection

The first method developed to generate transgenic mice was microinjection of cloned DNA into the pronuclei of fertilized mouse eggs *in vitro*. The DNA injected by this method is stably integrated into the mouse genome, including that of the germ line. The eggs are then transferred into the oviducts of a pseudo-pregnant foster mother, where they develop normally to birth and are capable of transmitting the foreign DNA to the next generation. The advantage of this technique is that there is no practical limit on the size of DNA fragment used; indeed, there is no apparent correlation between the efficiency of integration and the length of the DNA used. The transgenic construct (that is, the DNA injected into the egg) is generally designed to contain a specific promoter, the gene of interest, and a polyadenylation signal.

A schematic overview illustrating the basic methodology for generating transgenic mice is shown in Figure 6.2. The transgenic DNA usually inserts randomly as head to tail concatamers of variable lengths at a single integration site. Most often, this results in the expression of the foreign (injected) DNA in the developing mouse. It can also lead to the disruption of a host gene at the site of insertion. Note that all individual founder mice obtained after injection of the same genetic material are different: the foreign DNA will be inserted into the genome at different sites, and as different-length concatamers. Consequently, several transgenic founders are required to validate a phenotype as being associated with expression of a transgene. Mice that develop from injected eggs are called founders, and these animals can be used to establish a transgenic line by mating with normal mice and selecting the affected progeny in the F_1 or F_2 generation. The pronuclear injection technology and the efficiency have not changed substantially over the years. There remain some problems and limitations. In particular, it is not possible to control either the number of copies of the transgene integrated following injection of the eggs or the site of transgene integration into the genome.

One crucial issue in transgenic technology is how to obtain the desired level of expression of a particular transgene. Transgene expression is influenced by the transcriptional regulatory elements within the cassette (the DNA fragment injected) and the site of integration; transgene copy number seems to have little effect on the level of transcription. The transcriptional regulatory elements may include the gene's promoter (the few hundred base pairs 5′ of the structural gene), its enhancer(s) (which may normally be found a few thousand base pairs away, 5′ or 3′ of the gene or in an intron), and distal elements (normally found even further away), inclusion of which may confer position-independence to expression from the cassette.

One example of such distal elements is the *tyrosinase* distal regulatory element (DRE) located at −15 kb to the transcription start site. It acts as a melanocyte-specific enhancer and also protects against spreading of condensed chromatin (Ganss et al. 1994; Porter & Meyer 1994; Giménez & Montoliu 2001). This DRE behaves differently in the two different pigment cell lineages; that is, neural crest-derived pigment cells compared with retinal pigment epithelium (RPE) cells that originate from the neural tube. The DRE is fully active in melanocytes that are developmentally of neural crest origin, but is only weakly active in the RPE, which originates from the neural tube (Camacho-Hubner & Beermann 2001; Murisier & Beermann 2006). Another distal element, located at −47 kb, has been identified recently (Murisier et al. 2007) and is required for appropriate spatial and temporal expression of tyrosinase in the RPE. No DRE has been found upstream from the *Dct* (also known as *Trp2*) gene: a minimal promoter of 3.2 kb is sufficient for both melanocyte- and RPE-specific expression (Murisier & Beermann 2006; Murisier et al. 2008). However, the *Dct* promoter is not always active in transgenic mice, suggesting that it is sensitive to the site of integration into the mouse genome. Therefore, the transcription regulatory elements most commonly used to drive transgene expression in melanocytes and their precursor, melanoblasts, are from the *Tyrosinase* locus.

Transgenic mouse technology can be used to evaluate the functions of specific gene products *in vivo*. Transgenic mice can also be used for studying transcriptional regulatory elements as was the case for the tyrosinase gene family, namely *Tyrosinase* (*Tyr*), *Tyrosinase-related protein 1* (*Tyrp1*), and *Dopachrome tautomerase* (*Dct*) (Murisier & Beermann 2006).

More recently, a new technology has been developed to diminish the expression of particular endogenous genes in a specific manner, namely by microinjection of short hairpin RNA (shRNA) into mouse eggs. This approach will undoubtedly become widely used to downregulate gene expression without the need to use homologous recombination.

The generation of transgenic mice by pronucleus injection, or additive transgenesis, involves addition of genetic material into the mouse genome in which the endogenous gene is still present. Other strategies use gene replacement, or targeted transgenesis, where an engineered complementary DNA (cDNA) replaces the endogenous gene. This strategy is now favored because the engineered cDNA can be integrated into the mouse genome at a precise and defined location, at the appropriate site with respect to its natural surrounding sequences; furthermore, the effect of the mutation introduced into the coding region can be analyzed without the expression of the wild-type allele.

6.2.2 Homologous recombination

A major advantage of gene-targeting approaches over transgenic techniques is that they involve homologous recombination in ES cells in cell culture (Fig. 6.3): this allows the site of integration to be determined and the genetic change to be introduced in a predetermined and accurate way. The

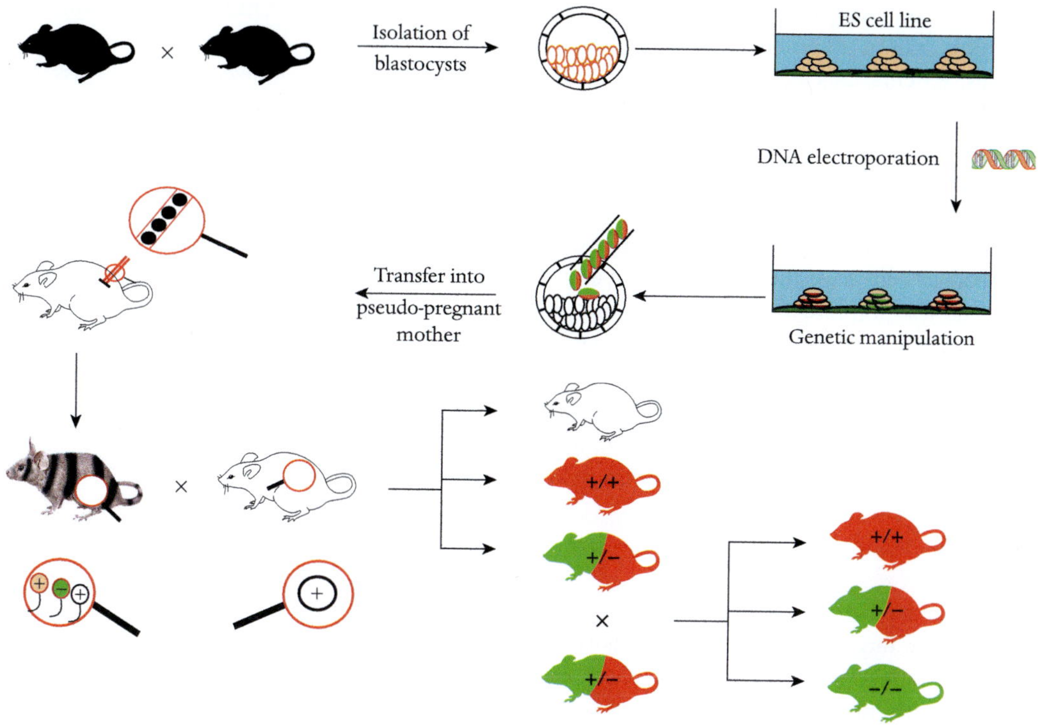

Figure 6.3 Generation of mutant mice using homologous recombination in embryonic stem (ES) cells. ES cells are pluripotent cells derived from the inner cell mass of the blastocyst. The inner cell mass gives rise to the embryo proper and a part of the placenta. A targeting vector can be electroporated into ES cells and *in vitro* selection is carried out to isolate recombined ES cell clones in which the foreign DNA construct has properly recombined with the homologous gene. These ES cells are microinjected into a recipient blastocyst, which is transferred into the uterus of a pseudo-pregnant recipient mother. ES cells contribute to all tissues and chimeric (allophonic) mice are generated. Coat color (usually agouti or albino) is used to identify successful germline transmission. Only one locus is targeted in the cell, and the mutated gene is transmitted in 50% of the progeny. The generated mutation is detected by standard molecular biology approaches (PCR or Southern blot analysis). Heterozygous mice are intercrossed to produce homozygous descendants.

establishment of totipotent/pluripotent ES cell lines from mouse embryos in 1981 has led to a new direction in transgenic technology (Martin 1981; Kaufman et al. 1983). ES cells allow genetic manipulation and modification *in vitro* using standard molecular and cellular techniques, thereby avoiding the problem of capturing and injecting individual oocytes, which is the main limitation of pronucleus injection.

ES cells

ES cells derived from preimplantation embryos are totipotent: that is, they have the unique capacity to contribute to all of the tissues that make up the adult animal from which they were derived. ES cells can be maintained *in vitro* as totipotent cells and may rapidly become multipotent in culture. The totipotent/multipotent capacity of ES cells can be exploited for the creation of transgenic animals. ES cells injected into a embryo at the blastocyst stage can colonize all embryo tissues including the germ line, thus giving rise to chimeric offspring that have tissues from the original embryo mixed with tissues from the ES cell. If, for example, the ES cell line becomes part of the testes and gives rise to sperm cells, then the chimeric mouse can be mated and genes from the ES cells will be transmitted to the next generation.

Studies in the 1980s, particularly in the laboratories of Smithies and Capecchi, demonstrated that mammalian cells have the enzymatic apparatus necessary for recombination between incoming DNA and the homologous sequence in the genome (Capecchi 1989; Bronson & Smithies 1994). Nevertheless, such homologous recombination is less frequent than random integration of the same DNA. Consequently, targeted vectors carrying a desired mutation can be used to transfect ES cells and cells in which there were homologous recombination events can be isolated. Today, hundreds of genes have been inactivated in this manner (gene knockouts, as described below) and the corresponding mutant mice have been created (see www.jax.org/).

Targeted constructs

Targeting constructs designed for homologous recombination in ES cells are generally complex. A basic gene-targeting construct consists of a core region containing the desired genetic structure and a positive selection cassette (conferring resistance to antibiotics, typically neomycin or hygromycin). This core is flanked by two regions of absolute sequence identity with the targeted region (homology arms) required for homologous recombination. ES cells can be transfected with the targeting construct by most of the gene transfer methods. However, to introduce a limited number of copies, electroporation is widely and commonly used. Clonal transgenic ES cell lines that carry the transgene in the genome can then be selected by exploiting the selection markers. As homologous recombination in ES cells is a rare event relative to random integration of DNA, strategies to enrich for recombined ES cell clones have been developed. Many of these enrichment strategies make use of negative selection markers such as the *Herpes simplex virus type 1 thymidine kinase* (*Hsv-tk*) gene or the *Diphtheria toxin chain*-coding region (*Dt-a*). These elements are inserted into the targeting construct outside the homology region. ES cell clones with the desired homologous recombination will not contain these negative selection markers, because nonhomologous flanking sequences (those not between the pair of homology arms) will be eliminated prior to homologous integration. By contrast, most randomly integrated targeting constructs still retain the negative selection marker, thus permitting counter-selection against randomly integrated clones. These selected cells are then injected into the inner cell mass of a blastocyst embryo and replicate as a normal component of the developing embryo. The resulting new mice will consist of cells and tissues from both the normal embryo and the genetically modified ES cells. These chimeric offspring are then mated to determine whether the mutation is transmissible in the germline.

Generation of chimeric mice: the use of coat color

There are two main techniques to create chimeric mice from ES cells: (1) microinjection into early-stage embryos or (2) co-culture aggregation. Briefly, microinjection involves the placement of a small number of ES cells into either a morula or blastocyst stage embryo and then transferring the injected embryo to the uterus of a pseudo-pregnant surrogate mother. The injected embryos are allowed to go to term, resulting in the birth of a variable proportion of chimeric offspring.

Co-culture aggregation involves the removal of the zona pellucida from morula stage embryos and then placing the embryos on a lawn of ES cells. The ES cells aggregate with the embryos and the embryo/ES-cell aggregates are removed and cultured overnight to the blastocyst stage. In both cases, the blastocysts are then transferred to the uterus of a pseudo-pregnant mother, resulting in the birth of a variable number of chimeric offspring. The ES cells contribute to the formation of a chimeric mouse, composed of cells from a recipient blastocyst and ES-cell-derived cells. So that the chimeric offspring can be easily identified by their colors, the ES cells and recipient blastocysts or morulae are derived from mice with distinguishable coat-color alleles (often strain 129/Sv, which are agouti, and C57BL/6, which are nonagouti black).

Chimeras are bred with wild-type mice to assess germline transmission of the ES cell genome by observing coat color and targeted mutations by molecular techniques (PCR, Southern-blot analysis). The chimeric mice obtained by either technique are then used as founder animals to generate new transgenic animal lines. The extent of the contribution of ES cells to the formation of the chimeric mouse can be evaluated by visual assessment of coat-color chimerism (extent of, say, agouti hair from 129/Sv-derived ES cells in the nonagouti black coat of C57BL/6). Although the coat-color patterns of allophenic (generally termed chimeric) mice do not necessarily reflect the relative proportions of each strain in the other organs, they are widely used as an indication of the overall contribution of the ES cells to the germ line. Thus, the generation of chimeric mice by aggregation of morulae from lines of different coat color is a convenient tool for transgenic experiments.

However, experiments of this type as early as the late 1960s allowed B. Mintz to make fundamental and significant observations about the development of melanocytes: chimeric mice displayed broad transverse bands of color, with each side of the mouse being patterned independently. These phenotypes were interpreted as reflecting the migratory history of melanoblasts during development. By assessing the patterns of many chimeric coats, Mintz proposed the existence of small numbers of melanoblast progenitors, each of which generates a discrete unilateral transverse band of color (Mintz 1967). Mintz's study, discussed extensively by Silvers (1979), and reviewed in Chapter 3 of this volume, was the first lineage analysis of mouse pigmentation and proposed the existence of about 34 founder melanoblasts arranged in two longitudinal mid-dorsal chains of 17 each. Several groups suggested that this number is probably underestimated and described the extensive mixing at the boundaries of the transverse bands. A recent and elegant lineage analysis by I. Jackson's group suggested that the number of founder melanoblasts is indeed much larger and that it should be measured directly in the embryo and not estimated from an adult chimera coat-color pattern (Wilkie et al. 2002). Analysis of the coat-color patterns of transgenic mice generated by targeting melanoblast development at a series of times during development would undoubtedly bring light to this important issue.

6.2.3 Conditional mutations

Conventional knockout and transgenic experiments have provided substantial insight into gene function. However, limitations intrinsic to conventional gene-targeting and transgenic approaches have become apparent. Examples of these limitations include embryonic-lethal knockout phenotypes, which cause death of the developing embryo, highly complicated cumulative phenotypes that result from some mutant genes and involve multiple tissues, and cytotoxic effects in response to novel patterns of gene expression. For many research aims, it may be necessary to analyze the function of a gene within a particular developmental window or cell lineage. To overcome some of the limitations and to control gene expression or extinction of gene function in a precise spatial and temporal fashion, more sophisticated conditional mouse models have been developed. These permit expression of mutant genes at specific times and/or in specific tissues during development. They can be used to avoid embryonic lethality, preventing unwanted pleiotropic side effects. In particular, the use of site-specific recombinases of the Cre and Flp integrase family have opened exciting new possibilities for creating conditional loss- or gain-of-function mouse models (Sauer 1998; Babinet & Cohen-Tannoudji 2001).

The Cre/LoxP system

Site-specific recombinases are enzymes that promote recombination of DNA through recognition of specific short nucleotide sequences of the chromosomes. Two recombinases have been used for constructing mouse mutants: Cre recombinase is an enzyme found in bacteriophage P1, a prokaryotic organism. Cre recombinase recognizes LoxP sites. Flp is a recombinase that was originally found in *Saccharomyces cerevisae*, a one-celled yeast organism, and recognizes FRT sites. Both LoxP and FRT sites are 34 bp-long specific sequences of nucleotides that consist of an 8 bp core region flanked by palindromic sequences of 13 bp. Thus, their recognition sequences are large enough that they are not expected to occur by chance in the mammalian genome. Depending on the orientation of pairs of these LoxP or Flp recognition sites, the recombinases catalyse excision (same orientation) or inversion (opposite orientation) of the DNA segments between them. Cre- and Flp-induced changes include sequence excision, duplication, integration, inversion, and chromosomal translocation. Both recombinases work equally well in ES cells. However, the Cre/LoxP system has been far more widely used and will be described in more detail in this chapter, particularly as concerns its use in knockout strategies (Sauer & Henderson 1988; Dymecki 1996; Garcia-Otin & Guillou 2006).

Conditional knockouts were developed in the laboratory of Klaus Rajewsky by combining ES cell technology with Cre recombinase methodology (Rajewsky et al. 1996). Genes are conditionally targeted by crossing two strains of transgenic mice. The first strain of mice is generated by homologous recombination in ES cells. These mice are normal except that the gene under study is flanked on both sides by extra DNA that is the specific sequence known as LoxP (a gene that is flanked by LoxP sequences is called 'floxed'). The gene will function normally because normal mice do not express the Cre recombinase enzyme that would affect the floxed sequences. The second strain of transgenic mice does contain the Cre recombinase gene and the gene transcription is controlled by a cell-type-specific promoter.

When mice of the floxed strain and mice of the specially designed recombinase strain are mated together, the offspring contain both the floxed gene and the recombinase gene. Because the recombinase gene has been designed with specific regulatory elements, it will produce the enzyme at a time in development when those regulatory elements normally begin to function. For example, if the recombinase gene has been designed with a promoter from the *Tyrosinase* locus, then the recombinase gene would 'turn on' only in pigment cells, and would recognize the floxed DNA in the pigment cells. This is known as conditional targeting because the gene is 'targeted' to be mutated (in this example) only in pigment cells.

Two points must be respected for optimal conditional targeting strategies: (1) the floxed strain should not have any mutant phenotype, and (2) the expression of Cre in the transgenic mice must be tightly controlled such that it is expressed only in particular cells. To create floxed mice with no mutant phenotype, it is necessary to introduce recombinase-recognition sites around an essential part of the gene of interest without compromising its function. Technically, it is difficult to flank the entire coding region with LoxP sites. Therefore an essential exon is often targeted by introducing LoxP into intronic sequences. The LoxP sites should be carefully positioned in the targeted locus and should not interfere with regulatory elements. The expression of the gene must be normal until it is induced by the Cre enzyme to change. Several techniques are available to accomplish this goal.

The tissue specificity of Cre expression is dependent on the transcriptional elements used in the cassette and on the site of insertion of the transgene into the mouse genome (see section 6.2.1). Position effects can lead to mosaicism of transgene expression in the target tissue and also to unexpected expression in other tissues: either the cells in which there is recombination express *Cre* transgene or Cre is expressed in the cells of the lineage leading to it. An extreme example illustrating this difficulty is the platelet endothelial cell adhesion molecule-1 (PECAM-1) promoter which was initially described to be active mainly in endothelial cells but when used to drive Cre expression recombined the floxed allele ubiquitously (Terry et al. 1997). In this case, Cre was, unexpectedly, expressed very early in development, resulting in the transmission of the deleted allele to most cells of the embryo. To control the window of Cre expression better and to exogenously induce Cre activity only when desired, inducible systems have been developed.

One common inducible system involves the use of steroid receptors. Sequences encoding Cre proteins have been fused to those encoding ligand-binding domains of steroid receptors, a mutated ligand-binding domain of the human estrogen receptor being most often used. Consequently, the Cre activity of the resulting fusion protein is hormone dependent and LoxP sites only recombine after administration of the appropriate hormone (tamoxifen in the case of the mutated estrogen receptor). This strategy allows the spatiotemporal control of gene modification and it has been developed for the analysis of pigment cell lineages (Bosenberg et al. 2006; Yajima et al. 2006).

Cre-transgenic lines designed to target pigment cells

To induce specific recombination in pigment cells, several lines of transgenic mice expressing Cre have been generated. Each of them has strengths and weaknesses and they should be used for what they are good at. One of the main challenges to this type of experimental biology is to find, among the various available tools, the best approach to answering to the biological question asked with the various tools available. The major limitations are mosaics, and the lack and/or ectopic expression of Cre in the mice (Montoliu et al. 2004). Four promoters have been used to target pigment cells, and they are those of the genes encoding TYRP1, DCT, microphthalmia-associated transcription factor (MITF), and tyrosinase (TYR).

Cre can either be inserted next to the normal genomic promoter using homologous recombination, in which case it disrupts expression of a copy of the normal gene, or it can be attached to an extra copy of the chosen promoter and inserted randomly by classical transgenesis.

The first Cre line generated to target pigment cells used the TYRP1 promoter as transcriptional elements to drive Cre expression in a transgene construct inserted randomly into the genome (Tyrp1::Cre) (Mori et al. 2002). This line induces efficient Cre-mediated excision of DNA selectively in the RPE but also in the ciliary margin of the retina and in some cells in the neural retina, but not (or very rarely) in neural crest-derived melanocytes in skin, choroid, and hair bulbs.

Expression of Cre driven by the DCT promoter has been achieved both by pronucleus injection (using a exogenous DCT promoter; Dct::Cre) and by homologous recombination by the insertion of Cre coding sequences into the normal *Dct* locus ($Dct^{tm1(cre)Bee}$) (Guyonneau et al. 2002, 2004). In these mice, Cre expression is found in RPE cells, melanoblasts, and melanocytes, and in the telencephalon (all of these sites reflecting endogenous DCT expression), but unexpectedly, also in dorsal root ganglia and caudal nerves. To detect the pattern of Cre activity, two independent lines of reporter mice were used: R26R (also called Rosa26R), constructed by P. Soriano, and the Z/EG mouse produced by C. Lobe, expressing LacZ or green fluorescent protein (GFP), respectively, after recombination (Lobe et al. 1999; Soriano 1999). Surprisingly, these reporter lines do not reveal exactly the same pattern of expression. *LacZ* reporter gene activity is seen in the presumptive RPE, in migrating melanoblasts, telencephalon, and caudal nerves and dorsal root ganglia, whereas there is no GFP reporter expression in the last two of these sites. This observation highlights an essential aspect of genetic recombination using the Cre/LoxP system: LoxP-targeted different loci do not always recombine with the same efficiency or in the same target cells with the same Cre-expressing mouse. Epigenetic regulation of the genomic conformation at an given locus may account for this diversity of recombination efficiency. Therefore, the pattern of recombination with the locus of interest must be analyzed not only with reporter constructs, because such constructs can display particular features. Other explanations are that LacZ staining may generate more artefacts compared to the GFP fluorescent signal or that the endogenous Rosa26 promoter is more active than the chicken β-actin promoter present in the Z/EG construct.

When Cre was targeted into the *Dct* locus ($Dct^{tm1(cre)Bee}$), the pattern of LacZ staining after crossing with the R26R line was similar to that obtained with transgenic Cre, and melanoblasts/cytes and RPE were clearly labeled. Here again, the efficiency of recombination appears to vary according to the targeted locus. Targeting the *Rosa26* locus using the $Dct^{tm1(cre)Bee}$ mouse line resulted in a relatively low and variable efficiency of recombination with only a few hair follicles stained in the adult mice. Targeting the *Pten* locus seemed more efficient, with recombination in 75% of the melanocytes isolated (Inoue-Narita et al. 2008). Interestingly the $Dct^{tm1(cre)Bee}$-knockin mouse line has been successful in targeting melanocyte stem cells (Osawa et al. 2005; Wong et al. 2006; Aoki et al. 2009).

When Cre recombinase was placed under the control of regulatory elements from *Mitf-M* using a bacterial artificial chromosome, recombination could be detected in melanoblasts or melanocytes (Alizadeh et al. 2008). In this case, two of nine founder mice exhibited patterns of X-gal (LacZ) staining characteristic of melanocyte lineage cells after crossing with the R26R strain, and there was no ectopic expression (outside the melanocyte lineage). A detailed characterization of embryonic expression of the Mitf::Cre transgene has not been performed, but preliminary experiments suggested that the intensity of staining and the number of stained cells were both lower than what has been described for the Dct::LacZ during development. Therefore, considering Dct::LacZ/° mice as the standard for the detection of the number of melanoblasts during development, the Mitf::Cre/+ display a certain degree of mosaicism. However, after crossing Mitf::Cre/+ with R26R/+ mice, all hair follicles analyzed exhibited X-gal staining on P10 (day 10 after birth) indicating a generally better efficiency for targeting melanocytes than can be obtained with Dct::Cre mice. Another difference between the Mitf::Cre/+ mice and the Dct::Cre/° mice was that Cre was not expressed in the RPE of the Mitf::Cre/+ mice.

The regulatory elements of mouse tyrosinase have been used to drive Cre expression in a constitutive (Delmas et al. 2003; Tonks et al. 2003) or inducible (Bosenberg et al. 2006; Yajima et al. 2006) manner in pigment cells. Robust expression is detected in melanoblasts or melanocytes without apparent mosaicism (Delmas et al. 2003). Tyr::Cre mice produce Cre recombinase in melanocytes and in many embryonic tissues that normally express tyrosinase but also in the brain and peripheral nerves.

A recent cell lineage tracing experiment revealed that the transcriptional elements of tyrosinase used in this transgenic line are active in a subset of vagal neural crest cells, targeting genes much earlier during development than expected (Puig et al. 2009). The targeted cells are a subset of smooth muscle cell precursors forming the ductus arteriosus and neuronal precursors associated with the intestine. The artificial combination of proximal and distal regulatory elements of the *Tyrosinase* locus used in these constructs may result in unexpected expression of Cre recombinase at ectopic sites or in temporal misregulation during development. In contrast, one can use Tyr::Cre-ERt2/°

inducible expression, where Cre is physically attached to an estrogen receptor, which makes it non-functional until an estrogen is added, when the conformation changes. This is highly specific for melanoblasts/cytes. In these mice, Cre activity can be induced by giving tamoxifen to the pregnant mother at the desired time of mouse development. Like Mitf::Cre/+ mice, no X-gal-stained cells have been observed in the RPE of the developing eyes. The correct dose of tamoxifen must be used for optimal induction of Cre-ERt2-mediated recombination. Yajima et al. (2006) proposed that a single injection of 2 mg of tamoxifen/40 g of body weight at embryonic day (E) 10.5 leads to detectable blue staining of melanoblasts, but the number of stained cells did not reach the maximum observed with Dct::LacZ/° mice. Induced recombination can be performed after birth by tamoxifen injection or topical application directly to the skin, but this approach is associated with the problem of accessibility of the stem cells. Note that among the inducible systems, Cre-ERt2 shows the most stringent control of uninduced recombinase activity (Yajima et al. 2006). Further work will provide more details about the efficiency of recombination using this approach, and can be expected to open new avenues of research, in particular concerning melanocyte development, homeostasis, and oncogenesis.

In conclusion, several lines of Cre mice have been generated using different transcriptional elements to target recombination in pigment cells and they constitute an extremely valuable set of tools for following cell lineages and examining gene function in melanocytes at various times during development and in the adult. Some of them can be used also to target RPE. It is important to remember that ectopic expression of Cre detected with one mouse reporter is not necessarily detected with another reporter or in the conditional floxed mouse of interest. Ectopic Cre expression can be deleterious or have no influence on the subsequent interpretation of the phenotype. At this point there is no 'perfect' Cre driver: all of them are useful, but each has its own specific limitations. In principle, there is almost no restriction to the experiments that can be designed and performed using this system, although the number of melanoblasts produced during development is limited and may therefore be difficult to analyze at the molecular level in the engineered mutant mice.

6.3 Coat-color transgenic mice

Transgenic mice have been generated by the microinjection of cDNA or bacterial artificial chromosome (or BAC) clones into mouse eggs, leading to the modification of coat-color phenotype. Such mice express the transgene in either melanocytes or keratinocytes, depending on the transcriptional regulatory element used in the construct. This clearly illustrates how genetic effects on melanocyte development and function may not be cell-autonomous, depending upon the specific gene locus, and that they can be highly dependent on the surrounding context. Table 6.1 lists transgenic mice with coat-color phenotypes already obtained. These phenotypes can be divided into three categories: hypopigmentation (Chapters 3 and 4), hyperpigmentation, and pigment-type switching (Chapter 5). Each category can be subdivided according to the stability of the phenotype observed (whether there are changes with time or during the life of the animals). If the phenotype is apparent within the first week after birth, a defect during embryonic development can be envisaged (as in white spotting, discussed in Chapter 3). If the phenotype changes with time, a melanocyte follicular stem cell defect is more likely, as in 'fading with age' discussed in Chapter 3. Here, we will consider in particular mouse mutants generated by transgenesis or gene targeting which display a visible coat-color mutant phenotype. We will not consider mutants which are embryonic lethal or which present an induced phenotype (either by chemical induction or targeting). In addition, we

Table 6.1 Transgenic mice with altered color phenotypes

Transgene symbol	Promoter used	Transgene name, gene information	Reference
Hypopigmented/white spotting			
Tg(PGK1-KITLG*220)441Daw	Pgk	Human *KITLG* (SCF), 220 aa isoform	Majumdar et al. (1996)
Tg(Tyr-ctnnb1)Lrue	Tyrosinase	β-Catenin	Delmas et al. (2007)
Tg(Tyr-SV40E)Mi	Tyrosinase	SV40 early region including large T antigen	Klein-Szanto et al. (1991)
Tg(Tyr-DT)3Bee	Tyrosinase	Diphtheria toxin	Camacho-Hubner and Beermann (2001)
Hyperpigmented			
Tg(Tyr-HRAS)60Lc	Tyrosinase	*Hras*	Chin et al. (1997)
Tg(Tyr-NRAS*Q61K)1Bee	Tyrosinase	*Nras*	Ackermann et al. (2005)
Tg(KRT14-Edn3)1Takk	Keratin 14	*Edn3*	Kurita et al. (2005)
Tg(KRT14-HGF)43Takk	Keratin 14	*Hgf*	Kunisada et al. (2000)
Tg(KRT14-Kitl)1Takk	Keratin 14	*Kitl* (SCF)	Kunisada et al. (1998a, 1998b)
Tg(KRT5-Edn3)Koss	Keratin 5	*Edn3*	Garcia et al. (2008)
Tg(Mt1-RET)192Ina	Metallothionein	*Ret*	Iwamoto et al. (1991)
Tg(MT-HGFSF)18Lmb	Metallothionein	*Hgf*	Takayama et al. (1996)
Tg(KRT5-Foxn1)Jlb	Keratin 5	*Foxn1*	Weiner et al. (2007)
Tg(Tyr-BrafV600E)Hal	Tyrosinase	*Braf*	Goel et al. (2009)
Pigment-type switching			
Tg(BACT-A)Ngc	β-Actin	*Agouti*	Perry et al. (1995)
Tg(TRP1iA)Rbg	Tyrosinase-related protein-1	*Agouti*	Kucera et al. (1996)
Tg(KRT14-A)Rbg	Keratin 14	*Agouti*	Kucera et al. (1996)
Tg(KRT14-A)Gsb	Keratin 14	*Agouti*	Millar et al. (1995)
Tg(Mc1r)Jkn	BAC clone	*Mc1r*	Healy et al. (2001)
Tg(MC1R)1Jkn	Human/BAC clone	*Mc1r*	Jackson et al. (2007)

BAC, bacterial artificial chromosome; PGK, phosphoglycerate kinase. Transgene symbols in roman type are named according to JAX nomenclature; transgene symbols in italic type are shown by their proposed name.

will not address mutants which present abnormal hair follicles which can lead to pigmentary abnormalities not due to melanocyte dysfunction or location but rather to disorganization of the whole skin structure. We also exclude mutants which have a coat-color phenotype generated by DNA insertions without clear characterization of the contribution of the gene altered in the phenotype observed.

6.3.1 Transgenic mice generated with melanocyte targeting

The rescue of a *Tyrosinase*-locus null mutation in the mouse by the expression of a functional *Tyrosinase* gene introduced by transgenic technology clearly opened new possibilities for research into pigmentation (Beermann et al. 1990). First, it demonstrated that the *C* (*Tyrosinase, Tyr*) locus encodes the structural gene for tyrosinase and not a regulator for tyrosinase expression. Second, the expression of the minigene introduced is apparently restricted to melanocytes, which suggested the possibility of using the regulatory elements of the *Tyrosinase* gene to drive transgene expression specifically in this cell type (Beermann et al. 1992). Constructs involving the tyrosinase promoter driving the SV40 early region oncogenic sequences have been generated and used to show that this promoter was not entirely specific for melanocytes: tumors developed both in cells producing and those not producing melanin. The specificity of the various promoters used to target pigment cells is discussed previously (see the section 'Cre-transgenic lines designed to target pigment cells' above). Interestingly, these transgenic mice expressing SV40 displayed hypopigmentation. The authors suggested that this hypopigmentation resulted from diminished differentiation of melanized pigment granules in the melanocytes of the hair bulbs *in vivo*. These transgenic mice were the precursors of a series of mice designed to transform melanocytes into melanomas by expressing various oncogenes (see the recent review; Larue & Beermann 2007). These experiments highlighted the oncogenic roles of the G-protein-coupled receptor GRM1, of *Hras* and *Nras*, and the WNT/β-catenin signaling pathway. Hyperpigmentation was observed with Tyr::H-Ras^{V12G}, Tyr::N-Ras^{Q61K}, and Tyr::B-RAF^{V600E} mice (these are oncogenic mutant alles found in cancers) (Chin et al. 1997; Ackermann et al. 2005; Goel et al. 2009). In contrast, hypopigmentation was observed with activation of the WNT/β-catenin pathway, although again an oncogenic mutant allele was used (Delmas et al. 2007). Thus, these transgenic mice obtained with transgenes expressing candidate oncogenes for melanoma were extremely useful for demonstrating the causality of these genes in melanomagenesis. However, the molecular and cellular mechanisms responsible for their coat-color phenotype remain to be fully characterized.

6.3.2 Transgenic mice that target cells other than melanocytes

Numerous factors have been suggested to be produced by keratinocytes or fibroblasts and to regulate the differentiation and proliferation of mammalian melanocytes *in vitro*. These include hepatocyte growth factor (HGF), nerve growth factor (NGF), stem cell factor/KIT ligand (SCF, KITL), endothelins 1 and 3 (EDN1, EDN3), α-melanocyte-stimulating hormone (α-MSH), and basic fibroblast growth factor (bFGF, FGF2). Several transgenic mice overexpressing various keratinocyte-derived factors in keratinocytes have been generated to evaluate the significance and function of these factors *in vivo*. Kunisada and coworkers use the epidermis-specific cytokeratin K14 promoter to deliver the hepatocyte growth factor in epidermis. Genes for this growth factor and its receptor MET are not among the loci that affect coat color of mice, as assessed by the analysis of natural or chemically induced mouse mutants. However, two observations suggest that the HGF/MET signaling pathway may stimulate melanocyte growth *in vivo*.

First, HGF is produced in the developing skin, and second transgenic mice expressing HGF under the ubiquitously expressed metallothionein promoter (MT1) show dermal and epidermal hyperpigmentation (Takayama et al. 1996). The phenotype of the MT1::RET mouse suggest that this promoter may give highest expression in melanocytes. Transgenic mice expressing HGF under the control of the cytokeratin K14 promoter (active throughout life in the basal layer of the epidermis) showed increased numbers of melanoblasts during development followed by an explosive increase in the mature melanocyte count at birth (Kunisada et al. 2000). Despite the expression of the growth factor in the epidermis, these abnormally abundant melanocytes were restricted to the dermis. Transgenic HGF suppressed the expression of adhesive molecules, including E-cadherin, that may be required for the penetration of melanocytes into the epidermis.

Transgenic mice have been obtained expressing a constitutively active form of the tyrosine kinase receptor RET (receptor for glial-derived neutrophic factor, GDNF), encoded by a transgene consisting of a fusion between a ring finger protein (*Rfp*) gene and *Ret* under the transcriptional control of the metallothionein promoter. These transgenic mice showed hyperpigmented skin and developed melanocytic tumors (Iwamoto et al. 1991). It was proposed that RET expression confers a growth advantage to melanocytes.

Transgenic mice targeting the KIT/KITL signaling pathway were also generated, although numerous natural mutants exist. In mice, KITL is expressed transiently in the mesenchyme underlying skin during development, then in epidermis during late embryonic development (Chapter 3), and from soon after birth is confined to the dermal papillae of hair follicles where melanoblasts mature into melanocytes. Transgenic mice producing KITL under the regulatory sequences of human cytokeratin K14 have been generated. The production of KITL in basal epidermal keratinocytes increases the number of interfollicular melanocytes present after birth and reproduces the phenotype of human cutaneous mastocytosis (Kunisada et al. 1998a, 1998b). These mice thus differ from wild-type mice, which present a reduction in the number of interfollicular melanocytes after birth; indicating that the survival/proliferation of melanocyte precursors was extended by locally produced KITL. Interestingly, melanocyte stem cells independent of the KIT signal, which are present only in the hair follicles of normal mice, were maintained in the epidermis of these transgenic mice. There are several isoforms of KITL in both mice and humans due to alternative splicing of transcripts; one of these isoforms is a membrane-associated protein of 220 amino acids that lacks the proteolytic cleavage site of the full-length KITL. Transgenic mice expressing the human SCF220 isoform (which cannot be cleaved in murine cells) have coat-color deficiencies (Majumdar et al. 1996). This phenotype mimics heterozygous W/+ mice, probably due to occupancy of murine KIT by human KITL and less than optimal activation of the receptor. Alternatively, since the human KITL was expressed from a human phosphoglycerate kinase (PGK) promoter, likely to be ubiquitously active, many embryonic melanoblasts may migrate to incorrect sites, rather than the epidermis where they are normally attracted by KITL in late development (Chapter 3). Thus, the expression of normal membrane-associated KITL in transgenic mice mimics a codominant mutation of *Kit*.

Mice expressing EDN3 have been produced using the K14 and K5 promoters (Kurita et al. 2005; Garcia et al. 2008). *Edn3* encodes a ligand important for neural crest cell development but the full details of its involvement during melanoblast development have not been fully characterized. *In situ* hybridization suggests that EDN3 is expressed between E8.0 and E15.0 in the mesenchymal cells that surround the melanocyte pathway during mouse development but its contribution during all developmental stages has not been demonstrated. Mice expressing *Edn3* under the control of K14 and an inducible K5 promoter display hyperpigmented skin (Kurita et al. 2005; Garcia et al. 2008). Using the inducible system for the production of EDN3 at different times during development, Kos and colleagues demonstrated that EDN3 needs to be present at E10.5 for the generation of the dark-skin phenotype and that it is also required to maintain an extensive population of dermal melanocytes after birth. Transgenic K5::Edn3 mice that are also genetically yellow (A^y/a) show marked darkening of the coat compared to littermates, suggesting unexpectedly that excessive EDN3 signaling can also modify pigment-type switching (Garcia et al. 2008).

Transgenic mice expressing FOXN1, a transcription factor mutated in nude mice, under the transcriptional control of the K5 promoter exhibit ectopic pigmentation and abnormally abundant epidermal melanocytes similar to what has been observed in SCF transgenic mice (Weiner et al. 2007). It was suggested that pigmentation patterns reflected the distribution of FOXN1 in the skin. This led to a new idea – which remains to be proved – that epithelial cells use FOXN1 to recruit melanocytes and induce their own pigmentation. FOXN1 may thus define a distinct cell population that controls the targeting of pigment in the skin.

As part of the work to characterize genes involved in the growth, survival, and differentiation of melanocytes, transgenic mice have been obtained that improve our understanding of pigment production and regulation in melanocytes. For example, transgenic mice were generated to identify the signal controlling the production of the two types of melanin pigments, eumelanin and pheomelanin. Eumelanin synthesis is stimulated in melanocytes by activation of the α-MSH receptor, melanocortin 1 receptor (MC1R). Loss of function of MC1R or production of agouti protein results in pheomelanin production (Robbins et al. 1993; Ollmann et al. 1998). Despite the large number of mutant alleles available, molecular characterization of *Nonagouti* gene mutations has shed little light on the major functional elements of the agouti protein. By introducing new mutations by site-directed mutagenesis followed by generation of transgenic mouse lines, several laboratories have made progress in the analysis of the function of the agouti protein and its pulse production during hair cycle; in particular, they have found that it acts in a paracrine manner (Millar et al. 1995; Perry et al. 1995; Kucera et al. 1996). The study of the *Nonagouti* gene is a good example of how transgenic technology can bring additional information concerning the function of genes and the characterization of important domains within the encoded protein. Subsequently, the group of I. Jackson demonstrated that a mouse bacterial artificial chromosome which contains *MC1R* efficiently rescued loss of MC1R in transgenic mice and that its overexpression suppresses the effect of the endogenous inverse agonist, agouti protein (see Chapter 5) (Healy et al. 2001).

Much of the currently available information about all these phenotypes could not have been obtained without the use of transgenic mice that have allowed the visualization of melanoblasts during development (Mackenzie et al. 1997). These transgenic mouse lines, carrying the LacZ reporter under the control of the DCT promoter, allow the detection of individual melanoblasts as early as E10.5. Also, a knockin of the *Kit* locus by LacZ has been used to mark melanoblasts during development (Bernex et al. 1996).

These various approaches certainly help elucidate the function of diverse genes and also the relation between sequence data and the phenotypes arising from alleles of genes known to be involved in pigmentation. The genotype–phenotype correlations emerging from these studies provide a complementary source of information to that obtained with natural and induced mutants.

The work with mice also complements and is complemented by studies of melanocytes *in vitro*. As stated previously, an important limitation in the analysis of melanocytes *in vivo* is the limited number of melanocytes present in the mouse skin and therefore the difficulty of isolating them for the purposes of cell biology and biochemistry. It is possible to culture melanocytes that are isolated from newborn skin explants. However, the establishment of melanocyte cell lines is a tedious task unless an *Ink4a-Arf*-null background is used to prevent cell senescence (Sviderskaya et al. 2002), a technical advance that also resulted from the development of transgenic mice of the appropriate genotype. Melanocyte cell lines obtained from mice that are mutant at specific color loci and also are null for *Ink4a-Arf* have been used to evaluate functions of numerous pigmentation loci (Suzuki et al. 2002; Gwynn et al. 2004; Chintala et al. 2005; Theos et al. 2005b, 2006; Wasmeier et al. 2006; Setty et al. 2008; Hida et al. 2009).

6.4 The coat-color mutants generated by gene targeting

Natural or induced coat-color mutants constitute a unique resource for studying molecules involved in pigmentation. Work with these mutants has revealed the involvement of hundreds of genes in the development and the homing of pigment cells. Recent advances in transgenic and gene-targeting technologies in mice make it possible now to manipulate the expression of these and other genes to study their roles in pigment cells *in vivo*. About 40 targeted mouse mutations that present a clear coat-color mutant phenotype have been generated to date (see Table 6.2).

6.4.1 New alleles of coat-color genes

In many cases, particular amino acids have been identified (for example, at sites of phosphorylation, ubiquitination, or others) as being important for the function of key proteins involved in the specification, the development, and the function of the melanocytes. For an *in vivo* assessment of amino acids that were so identified *in vitro*, new mutants were needed. Mutations of biomedical interest arise spontaneously only extremely rarely in mouse colonies, and therefore genetically engineered mice have been generated that carry mutations in these amino acids. Examples of such mutants are the Kit^{tm1Ber}, Kit^{tm2Ber}, and Kit^{tm3Ber} mice, which can be used to study the roles of the tyrosine residues at positions 567 and 569 in the juxtamembrane region of the KIT receptor tyrosine kinase (Kimura et al. 2004). Using Cre technology, mice were generated in which tyrosine residues 567 and 569 in exon 11 of *Kit* were mutated alone (or in combination) to phenylalanine (Kit^{tm1Ber}, Kit^{tm2Ber}, Kit^{tm3Ber} contain Y567F, Y569F, and both Y567F and Y569F mutations respectively). This work demonstrated that these two tyrosine residues, acting in concert, are essential for melanogenesis and mast cell development. Mutations in these residues diminish receptor autophosphorylation and attenuate activation of mitogen-activated protein (MAP) kinase signaling. However, these residues are not essential for the normal development of other cells including erythroid and interstitial cells of Cajal, and germ cells, tissues that also require KIT for their development, indicating a cell-type-specific role. This example illustrates how engineered mutations in a receptor tyrosine kinase can be used to understand the complex molecular events involved in the cell-specific activation of such receptors.

Another example of this approach to genotype–phenotype analysis is the $Mitf^{tm1.2Arnh}$ mouse line generated by the group of H. Arnheiter (Bismuth et al. 2008). The mutant was used to test the role of phosphorylation at serine 73 in exon 2 of MITF. This serine is phosphorylated after the activation of the MAP kinase pathway by KITL. None of the existing *Mitf* alleles (more than 30) was informative about this issue, so a novel mouse with a Ser73-to-Ala (S73A) mutation in *Mitf* was generated by gene targeting ($Mitf^{tm1.2Arnh}$). By chance, an allele in which the wild-type *Mitf* sequence was duplicated and inserted into an otherwise correctly targeted *Mitf* gene was obtained ($Mitf^{tm1Arnh}$). These mice display a patchwork coat with white or gray coat color, depending on the presence or absence of the neomycin resistance cassette used in the constructions ($Mitf^{tm1Arnh}$, $Mitf^{tm1.1Arnh}$). Revertants that lacked the duplicated wild-type sequence but retained the targeted codon were obtained and revealed that the serine-to-alanine mutation is not deleterious for melanocyte development

Table 6.2 Targeted genetic modifications giving altered color phenotypes

Allele symbol	Category	Transgenic used	Name, gene information	Reference
Hypopigmented/white spotting				
$Ap3b1^{tm1Sms}$	Targeted (reporter)		adaptor-related protein complex 3, β1 subunit	Yang et al. (2000)
$Atox1^{Gt(IRESBetageo)3Pgr}$	Gene trapped		antioxidant protein 1 homolog 1 (yeast)	Hamza et al. (2001)
$Bcl2^{tm1Dlo}$	Targeted (knockout)		B-cell leukemia/lymphoma 2	Nakayama et al. (1994)
$Bcl2^{tm1Sjk}$	Targeted (knockout)		B-cell leukemia/lymphoma 2	Veis et al. (1993)
$Bcl2^{tm1Tsu}$	Targeted (knockout)		B-cell leukemia/lymphoma 2	Kamada et al. (1995)
$Edn3^{tm1Ywa}$	Targeted (knockout)		endothelin 3	Baynash et al. (1994)
$Ednrb^{tm1(tTA)Tilg}/Ednrb^{tm3(tetO)Tilg}$	Targeted (knockin) Inducible		endothelin receptor type B	Shin et al. (1999)
$Ednrb^{tm2(rtTA)Tilg}/Ednrb^{tm3(tetO)Tilg}$	Targeted (knockin) Inducible		endothelin receptor type B	Shin et al. (1999)
$Ednrb^{tm1Myks}$	Targeted (reporter)		endothelin receptor type B	Lee et al. (2003)
$Ednrb^{tm1Ywa}$	Targeted (knockout)		endothelin receptor type B	Hosoda et al. (1994)
$Fzd4^{tm1Nat}$	Targeted (Reporter)		frizzled homolog 4 (*Drosophila*)	Wang et al. (2001)
$Gpc3^{tm1Arge}$	Targeted (knockout)		glypican 3	Chiao et al. (2002)
Kit^{tm1Alf}	Targeted (reporter)		kit	Bernex et al. (1996)
Kit^{tm1Ber}	Targeted (knockin)		kit	Kimura et al. (2004)
Kit^{tm2Ber}	Targeted (knockin) Inducible		kit	Kimura et al. (2004)
Kit^{tm3Ber}	Targeted (knockin)		kit	Kimura et al. (2004)
Kit^{tm1Bpr}	Targeted (knockin)		kit	Rubin et al. (2005)
Kit^{tm1Bsm}	Targeted (knockin) with neo cassette		kit	Kissel et al. (2000)
$Mitf^{mi\text{-}vga9}$	Transgenic (random, gene disruption)		microphthalmia-associated transcription factor	Hodgkinson et al. (1993)
$Mitf^{tm1Arnh}$	Targeted (knockin)		microphthalmia-associated transcription factor	Bismuth et al. (2008)
$Mitf^{tm1.1Arnh}$	Targeted (knockin)		microphthalmia-associated transcription factor	Bismuth et al. (2008)
$Mitf^{mi\text{-}tg}$	Transgenic (random, gene disruption)		microphthalmia-associated transcription factor	Krakowsky et al. (1993)
$Myc^{tm2Fwa}/Myc^{tm2Fwa}$; Tg(Wnt1-cre)11Rth/°	Targeted (floxed/Frt)	Wnt1::Cre	myelocytomatosis oncogene	Wei et al. (2007)

Table 6.2 (*Cont'd*)

Allele symbol	Category	Transgenic used	Name, gene information	Reference
$Notch1^{tm1Agt}/Notch1^{tm1Agt}$; Tg(Tyr-cre)2Lru/°	Targeted (floxed/Frt)	Tyr::Cre	notch gene homolog 1 (*Drosophila*)	Schouwey et al. (2007)
$Notch2^{tm1Frad}/Notch2^{tm1Frad}$; Tg(Tyr-cre)2Lru/°	Targeted (floxed/Frt)	Tyr::Cre	notch gene homolog 2 (*Drosophila*)	Schouwey et al. (2007)
$Ntrk1^{tm1Bbd}$	Targeted (knockout)		neurotrophic tyrosine kinase, receptor, type 1	Smeyne et al. (1994)
$Pax3^{tm1Buck}$	Targeted (reporter)		paired box gene 3	Relaix et al. (2003)
$Pax3^{tm3(Pax7)Buck}$	Targeted (knockin)		paired box gene 3	Relaix et al. (2004)
$Pcbd1^{tm1Grc}$	Targeted (knockout)		dimerization cofactor of hepatocyte nuclear factor 1α	Bayle et al. (2002)
$Rbpj^{tm1Hon}/Rbpj^{tm1Hon}$; Tg(Tyr-cre)2Lru/°	Targeted (floxed/Frt)	Tyr::Cre	recombination signal binding protein for immunoglobulin κ J region	Schouwey et al. (2007)
$Snai2^{tm2Grid}$	Targeted (knockout)		snail homolog 2 (*Drosophila*)	Pérez-Losada et al. (2002)
$Sox10^{tm3(Sox8)Weg}$	Targeted (knockin)		SRY-box-containing gene 10	Kellerer et al. (2006)
$Tcfap2a^{tm1Hsv}/Tcfap2a^{tm2Will}$; Tg(Wnt1-cre)11Rth/°	Targeted (floxed/Frt)	Wnt1::Cre	transcription factor AP-2α	Brewer et al. (2004)
Hyperpigmented				
$Sufu^{tm1Rto}$	Targeted (knockout)		suppressor of fused homolog (*Drosophila*)	Svárd et al. (2006)
Pigment-type switching				
$Drd2^{tm1Mok}$	Targeted (knockout)		dopamine receptor 2	Yamaguchi et al. (1996)
$Ggt1^{tm1Zuk}$	Targeted (knockout)		γ-glutamyltransferase 1	Lieberman et al. (1996)
$Sox2^{ysb}$	Transgenic (random, gene disruption)		SRY-box-containing gene 2	Dong et al. (2002)
$Pomc^{tm1Sora}$	Targeted (reporter)		pro-opiomelanocortin-α	Challis et al. (2004)
$Pomc^{tm2Ute}$	Targeted (knockout)		pro-opiomelanocortin-α	Yaswen et al. (1999)
$Prkaca^{tm3Gsm}/Prkaca^{tm3Gsm}$/; Tg(Mitf-cre)710-59Gsb/°	Targeted (floxed/Frt)	Mitf::Cre	catalytic α-subunit of protein kinase A	Alizadeh et al. (2008)

Grey shading indicates coat-color changes during the life of the mouse.

($Mitf^{tm1.2Arnh}$). These novel *Mitf* mutants thus revealed the role of serine 73 *in vivo*, the influence of the neomycin resistance cassette in the phenotype generated, and a system of reversion of an unstable targeted allele of *Mitf* with the highest spontaneous reversion rates in mammals. Note that without the rigorous and detailed molecular analysis of the $Mitf^{tm1Arnh}$ allele performed by the group of Arnheiter, incorrect conclusions could have been made about the role of this serine in MITF during development of the melanocyte lineage. This example also illustrates the importance of this type of approach for dissecting the *in vivo* roles of the multiple signaling pathways present in melanocytes and indicates that a phosphorylation site identified as crucial *in vitro* can be not so critical *in vivo*.

Another type of mutant has been designed to inactivate known coat-color genes at various times during development. Work with natural mutants of the *Endothelin receptor B* (*Ednrb*) locus established clearly that this protein is required for the development of melanocytes but the specific time at which it is required was not known. New mutant mice were generated in which the endogenous *Ednrb* locus is under the control of the tetracycline-dependent transactivators tTa or rtTA (an inducible system described in Prosser & Rastan 2003). Using this system, the group of S. Tilghman showed that EDNRB is required for melanoblast development during a restricted period, between E10 and E12.5 (Shin et al. 1999). Other knockins of known coat-color genes have been generated: they involve replacing the gene of interest with either a marker to trace the melanocyte cell lineage (Bernex et al. 1996) or a member of the same family, like Sox10 with Sox8 (Kellerer et al. 2006).

Pro-opiomelanocortin (POMC), its derived peptide α-MSH, and the α-MSH receptor MC1R, regulate the eumelanin/pheomelanin switch in mice. Numerous questions about the molecular mechanism underling this process remain unanswered (see above and Chapter 5). Several knockouts have been generated targeting genes of the POMC/α-MSH/MC1R axis and its downstream signaling. POMC-derived peptides and more specifically the pro-opiomelanocortin-α have been knocked out and the corresponding mouse lines obtained: $Pomc^{tm2Ute}$ and the $Pomc^{tm1Sora}$, respectively (Yaswen et al. 1999; Challis et al. 2004). Surprisingly, these mice, on a mixed, 129-based agouti background, have yellowish coats compared to the wild-type littermates but are not completely yellow. Thus, in the mouse, the lack of the ligand (POMC) does not result in a phenotype congruent with the lack or antagonism of MC1R; this suggests the existence of other ligands for this melanocortin receptor. Alternatively, the receptor may have a ligand-independent constitutive activity. To analyse this axis and the signaling associated with it in more detail, the Cre/LoxP system was used to mutate the catalytic α-subunit of protein kinase A (PKA) to cause constitutive activation of PKA in melanocytes (Alizadeh et al. 2008). This kinase is the classical effector for MC1R; MC1R couples to adenylate cyclase through G_s in melanocytes. The group of G. Barsh used this model to demonstrate that constitutive activation of PKA in an agouti background leads to a darker coat and also to darker skin. The range of coat-color darkening varied from umbrous (black dorsum and dark agouti flanks) to a nonagouti phenotype in which pheomelanin pigmentation was replaced by eumelanin. Thus, the activation of PKA promotes synthesis of eumelanin instead of pheomelanin. cAMP does appear to be a key second messenger for pigment-type switching towards eumelanin, but not all the target molecules have yet been identified (see also Chapter 3), and a cAMP-independent pathway also appears necessary for agouti signal protein (ASP) signaling of pheomelanogenesis (Hida et al. 2009). Moreover, there is some evidence for melanogenic effects of cAMP that are independent of PKA (Khaled et al. 2002). This type of *in vivo* approach could be used to determine the specific roles of POMC and PKA in the balance between eumelanin and pheomelanin synthesis.

Transgenic technology has been used to elucidate the differences between human and mouse genes within this pathway. A good example is the *Melanocortin receptor* (*Mc1r*) gene and its sensitivity and requirement for the agonist α-MSH. Signaling by mouse MC1R is largely ligand-independent *in vivo*, as deletions of POMC, the precursor of α-MSH, have little effect on eumelanin synthesis (see above and Chapter 5). In contrast, most humans that lack POMC have red hair, suggesting that human MC1R requires α-MSH to signal. Jackson and collaborators generated transgenic mouse model in which coat pigmentation is mediated solely by the human *MC1R*: they integrated the human *MC1R* into the genome of *Mc1r*-mutant mice (*Recessive yellow*, *Mc1r^{e}* homozygous) (Jackson et al. 2007). They found that the human receptor is more sensitive than the mouse MC1R to exogenous ligand and demonstrated that the difference lies in the receptor itself rather than in the signaling of mouse melanocytes.

6.4.2 New pigmentation loci identified

The Notch signaling pathway regulates various processes including cell proliferation, cell-fate decisions, differentiation, and stem cell maintenance. Several members of the Notch signaling pathway are expressed in melanocytes and seem to be upregulated during oncogenic transformation (Hoek et al. 2004). Jagged2, one of the Notch ligands, is expressed in the basal layer of keratinocytes. These various findings suggested that Notch signaling is functional in melanocytes (Moriyama et al. 2006). To investigate the involvement of Notch signaling in melanocytes, several groups constructed mouse mutants. Conditional deletion of Notch's partner in the nucleus, the recombination binding protein-J (RBP-Jk), in melanocytes causes elimination of melanoblasts and melanocyte stem cells of the epidermis (Moriyama et al. 2006). Additional experiments suggested that Notch/RBP-J pathway signaling via HES1 plays a critical role in the maintenance of melanoblasts by preventing apoptosis (Moriyama et al. 2006). Using the same conditional RBP-Jk construct and the same Tyr::Cre inducer mice, another study showed that Notch signaling plays a role both in maintaining the immature status of melanoblasts (Aubin-Houzelstein et al. 2008), and in melanoblasts and melanocytes being appropriately located in the outer root sheath and in the hair matrix, respectively. Ablation of Notch1 and Notch2 receptors using the same Tyr::Cre mice induced hair graying (Schouwey et al. 2007). The number of DCT-expressing cells was not affected during embryonic stages, but melanocytes located within the hair matrix progressively disappeared during the first regeneration of the hair follicle. Interestingly, the severity of the phenotype depended on the dose of the two receptors. Dermal and choroidal melanocytes were unaffected in *Notch1* and *Notch2* double mutants, suggesting that Notch signaling is required only for epidermal melanocytes (Schouwey et al. 2007), or only for epidermal melanocyte stem cells as discussed in Chapter 3.

The transcription factor activating protein AP2α was long suspected to be important in neural crest cells, and the construction of null mutants genetically confirmed this view (Brewer et al. 2004). However, the involvement of AP2α in mouse pigmentation could not be directly addressed because AP2α-null mutant mice did not survive beyond birth. AP2α was therefore conditionally inactivated using the Wnt-1::Cre transgene; the resulting mutant mice displayed white patches of variable size on the belly and on the tail, and white paws (Brewer et al. 2004). This phenotype suggested a deficiency in the number of melanoblasts and/or a migratory defect (see Fig. 3.16). The enteric nervous system appeared to be unaffected, so it was concluded that AP2α mutants can be classified in the same category as *Kit* and *Mitf* mutants (melanocyte lineage-specific within the neural crest) rather than in the category containing *Pax3*, *Sox10*, *Edn3*, and *Ednrb* mutants (other neural crest cells also affected).

The significance of the anti-apoptotic molecule BCL2 in mouse coat color was revealed by the hair-graying phenotype of mice with homozygous *Bcl2*-null mutations (Veis et al. 1993; Kamada et al. 1995) (see section 3.4.7). The *Bcl2*-null mice are born pigmented but become white (or less pigmented, depending upon the background genome) from the onset of the first hair cycle. Nishikawa's group further demonstrated that there is a complete absence of melanocyte stem cells in these mice and that the melanocytes populating hair bulbs in the first hair cycle are not derived from melanocyte stem cells (Mak et al. 2006).

Transgenic mice designed to target the melanocyte lineage have only recently been constructed, and the number of new mutants will undoubtedly increase substantially in the near future. Even at this early stage of progress in the field, new actors in pigmentation have been identified and described.

6.5 Influence of the genetic background

An important point to consider when generating and analyzing mouse mutations is the issue of genetic background, which can have both subtle and profound effects on phenotype. The first decision to be made when generating genetically engineered mice is the strain to be used as donor of eggs. There is a limited choice of strains for pronuclear injection and certain strains or combinations of strains produce more robust eggs for this purpose. In any case, genes must be introduced into a defined genetic background so that phenotype can be compared with controls that have the same genetic background. Therefore, inbred mice should be used for transgenesis.

An inbred strain is defined as one that has been maintained for 20 or more generations of brother-to-sister mating and is essentially homologous at all genetic loci. One of the first inbred strains was DBA (named according to the mutant pigmentary loci, *Dilute* (*D*, *Myo5A*), *Brown* (*B*, *Tyrp1*), and *Nonagouti* (*A*) that it carries) and another was C57 (derived from the female 57 from the Granby mouse farm). The inbred strain most widely used for microinjection is C57BL/6. However, inbred mice are generally poor reproducers and when the genetic background is not critical, microinjections are performed with F_1 zygotes derived by crossing different inbred strains. For example, mice of the following inbred strains may be crossed to produce F_1 hybrids C57BL/6J, CBA/J, C3H/HeJ, and DBA/2J. With further interbreeding, however, beyond the F_1 generation, the offspring will become mixed and variable and will no longer provide uniform genetic controls for evaluating the transgenic mice.

Most mouse ES cells used for gene targeting are derived from the strain 129/Sv although C57BL/6 is the reference strain for most biomedical research, and in particular for that in the pigmentation field. Therefore, to compare the transgenic mouse with normal mice of the same genetic background it is necessary to backcross the transgenic mutation on to the preferred background. Backcrossing involves mating the mutant mouse to an inbred mouse of the appropriate inbred strain and selecting the offspring that carry the mutation to repeat the process. After 10 generations of backcross breeding, the resulting congenic strain is 99.9% similar to the reference inbred strain, except for the chromosomal region harboring the targeted (transgenic) gene; this region can encompass up to 300 genes that will be those of the starting strain.

The alternative to 129/Sv ES cells is C57BL/6 ES cell lines but their reliability has to be further demonstrated. A comparison between the performances of 129/Sv ES cells (CJ7) and a C57BL/6-ES cell line (BRUCE4) shows that 129/Sv ES cells are more efficient for chimera generation, but that C57BL/6-ES cells would have better germline transmission (Seong et al. 2004). Moreover, it appears that ES cells derived from F_1 129/Sv × 129/Ola blastocysts are stable in cell culture to allow an efficient germline transmission. In the future, genetically engineered mice will be certainly more often generated with C57BL/6-ES particularly for work in which genetic background and fertility are important factors.

It is critical for the interpretation of any knockout and transgenic genotype to determine whether the observed phenotypic difference is the consequence of the targeted mutation or is the result of background genetic variation. Unfortunately, for the technical reasons described above, the first genetically engineered mice were generated in hybrid strains; consequently, many published phenotypes are from genetically mixed backgrounds and are not always compared to appropriate controls with a genetically similar background. More effort and rigor must be used in analyzing phenotypes arising from genetic manipulation and researchers should not forget that the consequences of a single genetic modification can be strongly influenced by other genes. For example, among pigmentary mutants, the strong effect of the background genome on the observed phenotype is well illustrated by the *Nonagouti* locus that is discussed in Chapter 5. Background can similarly affect phenotypes at any level, including cell biology, where it may be important to medical applications.

The health status of the animal colony is also an important factor to consider when analyzing the phenotype of genetically engineered mice. Many environmental factors can influence expression of transgenes or ordinary mutations, although this issue is often underestimated or ignored. For example, reports in the literature have attributed skin disease to mutation at *Faded* (*Fe*; Chapter 4). However, in M.L. Lamoreux's hands, *Fe* was backcrossed on to C57BL/6J and no skin problems were observed. It is always difficult to determine exactly the contribution of environmental events in transgene-influenced phenotypic traits. It is as important to control the environmental parameters as the genetic variables. When possible, genetically engineered mice should be produced and analyzed in specific-pathogen-free (SPF) animal colonies. In addition, temperature, humidity, light cycle, and air supply should also be tightly controlled to ensure valid and reproducible results.

6.6 Conclusions

A revolution is under way concerning the generation of genetically engineered mice carrying new genetic traits. Since the production of chimeric mice in the 1970s followed by the development of successful pronuclear injection techniques in the early 1980s and then that of embryonic ES cell lines leading to the production of the first knockouts in the late 1980s, many new mutants have been created. Methods involving the recombination of huge portions of different genomes by nuclear transfer and transfer of chromosomal material have been developed but are currently less often used. It is difficult to predict how much we will learn in the pigmentation field from all these mutants.

New coat-color mutants have been generated mainly for four reasons:

1. the analysis of the relation between sequence data and the function of genes known to be involved in pigmentation,
2. the discovery of the function of genes suspected to act in pigmented cell lineages,
3. the alteration of the timing and site of gene expression to identify the window of action of these genes during development, and
4. the creation of mouse models for clinical pigmentary disorders and melanomas.

Numerous mouse models have been generated around the world but, unfortunately, many are still under careful analysis and have not yet been published. New color mutants will emerge and they will include mutants of gene products involved in general processes like cell signaling, cell cycle, adhesion, and transcription, playing a specific role in melanocytes. They will also include mutants of housekeeping-gene products that are necessary for all cell types. As we learn more about the extent to which different developmental processes interact with each other, and under what conditions of genetic background, it will be increasingly difficult to determine whether or not all genes with coat-, eye-, or skin-color phenotypes should be referred to as color genes. Many developmental processes such as homing require the coordinate action of two or more genes, so the combination of different genetically engineered mice may be required to obtain certain phenotypes.

Following on from the new mutants generated with sophisticated transgenic and gene targeting techniques, new methods of analysis will be developed and will create new research possibilities. Until now, melanoblasts have been visualized during development using β-galactosidase as a marker and this technique requires fixation of the tissue for analysis. More recently, GFP detection has been used and this approach avoids the need for a substrate or indeed any pretreatment. A fusion gene that includes DNA from the protein of interest and from GFP can be used and will provide a marker *in vivo* to follow gene expression and real-time protein localization. The limitation is the sensitivity of GFP detection. GFP detection and visualization in late embryos is currently not feasible. The development of noninvasive techniques such as echography, positron emission tomography, or biphoton will allow longitudinal studies in individual mice and will lead ultimately to better understand dynamic developmental processes. These imaging technologies will allow accurate localization of single cells within developing skin and promise to be very powerful for investigating melanocyte stem cells.

In short, the use of modern technologies is now adding new parameters to the 100 year history of the pigmentary system of the mouse as the premier model for our understanding of mammalian genetics, and particularly for our understanding of the relationships between the function of the gene and its effect upon the phenotype at all levels of organization. It began with the question 'why is the mutant yellow mouse yellow, and obese?' (Cuénot 1902). For every answer to that and the many other questions of pigmentary research we have discovered new questions that lead us deeper in our understanding of the nature of life itself.

7

Other Species and Other Resources

Figure 7.1 A brindle cow, with a mutation at the *Agouti* locus (Girardot et al. 2006). Courtesy of the Britten Trust.

7.1 Introduction

Phenotypically, it has always been clear that the colors of mammals, birds, and other vertebrates share common features, and the field of mouse pigmentation genetics has always benefitted from species comparisons (for example Searle 1968). With the sequencing of the genomes and other modern molecular and cell biological techniques (Chapter 6) direct comparisons are now possible at all levels from the molecular to the phenotypic. Such studies have already strongly impacted technologies that include medicine (Nordlund et al. 2006, with regard to pigmentation), agriculture

The Colors of Mice: A Model Genetic Network, 1st edition. By M. Lynn Lamoreux, Véronique Delmas, Lionel Larue, and Dorothy C. Bennett. Published 2010 by Blackwell Publishing Ltd.

(Womack 1987; Crouch 1990), and animal breeding. Perhaps the latter is best exemplified by a conversation with the rancher, owner of the above cow, who exclaimed in awestruck tones that he can now pluck a few hairs from the tail of his calves, send them off to the university, and the results of their evaluation will tell him which of his beef cattle will grow up to produce the most tender steaks. If you want to breed for color of horses, cats, or dogs, similar information is available (see below). To a large extent, these new technologies are possible because of the science of genomics.

With the advent of modern molecular techniques and genome sequencing, quite a number of species are now available to evaluate relationships between phenotypes and genotypes in comparison with the basic information generated in inbred mice (Jackson et al. 2006). This includes primate (Bradley & Mundy 2008), including Neanderthal (in 2008); dog (Schmutz et al. 2003; Newton et al. 2000); cat (O'Brien et al. 2008); mice of several species; cattle (Womack 2006); horses (Choudhary & Raudsepp 2006); chickens (Andersson 2003; Burt 2006); fish (Logan et al. 2006; Postlethwait 2006; Mitani et al. 2006; Morizot et al. 1991; Shima et al. 2003; Lamoreux et al. 2005; Cerdá-Reverter et al. 2005; Kelsh et al. 2004; Delfgaauw et al. 2003); reptiles (Rosenblum et al. 2004; Matsumoto 2002); red-footed boobies (Baião et al. 2007) and other birds (Mundy 2005); and even wooly mammoths (Römpler et al. 2006) and other species (Gojobori et al. 2003).

The mouse and human genomes were the first sequenced and the new technologies are now applied interactively to study human disease, as shown by the content of Online Mendelian Inheritance in Man (OMIM; www.ncbi.nlm.nih.gov/) and Mouse Genome Informatics (MGI; www.informatics.jax.org/), both of which refer to both human and murine research.

As genomic sequences from additional species become available for comparison, the differences in phenotypic expression that we find among species with similar genotypes inform our understanding of gene functions and ontogenetic processes (Klungland & Vage 2003). More importantly, phylogenetic similarities among the species, including primates (Bradley & Mundy 2008) and extinct species (Lalueza-Fox et al. 2007; Römpler et al. 2006), are enhancing and confirming our understanding of the process of evolution as one of the most important and powerful components of the living ecosystem (Wayne 1993; Wayne & Ostrander 2007; D'Erchia et al. 1996; Frost-Mason & Mason 2000; O'Brien et al. 1997, 1999; Eizirik et al. 2003; Murphy et al. 2004; Yu et al. 2008).

Clearly, we cannot discuss the broad contribution of pigmentation genetics in these varied disciplines. However, we envision that comparative genetics will play a large role in the future of pigmentation genetics (O'Brien et al. 1999). Therefore we will attempt to skim the surface, stressing homologies among the species, with a primary goal to provide some key references and reliable internet sources of information about pigmentation genetics. Your basic text on the subject of comparative mammalian pigmentation genetics should be Searle (1968).

7.2 Resources

Another reason for including a section about other species, however brief it must be, is the ocean of misinformation that is available on the worldwide web. With the advent of self-publishing we can expect to see a yet greater increase in personal opinions presented as though they were facts. Of course, for analytical evaluation of inter-species homologies it is important to use scientifically accurate information based in reproducible data. Finding this information can be a problem, but with the creative use of keywords, peer-reviewed papers published approximately within the last 20 years can be found at Pubmed (www.ncbi.nlm.nih.gov/pubmed).

We mention other sources below as a minimal starting point. Obviously there are many good websites that we do not mention.

7.2.1 The pigment cell societies

Studies of pigmentation genetics of course are not limited to the mouse. Members of pigment cell societies study creatures from ascidians and amphioxus to mice and humans, and all of the species mentioned below. The International Federation of Pigment Cell Societies also publishes the premier journal on the subject of pigmentation research, *Pigment Cell and Melanoma Research* (www.pigment.org/).

International Federation of Pigment Cell Societies	www.ifpcs.org/
PanAmerican Society for Pigment Cell Research	http://paspcr.med.umn.edu/
European Society for Pigment Cell Research	www.espcr.org/
Japanese Society for Pigment Cell Research	www.soc.nii.ac.jp/jspcr/
Asian Society for Pigment Cell Research	www.aspcr.org/

Databases

Mouse Genome Informatics (MGI) is the central source for basic information about mutant mice. MGI is maintained by The Jackson Laboratory (JAX; www.informatics.jax.org/). The MGI website lists the various subunits of MGI.

The Mouse Genome Database (MGD) includes data on gene characterization, nomenclature, mapping, gene homologies among mammals, sequence links, phenotypes, allelic variants and mutants, and strain data.

The Gene Expression Database (GXD) integrates different types of gene-expression information from the mouse and provides a searchable index of published experiments on endogenous gene expression during development.

The Mouse Tumor Biology (MTB) database project integrates data on the frequency, incidence, genetics, and pathology of neoplastic disorders, emphasizing data on tumors that develop characteristically in different genetically defined strains of mice.

The Gene Ontology (GO) project at MGI: the MGI group is a founding member of the Gene Ontology Consortium (www.geneontology.org). MGI fully incorporates the GO project in the database and provides a GO browser.

MouseCyc Project at MGI: the MouseCyc database focuses on *Mus musculus* metabolism and includes cell-level processes such as biosynthesis, degradation, energy production, and detoxification. It is part of the BioCyc (www.biocyc.org/) collection of pathway databases created at SRI International (Stanford Research Institute). Pathway information in MouseCyc is integrated with MGI data.

The International Mouse Strain Resource (IMSR; www.findmice.org/) lists international facilities that can provide resources (mice, frozen embryos or sperm, tissues, etc.). It can be accessed through the MGI database or directly. Stocks that are available from the mouse strain resource institutions are available only to qualified research institutions that have demonstrated (by inspection) that they conform to Federal guidelines that define humane management and treatment of laboratory animals. Because of this requirement the stocks are not available to the general public.

The Mutant Mouse Regional Resource Centers (MMRRC) is mentioned in several figure legends in earlier chapters of this volume. It is one of the institutions referenced in IMSR and MGI. MMRRC preserves and maintains the unique mouse mutant stocks that were developed by M.L. Lamoreux and which are pictured throughout this book.

Important online databases available for other species are listed at www.ncbi.nlm.nih.gov/Genomes/. This website provides links to relevant books, OMIM, and Pubmed, and a list of species that are cloned and being cloned that ranges from moss to malaria and includes most of those touched on below.

Breed associations sometimes provide good information. They are not authoritative sources of scientific information because they are not peer-reviewed, but an exhibition sponsored by one of these organizations is an excellent place to look at phenotypes. These are only a few of the North American associations with websites. Similar organizations exist in other countries.

American Cavy Breeders' Association	www.acbaonline.com/
American Rabbit Breeders' Association	www.arba.net/
American Kennel Club	www.akc.org/
American Dairy Goat Association	www.adga.org/
Alpaca and Llama Show Association	www.alsashow.net/index.html
American Livestock Breeds Conservancy	www.albc-usa.org/

Another helpful website is that of the University of California at Davis Veterinary Genetics Laboratory, www.vgl.ucdavis.edu/services/.

7.3 Other species

7.3.1 Rats, guinea pigs, and rabbits

Mice, rats, guinea pigs, and rabbits participated in the orgins of the science of pigmentary genetics in Europe and the USA (prominently under W.E. Castle at the Bussey Institute, at the time a part of Harvard University), where the newly appreciated Mendelian principles were applied to small animals that were already domesticated and to a few other species. Somewhat later, The Jackson Laboratory and other institutions began their important contributions to the genetics of mice, other rodents, and lagomorphs, as well as dogs. During this time, just about 100 years ago, many of the most obvious or most common pigmentary mutants were recognized and described (Green et al. 1966). These we have termed the 'classical pigmentary mutants,' and it is interesting to observe their presence in other mammalian species and, in the case of the *Albino* and *Microphthalmia* loci, in as primitive an organism as the ascidian (Yajima et al. 2003; Toyoda et al. 2004), although we will not attempt to follow the evolutionary trail to its roots.

The early pigmentary scientists were soon joined by their students and compatriots who began by unravelling the Mendelian genetics of the classical pigmentary mutants and soon proceeded to study their functions. As well as developing an appreciation of the science, they preserved the mutants that we use today. For only one example, W.F. Hollander recognized and saved (in his garage when the colony was threatened) the *Microphthalmia* mutation that was, half a century later, used to clone the *Mitf* locus. Only rarely do we find such important contributions, intellectual or otherwise, listed in the modern databases, and it is worthwhile to look at a few of the first studies of pigmentation genetics; for example: Castle (1924), Castle and Allen (1903), Castle and Little (1909, 1910), Cuénot (1902), Little (1911, 1913a, 1913b), Wright (1925, 1928), and Wright and Chase (1936).

For more information about the early days of the science of pigmentary genetics we suggest the following references: Searle (1968), Green (1966), Silvers (1979), and Nordlund et al. (2006).

Figure 7.2 A series of guinea pigs, all of which are yellow with interacting modifying genes. Photograph courtesy of Harry Claus.

*Laboratory rat (*Rattus norvegicus*)*

According to the National Center for Biotechnology Information (NCBI; www.ncbi.nlm.nih.gov/) the rat genome has been sequenced (Smits & Cuppen 2006a, 2006b; Dwinell et al. 2008).

*Guinea Pig, commonly called the cavy (*Cavia porcellus*)*

According to NCBI the guinea pig genome is in the process of being sequenced. A good description of this and several other species appears on Wikipedia (http://en.wikipedia.org/wiki/Guinea_pig).

Figure 7.2 was provided by Harry Claus, breeder and judge of guinea pigs, who has written a reliable little description of guinea pig genetics for fanciers that is based on his study of Sewell Wright and others. For example, Wright (1925) described the interactions exemplified by these cavies. The pamphlet by Claus is distributed by The American Cavy Breeders Association, 16540 Hogan Avenue, Hastings, MN 55033-9576, USA, c/o Lenore Gergen.

Guinea pigs and rabbits are the two known species with heritable, autosomal yellow/nonyellow chimerism, referred to in the guinea pig as the *Tortoiseshell* allele and in the rabbit as the *Japanese* allele at the *Recessive yellow* (*Mc1r*) locus, introduced in Chapter 3.

*Rabbit (*Oryctolagus cuniculus*)*

Rabbit genome sequencing is also apparently in process. After the early years of pigmentation research, rabbits have been used extensively in ophthalmology, and have contributed to our understanding of albinism (Aigner et al. 2000; Jeffery 2001a, 2001b) (Fig. 7.3).

7.3.2 Carnivores

The dog and cat are the most popular pet animals, and they have also contributed a good deal to our studies of pigmentation, in part because of the ways in which their pigmentation deviates from that of the mouse.

Figure 7.3 Hoppy McDaniel. A tricolor rabbit that has white spotting in addition to autosomal chimerism. The white spotting does not appear to modify the pigment pattern, as it does in the cat (Chapter 3).

*Dog (*Canis lupus familiaris*)*

The genome of the dog has been sequenced (NCBI; www.ncbi.nlm.nih.gov/genome/guide/dog/). See also Switonski et al. (2004), Karlsson et al. (2007), and Ostrander et al. (2000). A useful website is http://homepage.usask.ca/~schmutz/dogcolors.html.

Because quite a lot of breeding with close relatives has been done within some dog breeds, interesting medical conditions have surfaced, along with a wide variety of color phenotypes. The dog is thus a good model for those specific diseases and also for pigmentation. Dr C.C. Little's early analysis of canine pigmentation, *The Inheritance of Coat Color in Dogs* (Little 1957), was very helpful, in spite of the fact that it contained a few hypotheses that were later proven to be inaccurate.

Recent reports more accurately describe canine pigmentation genetics: Schmutz et al. (2002), Berryere et al. (2005), Philipp et al. (2005), and Schmutz and Berryere (2007). Another good source is Ruvinsky and Sampson (2001), *The Genetics of the Dog*.

Little's introduction to our understanding of pigment-type switching in the dog was followed, when the technology became available, by identification of the *Recessive yellow* gene of the dog at the *Mc1r* locus (Everts et al. 2000; Newton et al. 2000) and further evaluation of the components of the eumelanin/pheomelanin switch in dogs (Schmutz et al. 2003; Berryerre et al. 2005; Kerns et al. 2003). This series of studies culminated in the important recent discovery of a new ligand of MC1R, CBD103, the *Canine β-defensin 103* gene, as discussed in Chapter 5.

Another phenotype of dogs that was thought to be unique (*Merle*), is actually a unique manifestation at a known locus and contributed to our understanding of the *Silver* locus, as discussed in Chapter 4.

*Cat (*Felis catus*)*

Useful articles on the cat genome include Eizirik et al. (2003), O'Brien et al. (2008), and Schmidt-Küntzel et al. (2009). A useful website is http://faculty.vetmed.ucdavis.edu/faculty/lalyons/Sites/color.htm.

The cat, of course, has contributed to a major advance in our understanding of development, the Lyon Hypothesis, which has long graduated from the status of hypothesis and is now generally recognized as a normal developmental process in mammals (Lyon 1961). The resulting phenotypes that involve the X-linked *Orange* locus are discussed in Chapter 3.

Cell-autonomous X-linked yellow has been identified only in cats and Syrian hamsters (Alizadeh et al. 2009). In a recent and rather surprising study, it was found that the X-linked *Orange* (yellow) feline gene may not be *Mc1r* (*Recessive yellow*) as some had postulated. The study of X-linked yellow in the hamster (which is referred to as *Sly*) suggests that some factor in the pathway of *Mc1r* may be the X-linked element in that species, and this might be informative of the situation in the cat as well.

Cat fanciers and scientists alike have long benefitted from Roy Robinson's book, *Genetics for Cat Breeders* (Robinson 1971; now Vella et al. 2003) with its careful analysis of feline Mendelian genetics. When many other books, notably genetics textbooks, were claiming that 'black' was a sex-linked gene, Robinson clearly and accurately defined the relationships among the three major loci that function cooperatively in the cat to determine the three 'primary' colors of mammalian melanin pigment. An unfortunate misrepresentation of this relationship has been explicitly and implicitly perpetuated in the literature with the incorrect statement that *Orange* is dominant to *Black* in the cat. This assumption has been rebutted by Lamoreux (1973), and by Miller and Hollander (1986), who state: 'There is general failure of the textbook authors to identify mutants by contrast with wild-type standard, thus leading to the erroneous conclusion that black is sex-linked and allelic with yellow.' However, it is difficult to undo inaccurate statements when they are published in textbooks. Robinson also clearly reported that the tabby phenotype is not encoded at the *Agouti* locus, although a cat must be agouti in order to fully express tabby. Confirming the Mendelian genetics, the *Tabby* locus has been mapped (Lyons et al. 2006) and is not related to other known mammalian genes.

More recently, several of these loci have been identified in cats. The black versus brown decision in the cat, as in the mouse, the dog, and most other species that have been studied, is encoded at *Tyrp1* (Lyons et al. 2005; Schmidt-Küntzel et al. 2005). However, an additional mutant allele is available in the cat, originally named 'red' by breeders, in the Abyssinian breed, and called caramel when it was imported to the USA from Britain by M.L. Lamoreux; the name was subsequently changed to cinnamon (Schmidt-Küntzel et al. 2005). This should not be confused with the standard use of the term cinnamon to describe *A/A b/b* (*Agouti*, *Brown*) in cats, rabbits, and mice. Schmidt-Küntzel et al. (2005) also confirmed that the feline allelic *Albino* series includes '*Siamese*,' that would be equivalent to *Himalayan* in mice, guinea pigs, and rabbits, as well as '*Burmese*' that may be specific to cats. The *Agouti* locus also is homologous among species in which it has been investigated.

Similar to the dog, and differing from the mouse, the cat 'dilute' mutation involves the *Melanophilin* (*Mlph*) locus (Chapter 4) (Philipp et al. 2005).

An apparently feline-specific locus is the *I* (*Inhibitor*) locus reported by Turner and Robinson (1980). The phenotype resulting from dominant mutation at this locus in cats is very similar to the recessive homozygous mutant phenotype of *Chinchilla* (an *Albino*-locus mutation) in mice (Fig. 4.5 shows an agouti cat that is mutant at *I*). Pheomelanic pigment is preferentially reduced in the presence of these mutations. This phenotype in cats is known to fanciers as silver, but it is not homologous to the murine *Silver* or *Albino* loci, and has been mapped to a unique location distinct from any other known mammalian 'hypopigmentation' genes (Menotti-Raymond et al. 2009). It is tempting to believe that this gene may have something important to tell us about the cellular regulation of pheomelanogenesis (Chapter 5).

7.3.3 Hoofed mammals

Again, the animals most likely to provide information about genotype/phenotype interactions in the pigmentary system are those that have been domesticated, as they are released from the selective pressures that limit pigmentation phenotypes in wild aimals, and the people who develop breeds of domestic animals generally enjoy the colors that surface, and keep them. In the cases of swine, sheep, goats, llamas, and alpacas, the pigmentary phenotypes seem to complement and confirm observations in the other species discussed above. However, to date the contribution of these species to our basic understanding of pigmentation has not been extensive, perhaps in part because the greater interest is in meat and fiber. However, see Sponenberg et al. (1988). It is interesting that the difference between 'white' and black sheep is apparently at the *Agouti* locus, suggesting that an assemblage of modifing genes may have been selected over time to reduce pheomelanin in the white sheep (Norris & Whan 2008).

Pigs have made a special contribution to studies of melanoma and vitiligo, as we discussed in Chapter 3, and have helped to confirm the long-held theory that the immune system may be involved in some kinds of vitiligo. White-spotted pigs fail to get melanoma, which is not surprising as white spotting is the absence of pigment cells (Borovansky et al. 2005). More curious is the observation in the Sinclair breed of swine that 'yellow' pigs (which are probably actually agouti) do not succumb to the melanoma nor the vitiligo that is very common in black pigs of this breed (Fig. 7.4). Unfortunately this phenomenon has not been studied adequately.

Cattle

A useful website for the genetics of coat color in cattle is http://homepage.usask.ca/~schmutz/colors.html. The colors of cattle seem to follow most of the basic parameters of pigmentation that we have described for other mammalian species (Ibsen 1933; Lauvergne 1999; Searle 1968; Olson 1982, 1999; Adalsteinsson et al. 1995; Klungland et al. 1995; Joerg et al. 1996; Seitz et al. 1999; Rouzaud et al. 2005; Berryere et al. 2003; Schmutz et al. 2004; Guibert et al. 2004), except that the wild-type pattern of cattle (as of some other species, including horses) is associated primarily with the distribution of pigmentation over the surface of the body, rather than the banding of the hairs as in rodents and lagomorphs. The whiteness of the Charolais breed of cattle, found in several breeds and crossbred cattle, is caused by mutation at *Silver*, as discussed in Chapter 4 (Fig. 4.13) (Kuhn & Weikard 2007).

The promoter region of the bovine *Agouti* locus contains at least three alternative promoters (Girardot et al. 2006). 'Brindle' is an unusual mutation affecting this complex promoter region (Giraradot et al. 2005). In 'brindle' mutant cattle a transposable element in this promoter region causes a eumelanin/pheomelanin chimerism that resembles *Viable yellow* in the mouse. The cause of brindle in cattle and mice is different from the cause of a similar phenotype in

Figure 7.4 The adult is a 'red' Sinclair breed of pig, although she may actually be agouti. The piglets do not belong to her and are not defined. The red Sinclairs purportedly do not succumb to the melanoma that is characteristic of the breed, nor the vitiligo that is often associated with the melanoma.

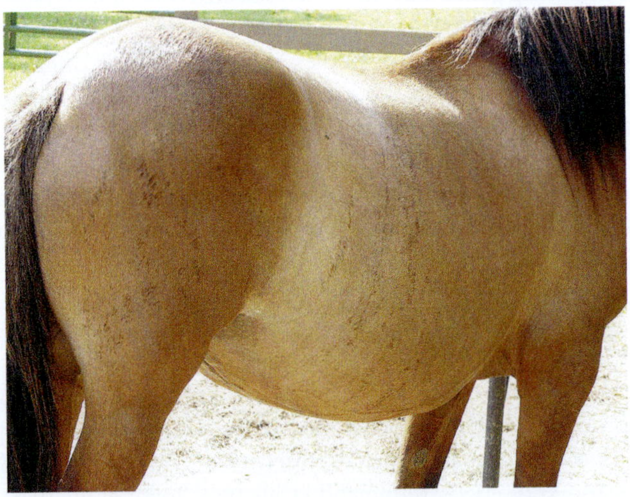

Figure 7.5 A horse with low-level brindling. The breed of the horse is Choctaw Colonial Spanish. Photograph courtesy of Phil Sponenberg (Sponenberg 2009).

dogs and humans (in humans they are called Blaschko stripes), and from 'brindling', which is very rarely seen in horses (Figs 7.1 and 7.5). Thus, in this case, the value of species comparisons lies in the fact that similar phenotypes seen in multiple species arose from different causes.

Horses

For information on the horse genome see Chowdhary and Raudsepp (2006). Horse breeders are fortunate to have at least two books that are reliably accurate: *Equine Color Genetics* (Sponenberg

2009), which focuses on pigmentation, with many pictures, and *The Genetics of the Horse* (Bowling & Ravinsky 2000).

Brindle horses, in appearance quite like the cow in Figure 7.1, or with more or less brindling of the darker and lighter colors (see Fig. 7.5), have been reported. Brindling in horses, however, is very rare and seems not to be heritable. Very likely it is caused by accidental fusion of two early embryos *in utero*.

The classical basic colors were found also in horses from the earliest reports of Castle and subsequent reports of Sponenberg et al. (1988b), for example, and many reports in the popular literature. As in cattle, the wild-type phenotype has more to do with distribution of eumelanin and pheomelanin over the body than within individual hairs. In addition, and more recently, a number of modifying genes have been identified in horses, which are quite well described at the website of the University of California at Davis Veterinary Genetics Laboratory (www.vgl.ucdavis.edu/services/).

One of the earliest horse genes studied was *Palomino* (Castle & King 1951). The website of UC Davis refers to this gene as 'cream;' it is homologous with *Underwhite* in the mouse, and therefore the correct genetic designation is *Matp* (Mariat et al. 2003). Another gene that reduces intensity of pigmentation is known as 'champagne' (Cook et al. 2008).

The unique contribution of the grey horse to our understanding of vitiligo is discussed in Chapter 3. Although a locus for appaloosa has been mapped (Terry et al. 2004), the appaloosa phenotype is such a complicated phenomenon that it must be polygenic, and the polygenes remain to be defined.

7.3.4 Birds

For information on bird genomes, see www.genome.gov/11008054 (see also Burt et al. 1995; Burt 2006). The classic reference on genetics of the fowl is Hutt (1949). Birds, particularly chicken and quail embryos, were the first model for study of the neural-crest derivation of melanocytes (Rawles 1940a, 1940b, 1948) and have continued an important component of pigment cell research (Le Douarin 1982; Dupin & Le Douarin 2003). Birds have also contributed significantly to our understanding of vitiligo, as we discussed in Chapter 3. Vitiligo is a multifaceted affliction with complex causation, and is therefore difficult to study in inbred mice. Better animal models for vitiligo have been the pig, horse, and chicken (Boissy & Lamoreux 1988; Wang & Erf 2003, 2004; Wick et al. 2006), especially the Smyth chicken model (originally known as the DAM chicken) developed by J.R. Smyth, Jr (Austin et al. 1992).

Beyond specific pigment cell function, we should not forget the multiple options provided by feather structure to study structural colors and other pigments that do not occur in hairy species. Irridescent colors, underlain by melanin and other sources of pigment, are found in parakeets (budgerigars), parrots, pigeons and doves, and chickens. Thus the phenotypes of birds are more varied than those of mammals. Enthusiastic groups of fanciers have studied and saved an impressive array of pigmentation genotypes, sometimes assisted, officially or unofficially, by scientists. The genetics of pigeon, chicken, and parakeet colors are well developed and understood within the literature of fanciers. We could not begin to discuss the genotypes and phenotypes they have saved and nurtured, except to say that homologies with the three 'primary colors' of melanin pigment – that is, controls over the black/brown alternative and eumelanin/pheomelanin – exist also in birds.

7.3.5 Fish and lower vertebrates

Information on fish genomes can be found at the following:

zebrafish	http://zfin.org/zf_info/zfbook/zfbk.html
platyfish	http://xiphophorus.org/xgsc.htm
medaka	National Bioresource Project (NBRP), www.shigen.nig.ac.jp/medaka/ (Kinoshita et al. 2009)

As is true of other vertebrate species, color variants of fish have been collected and maintained since the early days of the science of genetics, and in the case of the Japanese medaka since long before Mendel grew peas. The Tomita collection (Kelsh 2004; Kelsh et al. 2004; Lamoreux et al. 2005) of the color variants of medaka has already demonstrated homologies with mammalian genetics, as have zebrafish, a newer but more aggressively developed model (Postlethwait et al. 1998; Rawls et al. 2003) and the platyfish/swordtail model of inherited melanoma.

Fishes, as well as reptiles and amphibians, have two important differences from mammals and birds. First, their pigment cells can be be directly controlled by the nervous system, so that rapid color changes are possible (Logan et al. 2006). Second, there are at least three additional types of chromophores in these species. Each type of pigment cell produces pigment granules that are chemically distinct. Thus, their yellow/red sorts of colors are most often not produced in melanocytes.

Reptiles and amphibians have made equally important contributions to the study of pigmentation, especially in the early years of embryology.

Clearly, the comparative genetics of pigmentation is as infinitely fascinating as life itself.

But we must end our story somewhere.

Figure 7.6 The cow and calf are of the Highland breed. The calf is homozygous for *Silver*, its sire and dam are both heterozygous. Photograph courtesy of LEA-White Farms.

References

Abdel-Malek Z.A., Supp D. (2008) beta-Defensin 3: a novel and unexpected key that unlocks the melanocortin 1 receptor. *Pigment Cell and Melanoma Research* 21(1):7–8.

Abdel-Malek Z.A., Scott C.A., Furumura M., Lamoreux M.L., Ollmann M., Barsh G.S., Hearing V.J. (2001) The melanocortin 1 receptor is the principal mediator of the effects of agouti signaling protein on mammalian melanocytes. *Journal of Cell Science* 114:1019–1024.

Aberdam E., Bertolotto C., Sviderskaya E.V., de Thillot V., Hemesath T.J., Fisher D.E., Bennett D.C., Ortonne J.-P., Ballotti R. (1998) Involvement of microphthalmia in the inhibition of melanocyte lineage differentiation and of melanogenesis by agouti signal protein. *Journal of Biological Chemistry* 273(31):19560–19565.

Ackermann J., Frutschi M., Kaloulis K., McKee T., Trumpp A., Beermann F. (2005) Metastasizing melanoma formation caused by expression of activated N-RasQ61K on an INK4a-deficient background. *Cancer Research* 65(10):4005–4011.

Adalsteinsson S. (1974) Inheritance of the palomino color in Icelandic horses. *Journal of Heredity* 65:15–20.

Adalsteinsson S., Bjarnadottir S., Vage D.I., Jonmundsson J.V. (1995) Brown coat color in Icelandic cattle produced by the loci Extension and Agouti. *Journal of Heredity* 86(5):395–398.

Aigner B., Besenfelder U., Müller M., Brem G. (2000) Tyrosinase gene variants in different rabbit strains. *Mammalian Genome* 11(8):700–702.

Alberts B., Johnson A., Lewis J., Raff M., Roberts K., Walter P. (2008) *Molecular Biology of the Cell*, 5th edn. Garland, New York.

Alizadeh A., Fitch K.R., Niswender C.M., McKnight G.S., Barsh G.S. (2008) Melanocyte-lineage expression of Cre recombinase using Mitf regulatory elements. *Pigment Cell and Melanoma Research* 21(1):63–69.

Alizadeh A., Hong L.Z., Kaelin C.B., Raudsepp T., Manuel H., Barsh G.S. (2009) Genetics of sex-linked yellow in the Syrian hamster. *Genetics* 181(4):1427–1436.

Alkhateeb A., Fain P.R., Spritz R.A. (2005) Candidate functional promoter variant in the FOXD3 melanoblast developmental regulator gene in autosomal dominant vitiligo. *Journal of Investigative Dermatology* 125(2):388–391.

Alonso L., Fuchs E. (2003a) Stem cells of the skin epithelium. *Proceedings of the National Academy of Sciences USA* 100 Suppl 1:11830–11835.

Alonso L., Fuchs E. (2003b) Stem cells in the skin: waste not, Wnt not. *Genes & Development* 17:1189–1200.

Anagnostopoulos A.V. (2002) A compendium of mouse knockouts with inner ear defects. *Trends in Genetics* 18(10):499.

Ancans J., Tobin D.J., Hoogduijn M.J., Smit N.P., Wakamatsu W., Thody A.J. (2001) Melanosomal pH controls rate of melanogenesis, eumelanin/phaeomelanin ratio and melanosome maturation in melanocytes and melanoma cells. *Experimental Cell Research* 268:26–35.

Anderson M.G., Smith R.S., Hawes N.L., Zabaleta A., Chang B., Wiggs J.L., John S.W.M. (2002) Mutations in genes encoding melanosomal proteins cause pigmentary glaucoma in DBA/2J mice. *Nature Genetics* 30:81–85.

Anderson M.G., Libby R.T., Mao M., Cosma I.M., Wilson L.A., Smith R.S., John S.W. (2006) Genetic context determines susceptibility to intraocular pressure elevation in a mouse pigmentary glaucoma. *BMC Biology* 4:20.

Andersson L. (2003) Melanocortin receptor variants with phenotypic effects in horse, pig, and chicken. *Annals of the New York Academy of Sciences* 994:313–318.

Aoki H., Moro O. (2002) Involvement of microphthalmia-associated transcription factor (MITF) in expression of human melanocortin-1 receptor (MC1R). *Life Sciences* 71:2171–2179.

Aoki H., Motohashi T., Yoshimura N., Yamazaki H., Yamane T., Panthier J.J., Kunisada T. (2005) Cooperative and indispensable roles of endothelin 3 and KIT signalings in melanocyte development. *Developmental Dynamics* 233:407–417.

Aoki H., Yamada Y., Hara A., Kunisada T. (2009) Two distinct types of mouse melanocyte: differential signaling requirement for the maintenance of non-cutaneous and dermal versus epidermal melanocytes. *Development* 136:2511–2521.

Aoki Y., Saint-Germain N., Gyda M., Magner-Fink E., Lee Y.H., Credidio C., Saint-Jeannet J.P. (2003) Sox10 regulates the development of neural crest-derived melanocytes in *Xenopus*. *Developmental Biology* 259(1):19–33.

April C.S., Barsh G.S. (2006) Skin layer-specific transcriptional profiles in normal and recessive yellow ($Mc1r^{e}/Mc1r^{e}$) mice. *Pigment Cell Research* 19(3):194–205.

April C.S., Barsh G.S. (2007) Distinct pigmentary and melanocortin 1 receptor-dependent components of cutaneous defense against ultraviolet radiation. *PLoS Genetics* 3(1):0030–0043.

Argeson A.C., Nelson K.K., Siracusa L.D. (1996) Molecular basis of the pleiotropic phenotype of mice carrying the *hypervariable yellow* (A^{hvy}) mutation at the *agouti* locus. *Genetics* 142:557–567.

Arnheiter H. (1998) Evolutionary biology. Eyes viewed from the skin. *Nature* 391:632–633.

Artavanis-Tsakonas S., Rand M.D., Lake R.J. (1999) Notch signaling: cell fate control and signal integration in development. *Science* 284:770–776.

Aubin-Houzelstein G., Bernex F., Elbaz C., Panthier J.J. (1998) Survival of patchwork melanoblasts is dependent upon their number in the hair follicle at the end of embryogenesis. *Developmental Biology* 198(2):266–276.

Aubin-Houzelstein G., Panthier J.J. (1999) The patchwork mouse phenotype: implication for melanocyte replacement in the hair follicle. *Pigment Cell Research* 12(3):181–186.

Aubin-Houzelstein G., Djian-Zaouche J., Bernex F., Gadin S., Delmas V., Larue L., Panthier J.J. (2008) Melanoblasts' proper location and timed differentiation depend on Notch/RBP-J signaling in postnatal hair follicles. *Journal of Investigative Dermatology* 128(11):2686–2695.

Austin L.M., Boissy R.E., Jacobson B.S., Smyth J.R. Jr (1992) The detection of melanocyte autoantibodies in the Smyth chicken model for vitiligo. *Clinical Immunology and Immunopathology* 64(2):112–120.

Avantaggiato V., Dathan N.A., Grieco M., Fabien N., Lazzaro D., Fusco A., Simeone A., Santoro M. (1994) Developmental expression of the RET protooncogene1: differentiation of specific neural structures and the excretory system. *Cell Growth & Differentiation* 5:305–311.

Aybar M., Mayor R. (2002) Early induction of neural crest cells: lessons learned from frog, fish and chick. *Current Opinion in Genetics and Development* 12:452–458.

Babinet C., Cohen-Tannoudji M. (2001) Genome engineering via homologous recombination in mouse embryonic stem (ES) cells: an amazingly versatile tool for the study of mammalian biology. *Anais da Academia Brasileira de Ciencias* 73(3):365–383.

Bachner D., Schroder D., Gross G. (2002) mRNA expression of the murine glycoprotein (transmembrane) nmb (Gpnmb) gene is linked to the developing retinal pigment epithelium and iris. *Brain Research. Gene Expression Patterns* 1(3–4):159–165.

Bagher P., Jiao J., Owen Smith C., Cota C.D., Gunn T.M. (2006) Characterization of mahogunin ring finger-1 expression in mice. *Pigment Cell Research* 19(6):635–643.

Bagnara J.T., Matsumoto J. (2006) Comparative anatomy and physiology of pigment cells in non-mammalian tissues. In: Nordlund J.J. et al. (eds) *The Pigmentary System*, 2nd edn, pp. 11–59. Blackwell Publishing, Oxford.

Bagnara J.T., Matsumoto J., Ferris W., Frost S.K., Turner W.A. Jr, Tchen T.T., Taylor J.D. (1979) Common origin of pigment cells. *Science* 203(4379):410–415.

Baião P.C., Schreiber E.A., Parker P.G. (2007) The genetic basis of the plumage polymorphism in red-footed boobies (*Sula sula*): a melanocortin-1 receptor (MC1R) analysis. *Journal of Heredity* 98(4):287–292.

Barbosa M.D., Barrat F.J., Tchernev V.T., Nguyen Q.A., Mishra V.S., Colman S.D., Pastural E., Dufourcq-Lagelouse R., Fischer A., Holcombe R.F. et al. (1997) Identification of mutations in two major mRNA isoforms of the Chediak-Higashi syndrome gene in human and mouse. *Human Molecular Genetics* 6(7):1091–1098.

Bar-Eli M. (2001) Gene regulation in melanoma progression by the AP-2 transcription factor. *Pigment Cell Research* 14:78–85.

Baroffio A., Dupin E., Le Douarin N.M. (1988) Clone-forming ability and differentiation potential of migratory neural crest cells. *Proceedings of the National Academy of Sciences USA* 85:5325–5329.

Baroffio A., Dupin E., Le Douarin N.M. (1991) Common precursors for neural and mesectodermal derivatives in the cephalic neural crest. *Development* 112:301–305.

Barral D.C., Seabra M.C. (2004) The melanosome as a model to study organelle motility in mammals. *Pigment Cell Research* 17:111–118.

Barrallo-Gimeno A., Nieto M.A. (2005) The Snail genes as inducers of cell movement and survival: implications in development and cancer. *Development* 132:3151–3161.

Barrallo-Gimeno A., Holzschuh J., Driever W., Knapik E.W. (2003) Neural crest survival and differentiation in zebrafish depends on mont blanc / tfap2a gene function. *Development* 131:1463–1477.

Barsh G.S. (2006) Regulation of pigment type switching by agouti, melanocortin signaling, attractin and mahoganoid. In: Nordlund J.J. et al. (eds) *The Pigmentary System*, 2nd edn, pp. 395–409. Blackwell Publishing, Ocford.

Barsh G.S. (2007) How the dog got its spots. *Nature Genetics* 39(11):1304–1306.

Barsh G., Gunn T., He L., Schlossman S., Duke-Cohan J. (2000) Biochemical and genetic studies of pigment-type switching. *Pigment Cell Research* 13(Suppl 8):48–53.

Bataille V. (2003) Genetic epidemiology of melanoma. *European Journal of Cancer* 39(10):1341–1347.

Baxter L.L., Pavan W.J. (2002) The oculocutaneous albinism type IV gene Matp is a new marker of pigment cell precursors during mouse embryonic development. *Mechanisms of Development* 116(1–2):209–212.

Baxter L.L., Pavan W.J. (2003) Pmel17 expression is Mitf-dependent and reveals cranial melanoblast migration during murine development. *Gene Expression Patterns* 3(6):703–707.

Baxter L.L., Hou L., Loftus S.K., Pavan W.J. (2004) Spotlight on spotted mice: a review of white spotting mouse mutants and associated human pigmentation disorders. *Pigment Cell Research* 17:215–224.

Bayle J.H., Randazzo F., Johnen G., Kaufman S., Nagy A., Rossant J., Crabtree G.R. (2002) Hyper-phenylalaninemia and impaired glucose tolerance in mice lacking the bifunctional DCoH gene. *Journal of Biological Chemistry* 277(32):28884–28891.

Baynash A.G., Hosoda K., Giaid A., Richardson J.A., Emoto N., Hammer R.E., Yanagisawa M. (1994) Interaction of endothelin-3 with endothelin-B receptor is essential for development of epidermal melanocytes and enteric neurons. *Cell* 79(7):1277–1285.

Beaumont K.A., Shekar S.N., Cook A.L., Duffy D.L., Sturm R.A. (2008) Red hair is the null phenotype of MC1R. *Human Mutation* 29(8):E88–E94.

Bedell M.A., Cleveland L.S., O'Sullivan T.N., Copeland N.G., Jenkins N.A. (1996) Deletion and interallelic complementation analysis of Steel mutant mice. *Genetics* 142:935–944.

Beermann F., Ruppert S., Hummler E., Bosch F.X., Muller G., Ruther U., Schutz G. (1990) Rescue of the albino phenotype by introduction of a functional tyrosinase gene into mice. *EMBO Journal* 9(9):2819–2826.

Beermann F., Schmid E., Schutz G. (1992) Expression of the mouse tyrosinase gene during embryonic development: recapitulation of the temporal regulation in transgenic mice. *Proceedings of the National Academy of Sciences USA* 89(7):2809–2813.

Beermann F., Orlow S.J., Boissy R.E., Schmidt A., Boissy Y.L., Lamoreux M.L. (1995) Misrouting of tyrosinase with a truncated cytoplasmic tail as a result of the murine *platinum (c^p)* mutation. *Experimental Eye Research* 61(5):599–607.

Beermann F., Orlow S.J., Lamoreux M.L. (2004) The Tyr *(albino)* locus of the laboratory mouse. *Mammalian Genome* 15(10):749–758.

Bennett D.C. (1989) Mechanisms of differentiation in melanoma cells and melanocytes. *Environmental Health Perspectives* 80:49–60.

Bennett D.C. (1993) Genetics, development, and malignancy of melanocytes. *International Review of Cytology* 146:191–260.

Bennett D.C. (2008) How to make a melanoma: what do we know of the primary clonal events? *Pigment Cell and Melanoma Research* 21(1):27–38.

Bennett D.C., Huszar D., Laipis P.J., Jaenisch R. (1990) Phenotypic rescue of mutant brown melanocytes by a retrovirus carrying a wild-type tyrosinase-related protein gene. *Development* 110:471–475.

Benson K.F., Li F.Q., Person R.E., Albani D., Duan Z., Wechsler J., Meade-White K., Williams K., Acland G.M., Niemeyer G. et al. (2003) Mutations associated with neutropenia in dogs and humans disrupt intracellular transport of neutrophil elastase. *Nature Genetics* 35(1):90–96.

Berens W., Van Den Bossche K., Yoon T.-J., Westbroek W., Valencia J.C., Out C.J., Naeyaert J.M., Hearing V.J., Lambert J. (2005) Different approaches for assaying melanosome transfer. *Pigment Cell Research* 18(5):370–381.

Bernex F., De Sepulveda P., Kress C., Elbaz C., Delouis C., Panthier J.J. (1996) Spatial and temporal patterns of c-kit-expressing cells in WlacZ/+ and WlacZ/WlacZ mouse embryos. *Development* 122(10):3023–3033.

Berryere T.G., Schmutz S.M., Schimpf R.J., Cowan C.M., Potter J. (2003) TYRP1 is associated with dun coat colour in Dexter cattle or how now brown cow? *Animal Genetics* 34(3):169–175.

Berryere T.G., Kerns J.A., Barsh J.A., Schmutz S.M. (2005) Association of an Agouti allele with fawn or sable coat color in domestic dogs. *Mammalian Genome* 16:262–272.

Berryere T.G., Oulmouden A., Schmutz S.M. (2007) *Genetic studies of coat colors and length in Highland cattle.* Plant and Animal Genomes XV Conference, 13–17 January 2007, San Diego, CA.

Bertolotto C., Abbe P., Hemesath T.J., Bille K., Fisher D.E., Ortonne J.P., Ballotti R. (1998) Microphthalmia gene product as a signal transducer in cAMP-induced differentiation of melanocytes. *Journal of Cell Biology* 142(3):827–835.

Bharti K., Nguyen M.-T.T., Skuntz S., Bertuzzi S., Arnheiter H. (2006) The other pigment cell: specification and development of the pigmented epithelium of the vertebrate eye. *Pigment Cell Research* 19:380–394.

Bharti K., Liu W., Csermely T., Bertuzzi S., Arnheiter H. (2008) Alternative promoter use in eye development: complex role and regulation of the transcription factor MITF. *Development* 135(6):1169–1178.

Bismuth K., Skuntz S., Hallsson J.H., Pak E., Dutra A.S., Steingrimsson E., Arnheiter H. (2008) An unstable targeted allele of the mouse Mitf gene with a high somatic and germline reversion rate. *Genetics* 178(1):259–272.

Blake J.A., Ziman M.R. (2005) PAX3 transcripts in melanoblast development. *Development, Growth & Differentiation* 47(9):627–635.

Blewitt M.E., Vickaryous N.K., Paldi A., Koseki H., Whitelaw E. (2006) Dynamic reprogramming of DNA methylation at an epigenetically sensitive allele in mice. *PloS Genetics* 2(4)e49:0399–0405.

Blokzijl A., Dahlqvist C., Reissmann E., Falk A., Moliner A., Lendahl U., Ibanez C.F. (2003) Cross-talk between the Notch and TGF-β signaling pathways mediated by interaction of the Notch intracellular domain with Smad3. *Journal of Cell Biology* 163(4):723–728.

Boissy R.E., Lamoreux M.L. (1988) Animal models of an acquired pigmentary disorder – vitiligo. *Progress in Clinical and Biological Research* 256:207–218.

Boissy R.E., Spritz R.A. (2009) Frontiers and controversies in the pathobiology of vitiligo: separating the wheat from the chaff. *Experimental Dermatology* 18(7):583–585.

Boissy R.E., Zhao H., Oetting W.S., Austin L.M., Wildenberg S.C., Boissy Y.L., Zhao Y., Sturm R.A., Hearing V.J., King R.A., Nordlund J.J. (1996) Mutation in and lack of expression of tyrosinase-related protein-1 (TRP-1) in melanocytes from an individual with brown oculocutaneous albinism: a new subtype of albinism classified as 'OCA3.' *American Journal of Human Genetics* 58:1145–1156.

Boissy R.E., Sakai C., Zhao H., Kobayashi T., Hearing V.J. (1998) Human tyrosinase related protein-1 (TRP-1) does not function as a DHICA oxidase activity in contrast to murine TRP-1. *Experimental Dermatology* 7(4):198–204.

Boissy R.E., Huizing M., Gahl W.A. (2006) Biogenesis of melanosomes. In: Nordlund J.J. et al. (eds) *The Pigmentary System*, 2nd edn, pp. 155–170. Blackwell Publishing, Oxford.

Bolande R.P. (1997) Neurocristopathy: its growth and development in 20 years. *Podiatr. Pathol. Lab. Med.* 17:1–25.

Bolós V., Peinado H., Pérez-Moreno M.A., Fraga M.F., Esteller M., Cano A. (2002) The transcription factor Slug represses E-cadherin expression and induces epithelial to mesenchymal transitions: a comparison with Snail and E47 repressors. *Journal of Cell Science* 116:499–511.

Bonilla C., Boxill L.A., Donald S.A, Williams T., Sylvester N., Parra E.J, Dios S., Norton H.L., Shriver M.D., Kittles R.A. (2005) The 8818G allele of the agouti signaling protein (ASIP) gene is ancestral and is associated with darker skin color in African Americans. *Human Genetics* 116(5):402–406.

Bondurand N., Pingault V., Goerich D.E., Lemort N., Sock E., Le Caignec C., Wegner M., Goossens M. (2000) Interaction between SOX10, PAX3 and MITF, three genes implicated in Waardenburg syndrome. *Human Molecular Genetics* 9:1907–1917.

Borovansky J., Horak V., Elleder M., Fortyn K., Smit N.P., Kolb A.M. (2005) Biochemical characterization of a new melanoma model – the minipig MeLiM strain. *Melanoma Research* 13(6):543–548.

Bosenberg M., Muthusamy V., Curley D.P., Wang Z., Hobbs C., Nelson B., Nogueira C., Horner J.W. 2nd, Depinho R., Chin L. (2006) Characterization of melanocyte-specific inducible Cre recombinase transgenic mice. *Genesis* 44(5):262–267.

Botchkareva N.V., Botchkarev V.A., Gilchrest B.A. (2003) Fate of melanocytes during development of the hair follicle pigmentary unit. *Journal of Investigative Dermatology Symposium Proceedings* 8(1):76–79.

Bottcher J.T., Niehrs C. (2005) Fibroblast growth factor signaling in vertebrate development. *Endocrine Reviews* 26(1):63–77.

Bouillet P., Cory S., Zhang L.C., Strasser A., Adams J.M. (2001) Degenerative disorders caused by Bcl-2 deficiency prevented by loss of its BH3-only antagonist Bim. *Developmental Cell* 5:645–653.

Bowling A.T., Ruvinsky A. (ed.). (2000) *The Genetics of the Horse*. CABI Publishing, Wallingford.

Bradley B.J., Mundy N.I. (2008) The primate palette: the evolution of primate coloration. *Evolutionary Anthropology* 17:97–111.

Brannan C.I., Lyman S.D., Williams D.E., Eisenman J., Anderson D.M., Cosman D., Bedell M.A., Jenkins N.A., Copeland N.G. (1991) Steel-Dickie mutation encodes a c-kit ligand lacking transmembrane and cytoplasmic domains. *Proceedings of the National Academy of Sciences USA* 88:4671–4674.

Branza-Nichita N., Petrescu A.J., Dwek R.A., Wormald M.R., Platt F.M., Petrescu S.M. (1999) Tyrosinase folding and copper loading in vivo: a crucial role for calnexin and alpha-glucosidase II. *Biochemical and Biophysical Research Communications* 261(3):720–725.

Brewer S., Feng W., Huang J., Sullivan S., Williams T. (2004) Wnt1-Cre-mediated deletion of AP-2alpha causes multiple neural crest-related defects. *Developmental Biology* 267(1):135–152.

Brilliant M.H. (2001) The mouse p (pink-eyed dilution) and human P genes, oculocutaneous albinism type 2 (OCA2), and melanosomal pH. *Pigment Cell Research* 14(2):86–93.

Brilliant M.H. (2006) Molecular regulation of melanin formation: melanosome transporter proteins. In: Nordlund J.J. et al. (eds) *The Pigmentary System*, 2nd edn, pp. 230–241. Blackwell Publishing, Oxford.

Brito F.C., Kos L. (2008)Timeline and distribution of melanocyte precursors in the mouse heart. *Pigment Cell and Melanoma Research* 21:464–470.

Bronson S.K., Smithies O. (1994) Altering mice by homologous recombination using embryonic stem cells. *Journal of Biological Chemistry* 269(44):27155–27158.

Brooks B.P., Larson D.M., Chan C.-C., Kjellstrom S., Smith R.S., Crawford M.A., Lamoreux L., Huizing M., Hess R., Jiao X. et al. (2007) Analysis of ocular hypopigmentation in Rab38 *cht/cht* mice. *Investigative Ophthalmology & Visual Science* 48(9):3905–3913.

Brunberg E., Andersson L., Cothran G., Sandberg K., Mikko S., Lindgren G. (2006) A missense mutation in PMEL17 is associated with the silver coat color in the horse. *BMC Genetics* 7:46.

Budd P.S., Jackson I.J. (1995) Structure of the mouse tyrosinase-related protein-2 / dopachrome tautomerase (Tyrp2 / Dct) gene and sequence of two novel slaty alleles. *Genomics* 29(1):35–43.

Bultmann S.J., Michaud E.J., Woychik R.P. (1992) Molecular characterization of the mouse *agouti* locus. *Cell* 71(7):1195–1204.

Burdo J., Dargusch R., Schubert D. (2006) Distribution of the cystine / glutamate antiporter system xc- in the brain, kidney, and duodenum. *Journal of Histochemistry and Cytochemistry* 54(5):549–557.

Burstyn-Cohen T., Stanleigh J., Sela-Donenfeld D., Kalcheim C. (2004) Canonical Wnt activity regulates trunk neural crest delamination linking BMP/noggin signaling with G1/S transition. *Development* 131:5327–5339.

Burt D.W. (2006) The chicken genome. *Genome Dynamics* 2:123–137.

Burt D.W., Bumstead N., Bitgood J.J., Ponce de Leon F.A., Crittenden L.B. (1995) Chicken genome mapping: a new era in avian genetics. *Trends in Genetics* 11(5):190–194.

Buscà R., Ballotti R. (2000) Cyclic AMP a key messenger in the regulation of skin pigmentation. *Pigment Cell Research* 13:60–69.

Busch C., Oppitz M., Sailer M.H., Just L., Metzger M., Drews U. (2006) BMP-2-dependent integration of adult mouse subventricular stem cells into the neural crest of chick and quail embryos. *Journal of Cell Science* 119:4467–4474.

Butz S., Larue L. (1995) Expression of catenins during mouse embryonic development and in adult tissues. *Cell Adhesion and Communication* 3(4):337–352.

Byers H.R. (2006) Melanosome processing in keratinocytes. In: Nordlund J.J. et al. (eds) *The Pigmentary System*, 2nd edn, pp. 181–190. Blackwell Publishing, Oxford.

Cable J., Steel P. (1998) Combined cochleo-saccular and neuroepithelial abnormalities in the *Varitint-waddler-J* (Va^J) mouse. *Hearing Research* 123(1–2):125–136.

Cable J., Barkway C., Steel K.P. (1992) Characteristics of stria vascularis melanocytes of *viable dominant spotting* (W^v/W^v) mouse mutants. *Hearing Research* 64(1):6–20.

Cable J., Jackson I.J., Steel K.P. (1995) Mutations at the W locus affect survival of neural crest-derived melanocytes in the mouse. *Mechanisms of Development* 50:139–150.

Cadigan K.M., Nusse R. (1997) Wnt signaling: a common theme in animal development. *Genes & Development* 11:3286–3305.

Cadigan K.M., Liu Y.I. (2006) Wnt signaling: complexity at the surface. *Journal of Cell Science* 119:395–402.

Camacho-Hubner A., Beermann F. (2001) Increased transgene expression by the mouse tyrosinase enhancer is restricted to neural crest-derived pigment cells. *Genesis* 29(4):180–187.

Candille S.I., Van Raamsdonk C.D., Chen C., Kuijper S., Chen-Tsai Y., Russ A., Meijlink F., Barsh G.S. (2004) Dorsoventral patterning of the mouse coat by Tbx15. *PLoS Biology* (1):0030–0042.

Candille S.I., Kaelin C.B., Cattanach B.M., Yu B., Thompson D.A., Nix M.A., Kerns J.A., Schmutz S.M., Millhauser G.L., Barsh G.S. (2007) A β-Defensin mutation causes black coat color in domestic dogs. *Science* 318:1418–1423.

Cano A., Perez-Moreno M.A., Rodrigo I., Locascio A., Blanco M.J., del Barrio M.G., Portillo F., Nieto M.A. (2000) The transcription factor Snail controls epithelial-mesenchymal transitions by repressing E-cadherin expression. *Nature Cell Biology* 2:76–83.

Capecchi M.R. (1989) The new mouse genetics: altering the genome by gene targeting. *Trends in Genetics* 5(3):70–76.

Carim L., Sumoy L., Andreu N., Estivill X., Escarceller M. (2000) Cloning, mapping and expression analysis of VPS33B, the human orthologue of rat Vps33b. *Cytogenetics and Cell Genetics* 89(1–2):92–95.

Carmona-Fontaine C., Mathews H.L., Kuriyama S., Moreno M., Dunn G.A., Parsons M., Stern C.D., Mayor R. (2008) Contact inhibition of locomotion in vivo controls neural crest directional migration. *Nature* 456:957–961.

Carver E.A., Jiang R., Lan Y., Oram K.O., Gridley T. (2001) The mouse Snail gene encodes a key regulator of the epithelial-mesenchymal transition. *Molecular and Cellular Biology* 21:8184–8188.

Cassady J.L., Sturm R.A. (1994) Sequence of the human dopachrome tautomerase-encoding TRP-2 cDNA. *Gene* 143(2):295–298.

Castle W.E. (1924) On the occurrence in rabbits of linkage in inheritance between albinism and brown pigmentation. *Proceedings of the National Academy of Sciences USA* 10(12):486–488.

Castle W.E., Allen G.M. (1903) The heredity of albinism. *Proceedings of the National Academy of Sciences USA* 38:603–622.

Castle W.E., Little C.C. (1909) The peculiar inheritance of pink eyes among colored mice. *Science* 30:313–314.

Castle W.E., Little C.C. (1910) On a modified mendelian ratio among yellow mice. *Science* 32:868–870.

Castle W.E., King F.L. (1951) New evidence on the genetics of the Palomino horse. *Journal of Heredity* 42:60–64.

Castle W.E., Singleton W.R. (1961) The palomino horse. *Genetics* 46:1143–1150.

Cecchi C., Biasotto M., Tosi M., Avner P. (1997) The mottled mouse as a model for human Menkes disease: identification of mutations in the Atp7a gene. *Human Molecular Genetics* 6(5):829.

Cerdá-Reverter J.M., Haitina T., Schiöth H.B., Peter R.E. (2005) Gene structure of the goldfish agouti-signaling protein: a putative role in the dorsal-ventral pigment pattern of fish. *Endocrinology* 146(3):1597–1610.

Challis B.G., Coll A.P., Yeo G.S., Pinnock S.B., Dickson S.L., Thresher R.R., Dixon J., Zahn D., Rochford J.J., White A. et al. (2004) Mice lacking pro-opiomelanocortin are sensitive to high-fat feeding but respond normally to the acute anorectic effects of peptide-YY(3–36). *Proceedings of the National Academy of Sciences USA* 101(13):4695–4700.

Chang B., Smith R.S., Hawes N.L., Anderson M.G., Zabaleta A., Savinova O., Roderick T.H., Heckenlively J.R., Davisson M.T., John S.W. (1999) Interacting loci cause severe iris atrophy and glaucoma in DBA/2J mice. *Nature Genetics* 21(4):405–409.

Chase H.B. (1951) Critical stages of hair development and pigmentation in the mouse. *Physiological Zoology* 24:1–8.

Chase H.B., Rauch H. (1950) Greying of hair. II. Response of individual hairs in mice to variations in x-radiation. *Journal of Morphology* 87:381–392.

Chen K., Manga P., Orlow S.J. (2002) Pink-eyed dilution protein controls the processing of tyrosinase. *Molecular Biology of the Cell* 13(6):1953–1964.

Chen K., Minwalla L., Ni L., Orlow S.J. (2004) Correction of defective early tyrosinase processing by bafilomycin A1 and monensin in pink-eyed dilution melanocytes. *Pigment Cell Research* 17(1):36–42.

Chi N., Epstein J.A. (2002) Getting your Pax straight: Pax proteins in development and disease. *Trends in Genetics* 18(1):41–47.

Chiao E., Fisher P., Crisponi L., Deiana M., Dragatsis I., Schlessinger D., Pilia G., Efstratiadis A. (2002) Overgrowth of a mouse model of the Simpson-Golabi-Behmel syndrome is independent of IGF signaling. *Developmental Biology* 243(1):185–206.

Chin L., Pomerantz J., Polsky D., Jacobson M., Cohen C., Cordon-Cardo C., Horner J.W. 2nd., DePinho R.A. (1997) Cooperative effects of INK4a and ras in melanoma susceptibility in vivo. *Genes & Development* 11(21):2822–2834.

Chintala S., Li W., Lamoreux M.L., Ito S., Wakamatsu K., Sviderskaya E.V., Bennett D.C., Park Y.-M., Gahl W.A., Huizing M., Spritz R.A. et al. (2005) Slc7a11 gene controls production of pheomelanin pigment and proliferation of cultured cells. *Proceedings of the National Academy of Sciences USA* 102(31):10964–10969.

Chiu E., Lamoreux M.L., Orlow S.J. (1993) Postnatal ocular expression of tyrosinase and related proteins: disruption by the pink-eyed unstable (p^{un}) mutation. *Experimental Eye Research* 57(3):301–305.

Chowdhary B.P., Raudsepp T. (2006) The horse genome. *Genome Dynamics* 2:97–110.

Chow R.L., Lang R.A. (2001) Early eye development in vertebrates. *Annual Review of Cell and Developmental Biology* 17:255–296.

Chowdhary B.P., Raudsepp T. (2006) The horse genome. *Genome Dynamics* 2:97–110.

Ciciotte S.L., Gwynn B., Moriyama K., Huizing M., Gahl W.A., Bonifacino J.S., Peters L.L. (2003) *Cappuccino*, a mouse model of Hermansky-Pudlak syndrome, encodes a novel protein that is part of the pallidin-muted complex (BLOC-1). *Blood* 101(11):4402–4407.

Clark L.A., Starr A.N., Tsai K.L., Murphy K.E. (2006) Retrotransposon insertion in silver is responsible for merle patterning of the domestic dog. *Proceedings of the National Academy of Sciences USA* 103(5):1376–1381.

Clark L.A., Wahl J.M., Rees C.A., Murphy K.E. (2008) Genome-wide linkage scan localizes the harlequin locus in the Great Dane to chromosome 9. *Gene* 418(1–2):49–52.

Clark V.M. (1986) The cell biology of the retinal pigment epithelium. In: *The Retina: A Model for Cell Biology Studies*, part II, Ruben A., Farber D. (eds), pp. 129–168. Academic Press, Orlando, FL.

Clouthier D.E., Hosada K., Richardson J.A., Williams S.C., Yanagisawa H., Kuwaki T., Kumada M., Hammer R.E., Yanagisawa M. (1998) Cranial and cardiac neural crest defects in endothelin- A receptor-deficient mice. *Development* 125:813–824.

Cook A.L., Sturm R.A. (2008) POU domain transcription factors: BRN2 as a regulator of melanocytic growth and tumourigenesis. *Pigment Cell and Melanoma Research* 21:611–626.

Cook A.L., Donatien P.D., Smith A.G., Murphy M., Jones M.K., Herlyn M., Bennett D.C., Leonard H., Sturm R.A. (2003) Human melanoblasts in culture: expression of BRN2 and synergistic regulation by fibroblast growth factor-2, stem cell factor, and endothelin-3. *Journal of Investigative Dermatology* 121:1150–1159.

Cook D., Brooks S., Bellone R., Bailey E. (2008) Missense mutation in exon 2 of SLC36A1 responsible for champagne dilution in horses. *PLoS Genetics* 4(9):e1000195.

Correia A.C., Costa M., Moraes F., Bom J., Novoa A., Mallo M. (2007) Bmp2 is required for migration but not for induction of neural crest cells in the mouse. *Developmental Dynamics* 236(9):2493–2501.

Cortese K., Giordano F., Surace F.M., Venturi C., Ballabio A., Tacchetti C., Marigo V. (2005) The ocular albinism type 1 (OA1) gene controls melanosome maturation and size. *Investigative Ophthalmology & Visual Science* 46:4358–4364.

Costin G.E., Valencia J.C., Vieira W.D., Lamoreux M.L., Hearing V.J. (2003) Tyrosinase processing and intracellular trafficking is disrupted in mouse primary melanocytes carrying the *underwhite (uw)* mutation. A model for oculocutaneous albinism (OCA) type 4. *Journal of Cell Science* 116(Pt 15):3203–3212.

Costin G.-E., Valencia J.C., Wakamatsu K., Ito S., Solano F., Milac A.L., Vieira W.D., Yamaguchi Y., Rouzaud F., Petrescu A.-J., Lamoreux M.L., Hearing V.J. (2005) Mutations in dopachrome tautomerase (Dct) affect eumelanin/pheomelanin synthesis, but do not affect intracellular trafficking of the mutant protein. *Biochemical Journal* 391:249–259.

Crouch M.L. (1990) Debating the responsibilities of plant scientists in the decade of the environment. *Plant Cell* 2:275–277.

Cuénot L. (1902) La loi de Mendel et l'hérédité de la pigmentation chez les souris. *Arch. Zool. Exp. Gen. Series 3* 10:xxvii–xxx.

Cunliffe P., Reed V., Boyd Y. (2001) Intragenic deletions at Atp7a in mouse models for Menkes disease. *Genomics* 74(2):155–162.

Czyzyk T.A., Sikorski M.A., Yang L., McKnight G.S. (2008) Disruption of the RIIbeta subunit of PKA reverses the obesity syndrome of Agouti lethal yellow mice. *Proceedings of the National Academy of Sciences USA* 105(1):276–281.

Dahlqvist C., Blokzijll A., Chapman G., Falk A., Dannaeus K., Ibanez C.F., Lendahl U. (2003) Functional Notch signaling is required for BMP4-induced inhibition of myogenic differentiation. *Development* 130:6089–6099.

Dakubo G.D., Mazerolle C., Furimsky M., Yu C., St-Jacques B., McMahon A.P., Wallace V.A. (2008) Indian hedgehog signaling from endothelial cells is required for sclera and retinal pigment epithelium development in the mouse eye. *Developmental Biology* 320:242–255.

Das A.V., James J., Zhao X., Rahnenfuhrer J., Ahmad I. (2004) Identification of c-kit receptor as a regulator of adult neural stem cells in the mammalian eye: interactions with Notch signaling. *Developmental Biology* 273:87–105.

Das A.V., James J., Bhattacharya S., Imbalzano A.N., Antony M.L., Hegde G., Zhao X., Mallya K., Ahmad F., Knudsen E., Ahmad I. (2007) SWI/SNF chromatin remodeling ATPase, BRM regulates the differentiation of early retinal stem cells/progenitors by influencing Brn3B expression and Notch signaling. *Journal of Biological Chemistry* 282(48):35187–35201.

DasGupta R., Fuchs E. (1999) Multiple roles for activated LEF/TCF transcription complexes during hair follicle development and differentiation. *Development* 126:4557–4568.

Davidson E.H. (1990) How embryos work: a comparative view of diverse modes of cell fate specification. *Development* 108:365–389.

De Bie T., Tranchevent L.C., van Oeffelen L.M., Moreau Y. (2007) Kernel-based data fusion for gene prioritization. *Bioinformatics* 23(13):i125–i132.

Delanet M., Duband J.-L. (1992) Transforming growth factor-β control of cell-substratum adhesion during avian neural crest cell emigration in vitro. *Development* 116:275–287.

Delfgaauw J., Duschl J., Wellbrock C., Froschauer C., Schartl M., Altschmied J. (2003) MITF-M plays an essential role in transcriptional activation and signal transduction in *Xiphophorus* melanoma. *Gene* 320:117–126.

Dell'Angelica E.C., Klumperman J., Stoorvogel W., Bonifacino J.S. (1998) Association of the AP-3 adaptor complex with clathrin. *Science* 280(5362):431–434.

Delmas V., Martinozzi S., Bourgeois Y., Holzenberger M., Larue L. (2003) Cre-mediated recombination in the skin melanocyte lineage. *Genesis* 36(2):73–80.

Delmas V., Beermann F., Martinozzi S., Carreira S., Ackermann J., Kumasaka M., Denat L., Goodall J.,Luciani F., Viros A. et al. (2007) Beta-catenin induces immortalization of melanocytes by suppressing p16INK4a expression and cooperates with N-Ras in melanoma development. *Genes & Development* 21(22):2923–2935.

De Melker A.A., Desban N., Duband J.L. (2004) Cellular localization and signaling activity of beta-catenin in migrating neural crest cells. *Developmental Dynamics* 230(4):708–726.

Deol M.S. (1970) The relationship between abnormalities of pigmentation and of the inner ear. *Proceedings of the Royal Society of London Series B Biological Sciences* 175(39):201–217.

D'Erchia A.M., Gissi C., Pesole G., Saccone C., Arnason U. (1996) The guinea-pig is not a rodent. *Nature* 381(6583):597–600.

De Strooper B., Annaert W. (2001) Where Notch and Wnt signaling meet. The presenilin hub. *Journal of Cell Biology* 152(4):F17–20 [Comment on: *Journal of Cell Biology* (2001) 152(4):785–794].

Detter J.C., Zhang Q., Mules E.H., Novak E.K., Mishra V.S., Li W., McMurtrie E.B., Tchernev V.T., Wallace M.R., Seabra M.C., Swank R.T., Kingsmore S.F. (2000) Rab geranylgeranyl transferase alpha mutation in the gunmetal mouse reduces Rab prenylation and platelet synthesis. *Proceedings of the National Academy of Sciences USA* 97:4144–4149.

Dickie M.M. 1954. The tortoiseshell house mouse. *Journal of Heredity* 45:158–139.

Dickie M.M. (1964) *Underwhite, gunmetal. Mouse News Letter* 30:30.

Dickinson M.E., Selleck M.A.J., McMahon A.P., Bonner-Fraser M. (1995) Dorsalization of the neural tube by the non-neural ectoderm. *Development* 121:2099–2106.

Dinulescu D.M., Fan W., Boston B.A., McCall K., Lamoreux M.L., Moore K.J., Montagno J., Cone R.D. (1998) *Mahogany (mg)* stimulates feeding and increases basal metabolic rate independent of its suppression of agouti. *Proceedings of the National Academy of Sciences USA* 95(21):12707–12712.

Dolinoy D.C. (2008) The agouti mouse model: an epigenetic biosensor for nutritional and environmental alterations on the fetal epigenome. *Nutrition Reviews* 66(Suppl 1):S7–S11.

Dolinoy D.C., Weidman J.R., Waterland R.A., Jirtle R.L. (2006) Maternal genistein alters coat color and protects A^{vy} mouse offspring from obesity by modifying the fetal epigenome. *Environmental Health Perspectives* 114:567–572.

Dong S., Leung K.K., Pelling A.L., Lee P.Y., Tang A.S., Heng H.H., Tsui L.C., Tease C., Fisher G., Steel K.P., Cheah K.S. (2002) Circling, deafness, and yellow coat displayed by yellow submarine (ysb) and light coat and circling (lcc) mice with mutations on chromosome 3. *Genomics* 79(6):777–784.

Dorsky R.I., Moon R.T., Raible D.W. (1998) Control of neural crest cell fate by the Wnt signalling pathway. *Nature* 396(6709):370–373.

Dorsky R.I., Raible D.W., Moon R.T. (2000) Direct regulation of nacre, a zebrafish MITF homolog required for pigment cell formation, by the Wnt pathway. *Genes & Development* 14(2):158–162.

Doucet S.M., Schawkey M.D., Rathburn M.K., Mays H.L. Jr, Montgomerie R. (2004) Concordant evolution of plumage colour, feather microstructure and melanocortin receptor gene between mainland and island population of a fairy-wren. *Proceedings of the Royal Society of London Series B Biological Sciences* 27:1663–1670.

Du J., Fisher D.E. (2002) Identification of Aim-1 as the underwhite mouse mutant and its transcriptional regulation by MITF. *Journal of Biological Chemistry* 277(1):402–406.

Duhl D.M.J., Stevens M.E., Vrieling H., Saxon P.J., Miller M.W., Epstein C.J., Barsh G.S. (1994) Pleiotropic effects of the mouse *lethal yellow* (A^y) mutation explained by deletion of a maternally expressed gene and the simultaneous production of agouti fusion RNAs. *Development* 120:1695–1708.

Duncan A.W., Rattis F.M., DiMascio L.N., Congdon K.L., Pazianos G., Zhao C., Yoon K., Cook J.M., Willert K., Gaiano N., Reya T. (2005) Integration of Notch and Wnt signaling in hematopoietic stem cell maintenance. *Nature Immunology* 6:314–322.

Dunn K.J., Williams B.O., Li Y., Pavan W.J. (2000) Neural crest-directed gene transfer demonstrates Wnt1 role in melanocyte expansion and differentiation during mouse development. *Proceedings of the National Academy of Sciences USA* 97(18):10050–10055.

Dunn K.J., Brady M., Ochsenbauer-Jambor C., Snyder S., Incao A., Pavan W.J. (2005) WNT1 and WNT3a promote expansion of melanocytes through distinct modes of action. *Pigment Cell Research* 18:167–180.

Dunn L.C., Thigpen L.W. (1930) The silver mouse, a recessive color variation. *Journal of Heredity* 21:495–498.

Dupin E., Le Douarin N.M. (2003) Development of melanocyte precursors from the vertebrate neural crest. *Oncogene* 22(20):3016–3023.

Dupin E., Baroffio A., Dulac C., Cameron-Curry P., Le Douarin N.M. (1990) Schwann-cell differentiation in clonal cultures of the neural crest, as evidenced by the anti-Schwann cell myelin protein monoclonal antibody. *Proceedings of the National Academy of Sciences USA* 87:1119–1123.

Duttlinger R., Manova K., Chu T.Y., Gyssler C., Zelenetz A.D., Bachvarova R.F., Besmer P. (1993) W-sash affects positive and negative elements controlling c-kit expression: ectopic c-kit expression at sites of kit-ligand expression affects melanogenesis. *Development* 118:705–717.

Duttlinger R., Manova K., Berrozpe G., Chu T.Y., DeLeon V., Timokhina L., Chaganti R.S., Zelenetz A.D., Bachvarova R.F., Besmer P. (1995) The W^{sh} and Ph mutations affect the c-kit expression profile: c-kit misexpression in embryogenesis impairs melanogenesis in W^{sh} and Ph mutatn mice. *Proceedings of the National Academy of Sciences USA* 92:3754–3759.

Dwinell M.R., Worthey E.A., Shimoyama M., Bakir-Gungor B., DePons J., Laulederkind S., Lowry T., Nigram R., Petri V., Smith J. et al., and the RGD Team (2008) The Rat Genome Database 2009: variation, ontologies and pathways. *Nucleic Acids Research* 2008:1–6.

Dymecki S.M. (1996) Flp recombinase promotes site-specific DNA recombination in embryonic stem cells and transgenic mice. *Proceedings of the National Academy of Sciences USA* 93(12):6191–6196.

Easty D.J., Bennett D.C. (2000) Protein tyrosine kinases in malignant melanoma. *Melanoma Research* 10(5):401–411.

Eisinger M., Marko O. (1982) Selective proliferation of normal human melanocytes in vitro in the presence of phorbol ester and cholera toxin. *Proceedings of the National Academy of Sciences USA* 79:2018–2022.

Eizirik E., Yuhki N., Johnson W.E. Menotti-Raymond M., Hannah S.S., O'Brien S.J. (2003) Molecular genetics and evolution of melanism in the cat family. *Current Biology* 13:448–453.

Enshell-Seijffers D., Lindon C., Morgan B.A. (2008) The serine protease corin is a novel modifier of the agouti pathway. *Development* 135(2):217–225.

Erickson C.A. (1985) Control of neural crest cell dispersion in the trunk of the avian embryo. *Developmental Biology* 111:138–157.

Erickson C.A., Weston J.A. (1983) An SEM analysis of neural crest migration in the mouse. *Journal of Embryology and Experimental Morphology* 74:97–118.

Erickson C.A., Goins T.L. (1995) Avian neural crest cells can migrate in the dorsolateral path only if they are specified as melanocytes. *Development* 121(3):915–924.

Erickson C.A., Reedy M.V. (1998) Neural crest development: the interplay between morphogenesis and cell differentiation. *Current Topics in Developmental Biology* 40:177–209.

Erickson C.A., Duong T.D., Tosney K.W. (1992) Descriptive and experimental analysis of the dispersion of neural crest cells along the dorsolateral path and their entry into ectoderm in the chick embryo. *Developmental Biology* 151:251–272.

Everts R.E., Rothuizen J., van Oost B.A. (2000) Identification of a premature stop codon in the melanocyte-stimulating hormone receptor gene (MC1R) in Labrador and Golden retrievers with yellow coat colour. *Animal Genetics* 31(3):194–199.

Falconer D.S. (1956) Viable and fertile Brindled male, and new evidence for the allelism of the sex-linked genes, Brindled and Mottled. *Mouse News Letter* 15:24–25.

Falconer D.S., Isaacson J.H. (1972) Sex-linked variegation modified by selection in brindled mice. *Genetic Research (Cambridge)* 20:291–316.

Feng L., Seymour A.B., Jiang S., To A., Peden A.A., Novak E.K., Zhen L., Rusiniak M.E., Eicher E.M., Robinson M.S. et al. (1999) The beta3A subunit gene (Ap3b1) of the AP-3 adaptor complex is altered in the mouse hypopigmentation mutant *pearl*, a model for Hermansky-Pudlak syndrome and night blindness. *Human Molecular Genetics* 8(2):323–330.

Fitzpatrick T.B., Breathnach A.S. (1963) The epidermal melanin unit system. *Dermatologische Wochenschrift* 147:481–489 [in German].

Foster M. (1965) Mammalian pigment genetics. *Advances in Genetics* 13:311–339.

Frost-Mason S.K., Mason K.A. (1996) What insights into vertebrate pigmentation has the axolotl model system provided? *International Journal of Developmental Biology* 40(4):685–693.

Fu M., Lui V.C.H., Sham M.H., Pachnis V., Tam P.K.H. (2004) Sonic hedgehog regulates the proliferation, differentiation, and migration of enteric neural crest cells in gut. *Journal of Cell Biology* 166(5):673–684.

Fuchs E., Tumbar T., Guasch G. (2004) Socializing with the neighbors: stem cells and their niche. *Cell* 116:769–778.

Fukuda M, Kuroda TS. (2002) Slac2-c (synaptotagmin-like protein homologue lacking C2 domains-c), a novel linker protein that interacts with Rab27, myosin Va/VIIa, and actin. *Journal of Biological Chemistry* 277(45):43096–43103.

Fukuda M., Kuroda T.S., Mikoshiba K. (2002) Slac2-a/Melanophilin, the missing link between Rab27 and Myosin Va: implications of a tripartite protein complex for melanosome transport. *Journal of Biological Chemistry* 277(14):12432–12436.

Galbraith D.B. (1964) The agouti pigment pattern of the mouse: A quantitative and experimental study. *Journal of Experimental Zoology* 155:71–90.

Galceran J., Farias I., Depew M.J., Clevers H., Grosschedl R. (1999) Wnt3a -/- like phenotype and limb deficiency in Lef1 -/- Tcf1 -/- mice. *Genes & Development* 13(6):709–717.

Galli S.J., Tsai M., Wershil B.K. (1993) The c-kit receptor, stem cell factor, and mast cells. What each is teaching us about the others. *American Journal of Pathology* 142(4):965–974.

Galli S.J., Zsebo K.M., Geissler E.N. (1994) The kit ligand, stem cell factor. *Advances in Immunology* 55:1–96.

Gans C., Northcutt R. (1983) Neural crest and the origin of vertebrates, a new head. *Science* 220:268–274.

Ganss R., Montoliu L., Monaghan A.P., Schutz G. (1994) A cell-specific enhancer far upstream of the mouse tyrosinase gene confers high level and copy number-related expression in transgenic mice. *EMBO Journal* 13(13):3083–3093.

Garcia R.J., Ittah A., Mirabal S., Figueroa J., Lopez L., Glick A.B., Kos L. (2008) Endothelin 3 induces skin pigmentation in a keratin-driven inducible mouse model. *Journal of Investigative Dermatology* 128(1):131–142.

García-Borrón J.C., Sánchez-Laorden B.L, Jiménez-Cervantes C. (2005) Melanocortin-1 receptor structure and functional regulation. *Pigment Cell Research* 18(6):393–410.

Garcia-Otin A.L., Guillou F. (2006) Mammalian genome targeting using site-specific recombinases. *Frontiers in Bioscience* 11:1108–1136.

Geissler E.N., Russell E.S. (1983a) Analysis of the hematopoietic effects of new *dominant spotting (W)* mutations of the mouse. I. Influence upon hematopoietic stem cells. *Experimental Hematology* 11(6):452–460.

Geissler E.N., Russell E.S. (1983b) Analysis of the hematopoietic effects of new dominant spotting (W) mutations of the mouse. II. Effects on mast cell development. *Experimental Hematology* 11(6):461–466.

Geissler E.N., McFarland E.C., Russell E.S. (1981) Analysis of pleiotropism at the *dominant white-spotting (W)* locus of the house mouse: a description of ten new *W* alleles. *Genetics* 97(2):337–361.

Geschwind I.I. (1966) Change in hair color in mice induced by injection of alpha-MSH. *Endocrinology* 79(6):1165–1167.

Hearn C.J., Murphy M., Newgreen D. (1998) GDNF and ET-3 differentially modulate the numbers of avian enteric neural crest cells and enteric neurons *in vitro*. *Developmental Biology* 197:95–105.

Hecht A., Kemler R. (2000) Curbing the nuclear activities of beta-catenin. Control over Wnt target gene expression. *EMBO Reports* 11:24–28.

Hemesath T.J., Steingrímsson E., McGill G., Hansen M.J., Vaught J., Hodgkinson C.A., Arnheiter H., Copeland N.G., Jenkins N.A., Fisher D.E. (1994) Microphthalmia, a critical factor in melanocyte development, defines a discrete transcription factor family. *Genes & Development* 8(22):2770–2780.

Hemesath T.J., Price E.R., Takemoto C., Badalian T., Fisher D.E. (1998) MAP kinase links the transcription factor Microphthalmia to c-Kit signalling in melanocytes. *Nature* 391(6664):298–301.

Henion P.D., Weston J.A. (1997) Timing and pattern of cell fate restrictions in the neural crest lineage. *Development* 124:4351–4359.

Hershey C.L., Fisher D.E. (2004) Mitf and Tfe3: members of a b-HLH-ZIP transcription factor family essential for osteoclast development and function. *Bone* 34(4):689–696.

Heston W.E., Vlahakis G. (1962) Genetic obesity and neoplasia. *Journal of the National Cancer Institute* 29:197–209.

Heuckeroth R.O., Pachnis V. (2006) Getting to the guts of enteric nervous system development. *Development* 133:2287–2290.

Hida T., Wakamatsu K., Sviderskaya E.V., Donkin A.J., Montoliu L., Lamoreux M.L., Yu B., Millhauser G.L., Ito S., Barsh G.S., Jimbow K., Bennett D.C. (2009) Agouti protein, mahogunin and attractin in pheomelanogenesis and melanoblast-like alternation of melanocytes: a cAMP-independent pathway. *Pigment Cell and Melanoma Research* 22:623–634.

Hirai Y., Nose A., Kobayashi S., Takeichi M. (1989) Expression and role of E- and P-cadherin adhesion molecules in embryonic histogenesis. *Development* 105:271–277.

Hirobe T. (1984a) Histochemical survey of the distribution of the epidermal melanoblasts and melanocytes in the mouse during fetal and postnatal periods. *Anatomical Record* 208:589–594.

Hirobe T. (1984b) Effects of genic substitution at the brown locus on the differentiation of epidermal melanocytes in newborn mouse skin. *Anatomical Record* 209(4):425–432.

Hirobe T. (2001) Endothelins are involved in regulating the proliferation and differentiation of mouse epidermal melanocytes in serum-free primary culture. *Journal of Investigative Dermatology Symposium Proceedings* 6(1):25–31.

Hirobe T. (2004) Role of keratinocyte-derived factors involved in regulating the proliferation and differentiation of mammalian epidermal melanocytes. *Pigment Cell Research* 18:2–12.

Hirobe T. (2005) Role of keratinocyte-derived factors involved in regulating the proliferation and differentiation of mammalian epidermal melanocytes. *Pigment Cell Research* 18:2–12.

Hirobe T., Wakamatsu K., Ito S. (2003) Changes in the proliferation and differentiation of neonatal mouse pink-eyed dilution melanocytes in the presence of excess tyrosine. *Pigment Cell Research* 16:619–628.

Hirobe T., Osawa M., Nishikawa S.-I. (2004a) Hepatocyte growth factor controls the proliferation of cultured epidermal melanoblasts and melanocytes from newborn mice. *Pigment Cell Research* 17:51–61.

Hirobe T., Takeuchi S., Hotta E., Wakamatsu K., Ito S. (2004b) Pheomelanin production in the epidermis from newborn *agouti* mice is induced by expression of the *agouti* gene in the dermis. *Pigment Cell Research* 17:506–514.

Hitoshi S., Alexson T., Tropepe V., Donoviel D., Elia A.J., Nye J.S., Conlon R.A., Mak T.W., Bernstein A., van der Kooy D. (2002) Notch pathway molecules are essential for the maintenance but not the generation of mammalian neural stem cells. *Genes & Development* 16:846–858.

Hodgkinson C.A., Moore K.J., Nakayama A., Steingrimsson E., Copeland N.G., Jenkins N.A., Arnheiter H. (1993) Mutations at the mouse microphthalmia locus are associated with defects in a gene encoding a novel basic-helix-loop-helix-zipper protein. *Cell* 74(2):395–404.

Hoek K., Rimm D.L., Williams K.R., Zhao H., Ariyan S., Lin A., Kluger H.M., Berger A.J., Cheng E., Trombetta E.S. et al. (2004) Expression profiling reveals novel pathways in the transformation of melanocytes to melanomas. *Cancer Research* 64(15):5270–5282.

Hong L., Simon J.D., Sarna T. (2006) Melanin structure and the potential functions of uveal Melanosomes. *Pigment Cell Research* 19:465–466.

Höning S., Ignacio V., Sandoval I.V., von Figura K. (1998) A di-leucine-based motif in the cytoplasmic tail of LIMP-II and tyrosinase mediates selective binding of AP-3. *EMBO Journal* 17(5):1304–1314.

Honoré S.M., Aybar M.J., Mayor R. (2003) Sox10 is required for the early development of the prospective neural crest in *Xenopus* embryos. *Developmental Biology* 260(1):79–96.

Hosoda K., Hammer R.E., Richardson J.A., Baynash A.G., Cheung J.C., Giaid A., Yanagisawa M. (1994) Targeted and natural (piebald-lethal) mutations of endothelin-B receptor gene produce megacolon associated with spotted coat color in mice. *Cell* 79(7):1267–1276.

Hou L., Panthier J.-J., Arnheiter H. (2000) Signaling and transcriptional regulation in the neural crest-derived melanocyte lineage: interactions between KIT and MITF. *Development* 127:5379–5389.

Hou L., Pavan W.J., Shin M.K., Arnheiter H. (2004) Cell-autonomous and cell non-autonomous signaling through endothelin receptor B during melanocyte development. *Development* 131(14):3239–3247.

Hough R.B., Lengeling A., Bedian V., Lo C., Bucan M. (1998) Rump-white inversion in the mouse disrupts dipeptidyl aminopeptidase-like protein 6 and causes dysregulation of Kit expression. *Proceedings of the National Academy of Sciences USA* 95:13800–13805.

Hu D.-N., Simon J.D., Sarna T. (2008) Role of ocular melanin in ophthalmic physiology and pathology. *Photochemistry and Photobiology* 84:639–644.

Huang X., Saint-Jeannet J.-P. (2004) Induction of the neural crest and the opportunities of life on the edge. *Developmental Biology* 275:1–11.

Huber W.E., Price E.R., Widlund H.R., Du J., Davis I.J., Wegner M., Fisher D.E. (2003) A Tissue-restricted cAMP transcriptional response: SOX10 modulates α-melanocyte-stimulating hormone-triggered expression of microphthalmia-associated transcription factor in melanocytes. *Journal of Biological Chemistry* 278(46):45224–45230.

Huelsken J., Birchmeier W. (2001) New aspects of Wnt signaling pathways in higher vertebrates. *Current Opinion in Genetics and Development* 11(5):547–553.

Huelsken J., Vogel R., Brinkmann V., Erdmann B., Birchmeier C., Birchmeier W. (2000) Requirement for beta-catenin in anterior-posterior axis formation in mice. *Journal of Cell Biology* 148(3):567–578.

Huelsken J., Vogel R., Erdmann B., Cotsarelis G., Birchmeier W. (2001) β-Catenin controls hair follicle morphogenesis and stem cell differentiation in the skin. *Cell* 105(4):533–545.

Huizing M., Didier A., Walenta J., Anikster Y., Gahl W.A., Kramer H. (2001) Molecular cloning and characterization of human VPS18, VPS11, VPS16, and VPS33. *Gene* 264:241–247.

Huizing M., Helip-Wooley A., Westbroek W., Gunay-Aygun M., Gahl W.A. (2008) Disorders of lysosome-related organelle biogenesis: clinical and molecular genetics. *Annual Review of Genomics and Human Genetics* 9:359–386.

Hume A.N., Collinson L.M., Rapak A., Gomes A.Q., Hopkins C.R., Seabra M.C. (2001) Rab27a regulates the peripheral distribution of melanosomes in melanocytes. *Journal of Cell Biology* 152:795–808.

Hume A.N., Collinson L.M., Hopkins C.R., Strom M., Barral D.C., Bossi G., Griffiths G.M., Seabra M.C. (2002) The leaden gene product is required with Rab27a to recruit myosin Va to melanosomes in melanocytes. *Traffic* 3:193–202.

Hume A.N., Ushakov D.S., Tarafder A.K., Ferenczi M.A., Seabra M.C. (2007) Rab27a and MyoVa are the primary Mlph interactors regulating melanosome transport in melanocytes. *Journal of Cell Science* 120:3111–3122.

Hunt D.M. (1974) Primary defect in copper transport underlies mottled mutants in the mouse. *Nature* 249(460):852–854.

Hunt D.M., Johnson D.R. (1972a) An inherited deficiency in noradrenaline biosynthesis in the brindled mouse. *Journal of Neurochemistry* 19(12):2811–2819.

Hunt D.M., Johnson D.R. (1972b) Aromatic amino acid metabolism in brindled (Mobr) and viable-brindled (Movbr), two alleles at the mottled locus in the mouse. *Biochemical Genetics* 6(1):31–40.

Hurlbut G.D., Kankel M.W., Lake R.J., Artavanis-Tsakonas S. (2007) Crossing paths with Notch in the hyper-network. *Current Opinion in Cell Biology* 19(2):166–175.

Hustad C.M., Perry W.L., Siracusa L.D., Rasberry C., Cobb L., Cattanach B.M., Kovatch R., Copeland N.G., Jenkins N.A. (1995) Molecular genetic characterization of six recessive viable alleles of the mouse *agouti* locus. *Genetics* 140(1):255–265.

Huszar D., Sharpe A., Jaenisch R. (1991) Migration and proliferation of cultured neural crest cells in W mutant neural crest chimeras. *Development* 112:131–141.

Hutt F.B. (1949) *Genetics of the Fowl*. McGraw-Hill (re-published by Norton Creek Press, 2003).

Ibsen H.L. (1933) *Cattle inheritance*. I. *Color*. Kansas Agricultural Experiment Station.

Ikeya M., Lee S.M., Johnson J.E., McMahon A.P., Takada S. (1997) WNT signaling required for expansion of neural crest and CNS progenitors. *Nature* 389:966–970.

Inagaki K., Suzuki T., Shimizu H., Ishii N., Umezawa Y., Tada J., Kikuchi N., Takata M., Takamori K., Kishibe M. et al. (2004) Oculocutaneous albinism type 4 is one of the most common types of albinism in Japan. *American Journal of Human Genetics* 74(3):466–471.

Incerti B., Cortese K., Pizzigoni A., Surace E.M., Varani S., Coppola M., Jeffery G., Seeliger M., Jaissle G., Bennett D.C. et al. (2000) *Oa1* knock-out: new insights on the pathogenesis of ocular albinism type 1. *Human Molecular Genetics* 9(19):2781–2788.

Ingham P.W., McMahon A.P. (2001) Hedgehog signaling in animal development: paradigms and principles. *Genes & Development* 15:3059–3087.

Inman D.M., Sappington R.M., Horner P.J., Calkins D.J. (2006) Quantitative correlation of optic nerve pathology with ocular pressure and corneal thickness in the DBA/2 mouse model of glaucoma. *Investigative Ophthalmology & Visual Science* 47(3):986–996.

Inoue A., Seidel M.G., Wu W., Kamizono S., Ferrando A.A., Bronson R.T., Iwasaki H., Akashi K., Morimoto A., Hitzler J.K. et al. (2002) Slug, a highly conserved zinc finger transcriptional repressor protects hematopoietic progenitor cells from radiation-induced apoptosis *in vivo*. *Cancer Cell* 2(4):249–251.

Inoue-Narita T., Hamada K., Sasaki T., Hatakeyama S., Fujita S., Kawahara K., Sasaki M., Kishimoto H., Eguchi S., Kojima I. et al. (2008) Pten deficiency in melanocytes results in resistance to hair graying and susceptibility to carcinogen-induced melanomagenesis. *Cancer Research* 68(14):5760–5768.

Introne W., Boissy R.E., Gahl W.A. (1999) Clinical, molecular, and cell biological aspects of Chediak-Higashi syndrome. *Molecular Genetics and Metabolism* 68(2):283–303.

Ito K., Sieber-Blum M. (1991) *In vitro* clonal analysis of quail cardiac neural crest development. *Developmental Biology* 148:95–106.

Ito S., Wakamatsu K. (2003) Quantitative analysis of eumelanin and pheomelanin in humans, mice, and other animals: a comparative review. *Pigment Cell Research* 16(5):523–531.

Ito S., Wakamatsu K. (2006) Chemistry of melanins. In: Nordlund J.J. et al. (eds) *The Pigmentary System*, 2nd edn. Blackwell Publishing, Oxford.

Ito S., Wakamatsu K. (2008) Chemistry of mixed melanogenesis – pivotal roles of dopaquinone. *Photochemistry and Photobiology* 84(3):582–592.

Iwamoto T., Takahashi M., Ito M., Hamatani K., Ohbayashi M., Wajjwalku W., Isobe K., Nakashima I. (1991) Aberrant melanogenesis and melanocytic tumour development in transgenic mice that carry a metallothionein/ret fusion gene. *EMBO Journal* 10(11):3167–3175.

Jackson I.J. (1994) Molecular and developmental genetics of mouse coat color. *Annual Review of Genetics* 28:189–217.

Jackson I.J. (1999) The mahogany mouse mutation: further links between pigmentation, obesity and the immune system. *Trends in Genetics* 15(11):429–431.

Jackson I.J., Chambers D.M., Tsukamoto K., Copeland N.G., Gilbert D.J., Jenkins N.A., Hearing V. (1992) A second tyrosinase-related protein, TRP-2, maps to and is mutated at the mouse slaty locus. *EMBO Journal* 11(2):527–535.

Jackson I.J., Budd P.S., Keighren M., McKie L. (2007) Humanized MC1R transgenic mice reveal human specific receptor function. Human molecular genetics 16(19):2341–2348.

Jackson P.J., Douglas N.R., Chai B., Binkley J., Sidow A., Barsh G.S., Millhauser G.L. (2006) Structural and molecular evolutionary analysis of agouti and agouti-related proteins. *Chemical Biology* (12):1297–1305.

Jaenisch R. (1985) Mammalian neural crest cells participate in normal embryonic development on microinjection into postimplantation mouse embryos. *Nature* 318:181–183.

Jamora C., Lee P., Kocieniewski P., Ashar M., Hosokawa R., Chai Y., Fuchs E. (2005) A signaling pathway involving TGF-B2 and Snail in hair follicle morphogenesis. *PLOS Biology* 3(1):e11.

Jancic C., Savina A., Wasmeier C., Tolmachova T., El-Benna J., Dang P.M., Pascolo S., Gougerot-Pocidalo M.A., Raposo G., Seabra M.C., Amigorena S. (2007) Rab27a regulates phagosomal pH and NADPH oxidase recruitment to dendritic cell phagosomes. *Nature Cell Biology* 9(4):367–378.

Jara J.R., Aroca P., Solano F., Martinez J.H., Lozano J.A. (1988) The role of sulfhydryl compounds in mammalian melanogenesis: the effect of cysteine and glutathione upon tyrosinase and the intermediates of the pathway. *Biochimica et Biophysica Acta* 967(2):296–303.

Jeffery G. (2001a) The albino retina: an abnormality that provides insight into normal retinal development. *Trends in Neuroscience* 20(4):165–169.

Jeffery G. (2001b) Architecture of the optic chiasm and the mechanisms that sculpt its development. *Physiological Review* 81(4):1393–1414.

Jenkins N.A. (1998) The murine dilute suppressor gene dsu suppresses the coat-color phenotype of three pigment mutations that alter melanocyte morphology, d, ash and ln. *Genetics* 119(4):933–941.

Jenkins N.A., Copeland N.G., Taylor B.A., Lee B.K. (1981) *Dilute (d)* coat colour mutation of DBA/2J mice is associated with the site of integration of an ecotropic MuLV genome. *Nature* 293(5831):370–374.

Jiang R., Lan Y., Norton C.R., Sundberg J.P., Gridley T. (1998) The Slug gene is not essential for mesoderm or neural crest development in mice. *Developmental Biology* 198:277–285.

Jiménez M., Tsukamoto K., Hearing V.J. (1991) Tyrosinases from two different loci are expressed by normal and by transformed melanocytes. *Journal of Biological Chemistry* 266(2):1147–1156.

Jiménez-Cervantes C., Martínez-Esparza M., Solano F., Lozano J.A., García-Borrón J.C. (1998) Molecular interactions within the melanogenic complex: formation of heterodimers of tyrosinase and TRP1 from B16 mouse melanoma. *Biochemical and Biophysical Research Communications* 253(3):761–767.

Jin E.-J., Erickson C.A., Takada S., Burrus L.W. (2001) Wnt and BMP signaling govern lineage segregation of melanocytes in the avian embryo. *Developmental Biology* 233:22–37.

Joerg H., Fries H.R., Meijerink E., Stranzinger G.F. (1996) Red coat color in Holstein cattle is associated with a deletion in the MSHR gene. *Mammalian Genome* 7(4):317–318.

Johnson R., Jackson I.J. (1992) Light is a dominant mouse mutation resulting in premature cell death. *Nature Genetics* 1(3):226–229.

Jones T. (1979) *Encyclopedia of Pet Mice*. T.F.H. Publications, Neptune City, NJ.

Jordan S.A., Jackson I.J. (2000) A late wave of melanoblast differentiation and rostrocaudal migration revealed in patch and rump-white embryos. *Mechanisms of Development* 92:135–143.

Jouneau A., Yu Y.-Q., Pasdar M., Larue L. (2000) Plasticity of cadherin-catenin expression in the melanocyte lineage. *Pigment Cell Research* 13:260–272.

Kadekaro A.L., Abdel-Malek Z.A. (2007) Walking in the footsteps of giants: melanocortins and human pigmentation, a historical perspective. *Pigment Cell Research* 20:150–152.

Kageyama R., Ohtsuka T., Tomita K. (2000) The bHLH gene Hes1 regulates differentiation of multiple cell types. *Molecules and Cells* 10(1).

Kalcheim C., Burstyn-Cohen T. (2005) Early stages of neural crest ontogeny: formation and regulation of cell delamination. *International Journal of Developmental Biology* 49:105–116.

Kamada S., Shimono A., Shinto Y., Tsujimura T., Takahashi T., Noda T., Kitamura Y., Kondoh H., Tsujimoto Y. (1995) bcl-2 deficiency in mice leads to pleiotropic abnormalities: accelerated lymphoid cell death in thymus and spleen, polycystic kidney, hair hypopigmentation, and distorted small intestine. *Cancer Research* 55(2):354–359.

Kameya S., Hawes N.L., Chang B., Heckenlively J.R., Naggert J.K., Nishina P.M. (2002) Mfrp, a gene encoding a frizzled related protein is mutated in the mouse retinal degeneration 6. *Human Molecular Genetics* (16):1879–1886.

Kantheti P., Qiao X., Diaz M.E., Peden A.A., Meyer G.E., Carskadon S.L., Kapfhamer D., Sufalko D., Robinson M.S., Noebels J.L., Burmeister M. (1998) Mutation in AP-3 delta in the mocha mouse links endosomal transport to storage deficiency in platelets, melanosomes, and synaptic vesicles. *Neuron* 21:111–122.

Kaput J., Klein K.G., Reyes E.J., Kibbe W.A., Cooney C.A., Jovanovic B., Visek W.J., Wolff G.L. (2004) Identification of genes contributing to the obese yellow A^{vy} phenotype: caloric restriction, genotype, diet x genotype interactions. *Physiological Genomics* 18(3):316–324.

Karlsson E.K., Baranowska I., Wade C.M., Salmon Hillbertz N.H.C.S., Zody M.C., Anderson N., Biagi T.A., Patterson N., Rosengren Pielberg G., Kulbokas III E.J. et al. (2007) Efficient mapping of Mendelian traits in dogs through genome-wide association. *Nature Genetics* 39(11):1321–1328.

Karpac J., Czyzewska K., Kern A., Brush R.S., Anderson R.E., Hochgeschwender U. (2008) Failure of adrenal corticosterone production in POMC-deficient mice results from lack of integrated effects of POMC peptides on multiple factors. *American Journal of Physiology Endocrinology and Metabolism* 295:E446–E455.

Katoh M., Katoh M. (2006) Cross-talk of WNT and FGF signaling pathways at GSK3β to regulate beta-catenin and SNAIL signaling cascades. *Cancer Biology and Therapy* 5(9):1059–1064.

Katoh M., Katoh M. (2007) WNT signaling pathway and stem cell signaling network. *Clinical Cancer Research* 13(14):4042–4045.

Kaufman M.H. (1992) *The Atlas of Mouse Development*. Academic Press, San Diego, CA.

Kaufman M.H., Robertson E.J., Handyside A.H., Evans M.J. (1983) Establishment of pluripotential cell lines from haploid mouse embryos. *Journal of Embryology and Experimental Morphology* 73:249–261.

Kawa Y., Ito M., Ono H., Asano M., Takano N., Ooka S., Watabe H., Hosaka E., Baba T., Kubota Y., Mizoguchi M. (2000) Stem cell factor and/or endothelin-3 dependent immortal melanoblast and melanocyte populations derived from mouse neural crest cells. *Pigment Cell Research* 13(Suppl 8):73–80.

Kellerer S., Schreiner S., Stolt C.C., Scholz S., Bosl M.R., Wegner M. (2006) Replacement of the Sox10 transcription factor by Sox8 reveals incomplete functional equivalence. *Development* 133(15):2875–2886.

Kelsh R.N. (2004) Genetics and evolution of pigment patterns in fish. *Pigment Cell Research* 17:326–336.

Kelsh R.N., Inoue C., Momoi A., Kondoh H., Furutani-Seiki M., Ozato K., Wakamatsu Y. (2004) The Tomita collection of medaka pigmentation mutants as a resource for understanding neural crest cell development. *Mechanisms of Development* 121(7–8):841–859.

Kerje S., Sharma P., Gunnarsson U., Kim H., Bagchi S., Fredriksson R., Schutz K., Jensen P., von Heijne G., Okimoto R., Andersson L. (2004) The Dominant white, Dun and Smoky color variants in chicken are associated with insertion/deletion polymorphisms in the PMEL17 gene. *Genetics* 168:1507–1518.

Kerns J.A., Olivier M., Lust G., Barsh G.S. (2003) Exclusion of melanocortin-1 receptor (mc1r) and agouti as candidates for dominant black in dogs. *Journal of Heredity* 94(1):75–9.

Kerns J.A., Newton J., Berryere T.G., Rubin E.M., Cheng J.-F., Schmutz S.M., Barsh G.S. (2004) Characterization of the dog Agouti gene and a nonagouti mutation in German Shepherd Dogs. *Mammalian Genome* 15:798–808.

Khaled M., Larribere L., Bille K., Aberdam E., Ortonne J.P., Ballotti R., Bertolotto C. (2002) Glycogen synthase kinase 3beta is activated by cAMP and plays an active role in the regulation of melanogenesis. *Journal of Biological Chemistry* 277(37):33690–33697.

Kim B.Y., Olzmann J.A., Barsh G.S., Chin L.S., Li L. (2007) Spongiform neurodegeneration-associated E3 ligase Mahogunin ubiquitylates TSG101 and regulates endosomal trafficking. *Molecular Biology of the Cell* 18(4):1129–1142.

Kim J., Lo L., Dormand E., Anderson D.J. (2003) SOX10 maintains multipotency and inhibits neuronal differentiation of neural crest stem cells. *Neuron* 38(1):17–31.

Kim S.-Y., Kawashima Y., Takahashi J., Yasuda Y. (1995) Quantitative analysis of coat-color patterns in BALB/c<->C57BL/6 mouse chimeras by means of a computer-based color image processor system. *Journal of Reproduction and Development* 41:47–55.

Kimmel C.B., Kane D.A., Ho R.K. (1991) Lineage specification during early embryonic development of the zebrafish. In: *Cell-Cell Interactions in Early Development*, J. Gerhart (ed.), pp. 203–225. Wiley, New York.

Kimura Y., Nina Jones N., Kluppel M., Hirashima M., Tachibana K., Cohn J.B., Wrana J.L., Pawson T., Bernstein A. (2004) Targeted mutations of the juxtamembrane tyrosines in the Kit receptor tyrosine kinase selectively affect multiple cell lineages. *Proceedings of the National Academy of Sciences USA* 101(16):6015–6020.

King R.A., Pietsch J., Fryer J.P., Savage S., Brott M.J., Russell-Eggitt I., Summers C.G., Oetting W.S. (2003a) Tyrosinase gene mutations in oculocutaneous albinism 1 (OCA1): definition of the phenotype. *Human Genetics* 113(6):502–513.

King R.A., Willaert R.K., Schmidt R.M., Pietsch J., Savage S., Brott M.J., Fryer J.P., Summers C.G., Oetting W.S. (2003b) MC1R mutations modify the classic phenotype of oculocutaneous albinism type 2 (OCA2). *American Journal of Human Genetics* 73:638–645.

Kinnear P.E., Jay B., Witkop C.J. Jr (1985) Albinism. *Surveys in Ophthalmology* 30(2):75–101.

Kinoshita M., Murata K., Naruse K., Tanaka M. (2009) *Medaka: Biology, Management and Experimental Protocols.* Wiley, Chichester.

Kirby M.L., Gale T.F., Stewart D.E. (1983) Neural crest cells contribute to normal aorticopulmonary septation. *Science* 220:1059–1061.

Kissel H., Timokhina I., Hardy M.P., Rothschild G., Tajima Y., Soares V., Angeles M., Whitlow S.R., Manova K., Besmer P. (2000) Point mutation in kit receptor tyrosine kinase reveals essential roles for kit signaling in spermatogenesis and oogenesis without affecting other kit responses. *EMBO Journal* 19(6):1312–1326.

Klaus S.N. (2006) A history of the science of pigmentation. In: Nordlund J.J. et al. (eds) *The Pigmentary System*, 2nd edn, pp. 5–10. Blackwell Publishing, Oxford.

Klein-Szanto A., Bradl M., Porter S., Mintz B. (1991) Melanosis and associated tumors in transgenic mice. *Proceedings of the National Academy of Sciences USA* 88(1):169–173.

Klungland H., Vage D.I. (2003) Pigmentary switches in domestic animal species. *Annals of the New York Academy of Sciences* 994:331–338.

Klungland H., Vage D.I., Gomez-Raya L., Adalsteinsson S., Lien S. (1995) The role of melanocyte-stimulating hormone (MSH) receptor in bovine coat color determination. *Mammalian Genome* 6(9):36–39.

Kluppel M., Nagle D.L., Bucan M., Bernstein A. (1997) Long-range genomic rearrangements upstream of Kit dysregulate the developmental pattern of Kit expression in W57 and W-banded mice and interfere with distinct steps in melanocyte development. *Development* 124:65–77.

Kobayashi T., Hearing V.J. (2007) Direct interaction of tyrosinase with Tyrp1 to form heterodimeric complexes *in vivo*. *Journal of Cell Science* 120(Pt 24):4261–4268.

Kobayashi T., Urabe K., Winder A., Jiménez-Cervantes C., Imokawa G., Brewington T., Solano F., García-Borrón J.C., Hearing V.J. (1994) Tyrosinase related protein 1 (TRP1) functions as a DHICA oxidase in melanin biosynthesis. *EMBO Journal* 13(24):5818–5825.

Kobayashi T., Vieira W.D., Potterf B., Sakai C., Imokawa G. (1995) Modulation of melanogenic protein expression during the switch from eumelanogenesis to pheomelanogenesis. *Journal of Cell Science* 108:2301–2309.

Kobayashi T., Imokawa G., Bennett D.C., Hearing V.J. (1998) Tyrosinase stabilization by Tyrp1 (the brown locus protein). *Journal of Biological Chemistry* 273:31801–31805.

Körner A.M., Pawelek J. (1980) Dopachrome conversion: a possible control point in melanin biosynthesis. *Journal of Investigative Dermatology* 75(2):192–195.

Kos L., Aronzon A., Takayama H., Maina F., Ponzetto C., Merlino G., Pavan W. (1999) Hepatocyte growth factor/scatter factor-MET signaling in neural crest derived melanocyte development. *Pigment Cell Research* 12:13–21.

Kos R., Reedy M.V., Johnson R.L., Erickson C.A. (2001) The winged-helix transcription factor FoxD3 is important for establishing the neural crest lineage and repressing melanogenesis in avian embryos. *Development* 128:1467–1479.

Krakowsky J.M., Boissy R.E., Neumann J.C., Lingrel J.B. (1993) A DNA insertional mutation results in microphthalmia in transgenic mice. *Transgenic Research* 2(1):14–20.

Kraman M., McCright B. (2005) Functional conservation of Notch1 and Notch2 intracellular domains. *FASEB Journal* 20:1096–3407.

Krause K., Foitzik K. (2007) Biology of the hair follicle: the basics. *Seminars in Cell & Developmental Biology* 18:245–254.

Kripke M.L. (1977) Latency, histology, and antigenicity of tumors induced by ultraviolet light in three inbred mouse strains. *Cancer Research* 37(5):1395–1400.

Krude H., Biebermann H., Luck W., Horn R., Brabant G., Grüters A. (1998) Severe early-onset obesity, adrenal insufficiency and red hair pigmentation caused by POMC mutations in humans. *Nature Genetics* 19(2):155–157.

Krull C.E. (2001) Segmental organization of neural crest migration. *Mechanisms of Development* 105(1–2):37–45.

Kubic J.D., Young K.P., Plummer R.S., Ludvik A.E., Lang D. (2008) Pigmentation PAX-ways: the role of Pax3 in melanogenesis, melanocyte stem cell maintenance, and disease. *Pigment Cell and Melanoma Research* 21:627–645

Kucera G.T., Bortner D.M., Rosenberg M.P. (1996) Overexpression of an Agouti cDNA in the skin of transgenic mice recapitulates dominant coat color phenotypes of spontaneous mutants. *Developmental Biology* 173(1):162–173.

Kühn Ch., Weikard R. (2007) An investigation into the genetic background of coat colour dilution in a Charolais x German Holstein F2 resource population. *Animal Genetics* 38:109–113.

Kulesa P., Ellies D.L., Trainor P.A. (2004) Comparative analysis of neural crest cell death, migration and function during vertebrate embryogenesis. *Developmental Dynamics* 229:14–29.

Kunisada T., Lu S.Z., Yoshida H., Nishikawa S., Nishikawa S., Mizoguchi M., Hayashi S., Tyrrell L., Williams D.A., Wang X., Longley B.J. (1998a) Murine cutaneous mastocytosis and epidermal melanocytosis induced by keratinocyte expression of transgenic stem cell factor. *Journal of Experimental Medicine* 187(10):1565–1573.

Kunisada T., Yoshida H., Yamazaki H., Miyamoto A., Hemmi H., Nishimura E., Shultz L.D., Nishikawa S., Hayashi S. (1998b) Transgene expression of steel factor in the basal layer of epidermis promotes survival, proliferation, differentiation and migration of melanocyte precursors. *Development* 125(15):2915–2923.

Kunisada T., Yamazaki H., Hirobe T., Kamei S., Omoteno M., Tagaya H., Hemmi H., Koshimizu U., Nakamura T., Hayashi S.I. (2000) Keratinocyte expression of transgenic hepatocyte growth factor affects melanocyte development, leading to dermal melanocytosis. *Mechanisms of Development* 94(1–2):67–78.

Kuraguchi M., Wang X.P., Bronson R.T., Rothenberg R., Ohene-Baah N.Y., Lund J.J., Kucherlapati1 M., Maas R.L., Kucherlapati1 R. (2006) Adenomatous polyposis coli (APC) is required for normal development of skin and thymus. *PLoS Genetics* 2(9):1362–1374.

Kurita K., Nishito M., Shimogaki H., Takada K., Yamazaki H., Kunisada T. (2005) Suppression of progressive loss of coat color in microphthalmia-vitiligo mutant mice. *Journal of Investigative Dermatology* 125(3):538–544.

Kwon B.S., Kim K.K., Halaban R., Pickard R.T. (1994) Characterization of mouse Pmel 17 gene and silver locus. *Pigment Cell Research* 7(6):394–397.

Kwon B.S., Halaban R., Ponnazhagan S., Kim K., Chintamaneni C., Bennett D., Pickard R.T. (1995) Mouse silver mutation is caused by a single base insertion in the putative cytoplasmic domain of Pmel 17. *Nucleic Acids Research* 23(1):154–158.

Lalueza-Fox C., Römpler H., Caramelli D., Stäubert C., Catalano G., Hughes D., Rohland N., Pilli E., Longo L., Condemi S. et al. (2007) A melanocortin 1 receptor allele suggests varying pigmentation among Neanderthals. *Science* 318(5855):1453–1455.

Lamason R.L., Mohideen M., Mest J.R., Wong A.C., Norton H.L., Aros M.C., Jurynec M.J., Mao X.Y., Humphreville V.R., Humbert J.E. et al. (2005) SLC24A5, a putative cation exchanger, affects pigmentation in zebrafish and humans. *Science* 310:1782–1786.

Lamoreux M.L. (1973) Brown-orange tortoiseshell cat. *Journal of Heredity* 64(4):240.

Lamoreux M.L. (1981) White spotting allele alters color of yellow mice. *Journal of Heredity* 72(3):223–224.

Lamoreux M.L. (1983) Developmental interactions in the pigmentary system of mice. II. Interaction in genetically identical mice. *Journal of Heredity* 74:440–442.

Lamoreux M.L. (1986) Dopachrome conversion and dopa oxidase activities in recessive yellow mice. Catalytic activities of extracts from pheomelanic and eumelanic tissues. *Journal of Heredity* 77(5):337–340.

Lamoreux M.L. (1999) Strain-specific white-spotting patterns in laboratory mice. *Pigment Cell Research* 12(6):383–390.

Lamoreux M.L., Mayer T.C. (1975) Site of gene action in the development of hair pigment in recessive yellow (e/e) mice. *Developmental Biology* 46(1):160–166.

Lamoreux M.L., Russell E.S. (1979) Developmental interaction in the pigmentary system of mice. I. Interactions between effects of genes on color of pigment and on distribution of pigmentation in the coat of the house mouse (*Mus musculus*). *Journal of Heredity* 70(1):31–36.

Lamoreux M.L., Galbraith D.B. (1986) DK/Lm: a strain of laboratory mouse with an unusual expression of the *lethal yellow* (A^y) phenotype. *Genetic Research* 48(1):35–40.

Lamoreux M.L., Pendergast P. (1987) Genetic controls over melanocyte differentiation: interaction of *agouti*-locus and *albino*-locus genetic defects. *Journal of Experimental Zoology* 243(1):71–79.

Lamoreux M.L., Woolley C., Pendergast P. (1986) Genetic controls over activities of tyrosinase and dopachrome conversion factor in murine melanocytes. *Genetics* 113(4):967–984.

Lamoreux M.L., Zhou B.K., Rosemblat S., Orlow S.J. (1995) The pinkeyed-dilution protein and the eumelanin/pheomelanin switch: in support of a unifying hypothesis. *Pigment Cell Research* 8(5):263–270.

Lamoreux M.L., Wakamatsu K., Ito S. (2001) Interaction of major coat color gene function in mice as studied by chemical analysis of eumelanin and pheomelanin. *Pigment Cell Research* 14:23–31.

Lamoreux M.L., Kelsh R.N., Wakamatsu Y., Ozato K. (2005) Pigment pattern formation in the medaka embryo. *Pigment Cell Research* 18(2):64–73.

Lang D., Liu M.M., Huang L., Engleka K.A., Zhang M., Chu E.Y., Lipner S., Skoultchi A., Millar S.E., Epstein J.A. (2005) Pax3 functions at a nodal point in melanocyte stem cell differentiation. *Nature* 433:884–887.

Langtimm-Sedlak C.J., Schroeder B., Saskowski J.L., Carnahan J.F., Sieber-Blum M. (1996) Multiple actions of stem cell factor in neural crest cell differentiation *in vitro*. *Developmental Biology* 174:345–359.

Larue L., Beermann F. (2007) Cutaneous melanoma in genetically modified animals. *Pigment Cell Research* 20(6):485–497.

Larue L., Antos C., Butz S., Huber O., Delmas V., Dominis M., Kemler R. (1996) A role for cadherins in tissue formation. *Development* 122(10):3185–3194.

Larue L., Kunisaka M., Goding C.R. (2003) β-catenin in the melanocyte lineage. *Pigment Cell Research* 16:312–317.

Lauber P.G. (1971) *Of Man and Mouse; How House Mice Became Laboratory Mice*. The Viking Press, NY.

Lauvergne J.J. (ed.) (1999) *Mendelian Inheritance in Cattle*. Wageningen Academic Press.

Lavado A., Olivares C., Garcia-Boron J.C., Montoliu L. (2005) Molecular basis of the extreme dilution mottled mouse mutation. A combination of coding and noncoding genomic alterations. *Journal of Biological Chemistry* 280:(6)4817–4824.

Lavado A., Jeffery G., Tovar V., delaVilla P., Montoliu L. (2006) Ectopic expression of tyrosine hydroxylase in the pigmented epithelium rescues the retinal abnormalities and visual function common in albinos in the absence of melanin. *Journal of Neurochemistry* 96(4):1201–1211.

Lavker R.M., Miller S., Wilson C., Cotsarelis G., Wei Z.G., Yang J.S., Sun T.T. et al. (1993) Hair follicle stem cells: their location, role in hair cycle, and involvement in skin tumour formation. *Journal of Investigative Dermatology* 101:16s–26s.

Le Douarin N. (1982) *The Neural Crest*. Cambridge University Press, Cambridge.

Le Douarin N.M., Kalcheim C. (1999) *The Neural Crest*, 2nd edn. Cambridge University Press, Cambridge.

Le Douarin N.M., Cruezet S., Couly G., Dupin E. (2004) Neural crest cell plasticity and its limits. *Development* 131:4637–4650.

Lee H.O., Levorse J.M., Shin M.K. (2003) The endothelin receptor-B is required for the migration of neural crest-derived melanocyte and enteric neuron precursors. *Developmental Biology* 259(1):162–175.

Lehman A.L., Silvers W.K., Puri N., Wakamatsu K., Ito S., Brilliant M.H. (2000) The *Underwhite (uw)* locus acts autonomously and reduces the production of melanin. *Journal of Investigative Dermatology* 115:601–606.

Leibovici M., Safieddine S., Petit C. (2008) Mouse models for human hereditary deafness. *Current Topics in Developmental Biology* 84:385–429.

Le Pape E., Wakamatsu K., Ito S., Wolber R., Hearing V.J. (2008) Regulation of eumelanin/pheomelanin synthesis and visible pigmentation in melanocytes by ligands of the melanocortin 1 receptor. *Pigment Cell and Melanoma Research* 21(4):477–486.

Le Pape E., Passeron T., Giubellino A., Valencia J.C., Wolber R., Hearing V.J. (2009) Microarray analysis sheds light on the dedifferentiating role of agouti signal protein in murine melanocytes via the Mc1r. *Proceedings of the National Academy of Sciences USA* 106:1802–1807.

Lerner A.B., Shiohara T., Boissy R.E., Jacobson K.A., Lamoreux M.L., Moellmann G.E. (1986) A mouse model for vitiligo. *Journal of Investigative Dermatology* 87(3):299–304.

Levy C., Khaled M., Fisher D.E. (2006) MITF: master regulator of melanocyte development and melanoma oncogene. *Trends in Molecular Medicine* 12(9):406–414.

Lewis J.L., Bonner J., Modrell M., Ragland J.W., Moon R.T., Dorsky R.I., Raible D.W. (2003) Reiterated Wnt signaling during zebrafish neural crest development. *Development* 131:1299–1308.

Levinson B., Vulpe C., Elder B., Martin C., Verley F., Packman S., Gitschier J. (1994) The mottled gene is the mouse homologue of the Menkes disease gene. *Nature Genetics* 6(4):369–373.

Li W., Rusiniak M.E., Chintala S., Gautam R., Novak E.K., Swank R.T. (2004) Murine Hermansky–Pudlak syndrome genes: regulators of lysosome-related organelles. *BioEssays* 26:616–628.

Li X.-d., Ikebe R., Ikebe M. (2005) Activation of myosin Va function by melanophilin, a specific docking partner of myosin Va. *Journal of Biological Chemistry* 280(18):17815–17822.

Li W., Feng Y., Hao C., Guo X., Cui Y., He M., He X. (2007) The BLOC interactomes form a network in endosomal transport. *Journal of Genetics and Genomics* 34(8):669–682.

Lieberman M.W., Wiseman A.L., Shi Z.Z., Carter B.Z., Barrios R., Ou C.N., Chevez-Barrios P., Wang Y., Habib G.M., Goodman J.C. et al. (1996) Growth retardation and cysteine deficiency in gamma-glutamyl transpeptidase-deficient mice. *Proceedings of the National Academy of Sciences USA* 93(15):7923–7926.

Lindner G., Menrad A., Gherard E., Merlino G., Welker P., Handjiski B., Roloff B., Paus R. (2000) Involvement of hepatocyte growth factor / scatter factor and Met receptor signaling in hair follicle morphogenesis and cycling. *FASEB Journal* 14:319–332.

Little C.C. (1911) The 'Dilute' forms of yellow mice. *Science* 33:896–897.

Little C.C. (1913a) 'Yellow' and 'Agouti' factors in mice. *Science* 38:205.

Little C.C. (1913b) *Experimental Studies of the Inheritance of Color in Mice.* Carnegie Inst. Pub. No. 179, pp. 17–102.

Little C.C. (1957) *The Inheritance of Coat Color in Dogs.* Comstock, Ithaca, NY.

Liu R.R., Brown C.E., Murphy T.H. (2007) Differential regulation of cell proliferation in neurogenic zones in mice lacking cystine transport by xCT. *Biochemical and Biophysical Research Communications* 364:528–533.

Liu T., DeCostanzo A.J., Liu X., Wang H., Hallagan S., Moon R.T., Malbon C.C. (2001) G protein signaling from activated rat frizzled-1 to the beta-catenin-Lef-Tcf pathway. *Science* 292(5522):1718–1722.

Llamazares M., Cal S., Quesada V., Lopez-Otin C. (2003) Identification and characterization of ADAMTS-20 defines a novel subfamily of metalloproteinases-disintegrins with multiple thrombospondin-1 repeats and a unique GON domain. *Journal of Biological Chemistry* 278:13382–13389.

Llanos R.M., Ke B.X., Wright M., Deal Y., Monty F., Kramer D.R., Mercer J.F. (2006) Correction of a mouse model of Menkes disease by the human Menkes gene. *Biochimica et Biophysica Acta* 1762(4):485–493.

Lobe C.G., Koop K.E., Kreppner W., Lomeli H., Gertsenstein M., Nagy A. (1999) Z / AP, a double reporter for cre-mediated recombination. *Developmental Biology* 208(2):281–292.

Loftus S.K., Larson D.M., Baxter L.L., Antonellis A., Chen Y., Wu X., Jiang Y., Bittner M., Hammer III J.A., Pavan W.J. (2002) Mutation of melanosome protein RAB38 in chocolate mice. *Proceedings of the National Academy of Sciences USA* 99:4471–4476.

Loftus S.K., Antonellis A., Matera I., Renaud G., Baxter L.L., Reid D., Wolfsberg T.G., Chen Y., Wang C.W., NISC Comparative Sequencing Program, Prasad M.K. et al. (2008) Gpnmb is a melanoblast-expressed, MITF-dependent gene. *Pigment Cell and Melanoma Research* 22:99–110.

Logan C.Y., Nusse R. (2004) The Wnt signaling pathway in development and disease. *Annual Review of Cell and Developmental Biology* 20:781–810.

Logan D.W., Burn S.F., Jackson I.J. (2006) Regulation of pigmentation in zebrafish melanophores. *Pigment Cell Research* 19(3):206–213.

Lopes V.S., Wasmeier C., Seabra M.C., Futter C.E. (2007) Melanosome maturation defect in Rab38-deficient retinal pigment epithelium results in instability of immature melanosomes during transient melanogenesis. *Molecular Biology of the Cell* 18:3914–3927.

Lopez V.M., Decatur C.L., Stamer W.D., Lynch R.M., McKay B.S. (2008) L-DOPA is an endogenous ligand for OA1. *PLoS Biology* 6(9): e236.

Luo R., Gao J., Wehrle-Haller B., Henion P.D. (2003) Molecular identification of distinct neurogenic and melanogenic neural crest sublineages. *Development* 130:321–330.

Lyon, M.F. (1960) A further mutation of the mottled type in the house mouse. *Journal of Heredity* 51:116–120.

Lyon M.F. (1961) Gene action in the X-chromosome of the mouse (*Mus musculus* L.). *Nature* 190:372–373.

Lyon M.F., Glenister P.H. (1982) A new allele *sash* (W^{sh}) at the W-locus and a spontaneous recessive lethal in mice. *Genetic Research (Cambridge)* 39:315–322.

Lyons L.A., Foe I.T., Rah H.C., Grahn R.A. (2005) Chocolate coated cats: TYRP1 mutations for brown color in domestic cats. *Mammalian Genome* 16(5):356–366.

Lyons L.A., Bailey S.J., Baysac K.C., Byrns G., Erdman C.A., Fretwell N., Froenicke L., Gazlay K.W., Geary L.A., Grahn J.C. et al. (2006) The Tabby cat locus maps to feline chromosome B1. *Animal Genetics* 37(4):383–386.

Mackenzie M.A., Jordan S.A., Budd P.S., Jackson I.J. (1997) Activation of the receptor tyrosine kinase Kit is required for the proliferation of melanoblasts in the mouse embryo. *Developmental Biology* 192(1):99–107.

Majumdar M.K., Everett E.T., Xiao X., Cooper R., Langley K., Kapur R., Vik T., Williams D.A. (1996) Xenogeneic expression of human stem cell factor in transgenic mice mimics codominant c-kit mutations. *Blood* 87(8):3203–3211.

Mak S.S., Moriyama M., Nishioka E., Osawa M., Nishikawa S. (2006) Indispensable role of Bcl2 in the development of the melanocyte stem cell. *Developmental Biology* 291(1):144–153.

Malbon C.C., Wang H., Moon R.T. (2001) Wnt signaling and heterotrimeric G-proteins: strange bedfellows or a classic romance? *Biochemical and Biophysical Research Communications* 287(3):589–593.

Manga P., Kromberg J.G.R., Box N.F., Sturm R.A., Jenkins T., Ramsay M. (1997) Rufous oculocutaneous albinism in southern African blacks is caused by mutations in the TYRP1 gene. *American Journal of Human Genetics* 61:1095–1101.

Manga P., Boissy R.E., Pifko-Hirst S., Zhou B.K., Orlow S.J. (2001) Mislocalization of melanosomal proteins in melanocytes from mice with oculocutaneous albinism type 2. *Experimental Eye Research* 72(6):695–710.

Mariat D., Taourit S., Guérin G. (2003) A mutation in the MATP gene causes the cream coat colour in the horse. *Genetics Selection Evolution* 35(1):119–133.

Markert C.L., Silvers W.K. (1956) The Effects of Genotype and Cell Environment on Melanoblast Differentiation in the House Mouse. *Genetics* 41(3):429–450.

Martin G.R. (1981) Isolation of a pluripotent cell line from early mouse embryos cultured in medium conditioned by teratocarcinoma stem cells. *Proceedings of the National Academy of Sciences USA* 78(12):7634–7638.

Martina J.A., Moriyama K., Bonifacino J.S. (2003) BLOC-3, a protein complex containing the Hermansky-Pudlak syndrome gene products HPS1 and HPS4. *Journal of Biological Chemistry* 278(31):29376–29384.

Martínez-Esparza M., Jiménez-Cervantes C., Bennett D.C., Lozano J.A., Solano F., García-Borrón J.C. (1999) The mouse silver locus encodes a single transcript truncated by the silver mutation. *Mammalian Genome* 10(12):1168–1171.

Martínez-Morales J.R., Dolez V., Rodrigo I., Zaccarini R., Leconte L., Bovolenta P., Saule S. (2003) OTX2 activates the molecular network underlying retina pigment epithelium differentiation. *Journal of Biological Chemistry* 278(24):21721–21731.

Martínez-Morales J.R., Rodrigo I., Bovolenta P. (2004) Eye development: a view from the retina pigmented epithelium. *Bioessays* 26(7):766–777.

Martinez-Morales J.R., Wittbrodt J. (2009) Shaping the vertebrate eye. *Current Opinion in Genetics and Development* 2009 (epub ahead of print).

Matera I., Watkins-Chow D.E., Loftus S.K., Hou L., Incao A., Silver D.L., Rivas C., Elliott E.C., Baxter L.L., Pavan W.J. (2008) A sensitized mutagenesis screen identifies Gli3 as a modifier of Sox10 neurocristopathy. *Human Molecular Genetics* 17(14):2118–2131.

Matesic L.E., Yip R., Reuss A.E., Swing D.A., O'Sullivan T.N., Fletcher C.F., Copeland N.G., Jenkins N.A. (2001) Mutations in Mlph, encoding a member of the Rab effector family, cause the melanosome transport defects observed in leaden mice. *Proceedings of the National Academy of Sciences USA* 98:10238–10243.

Matsui Y., Zsebo K.M., Hogan B.L. (1990) Embryonic expression of a haematopoietic growth factor encoded by the *Sl* locus and the ligand for c-kit. *Nature* 34:667–669.

Matsumoto J. (2002) Brightly colored pigmentation in lower vertebrates: wonder searching its mechanisms and significance in the context of phylogeny. *Pigment Cell Research* 15(4):310–319.

Matsumoto K., Tajima H., Nakamura T. (1991) Hepatocyte growth factor is a potent stimulator of human melanocyte DNA synthesis and growth. *Biochemical and Biophysical Research Communications* 176:45–51.

Mayer T.C. (1964) An experimental investigation of pattern development in lethal spotting and belted embryos. *Developmental Biology* 9:269–286.

Mayer T.C. (1965) The development of piebald spotting in mice. *Developmental Biology* 11:319–334.

Mayer T.C. (1970) A comparison of pigment cell development in albino, steel, and dominant-spotting mutant mouse embryos. *Developmental Biology* 23(2):297–309.

Mayer T.C. (1973a) Site of gene action in steel mice: analysis of the pigment defect by mesoderm-ectoderm recombinations. *Journal of Experimental Zoology* 184(3):345–352.

Mayer T.C. (1973b) The migratory pathway of neural crest cells into the skin of mouse embryos. *Developmental Biology* 34(1):39–46.

Mayer T.C. (1977) Enhancement of melanocyte development from piebald neural crest by a favorable tissue environment. *Developmental Biology* 56(2):255–262.

Mayer T.C. (1979) Interactions between normal and pigment cell populations mutant at the dominant-spotting (W) and steel (Sl) loci in the mouse. *Journal of Experimental Zoology* 210(1):81–88.

Mayer T.C., Green M.C. (1968) An experimental analysis of the pigment defect caused by mutations at the *W* and *Sl* loci in mice. *Developmental Biology* 18:62–75.

Mayer T.C., Fishbane J.L. (1972) Mesoderm-ectoderm interaction in the production of the agouti pigmentation pattern in mice. *Genetics* 71(2):297–303.

McElwee K.J., Sinclair R. (2008) Hair physiology and its disorders. *Drug Discovery Today: Disease Mechanisms* 5(2):e163–e173.

McGarry M.P., Borchers M., Novak E.K., Lee N.A., Ohtake P.J., Lee J.J., Swank R.T. (2002) Pulmonary pathologies in *pallid* mice result from nonhematopoietic defects. *Experimental and Molecular Pathology* 72(3):213–220.

McGowan K.A., Jun Z.L., Park C.Y., Beaudry V., Tabor H.K., Sabnis A.J., Zhang W., Fuchs H., de Angelis M.H., Myers R.M., Attardi L.D., Barsh G.S. (2008) Ribosomal mutations cause p53-mediated dark skin and pleiotropic effects. *Nature Genetics* 40(8):963–970.

Ménasché G., Fischer A., de Saint Basile G. (2002) Griscelli syndrome types 1 and 2. *American Journal of Human Genetics* 71(5):1237–1238.

Menkes J.H., Alter M., Steigleder G.K., Weakley D.R., Sung J.H. (1962) A sex-linked recessive disorder with retardation of growth, peculiar hair, and focal cerebral and cerebellar degeneration. *Pediatrics* 29:764–779.

Menotti-Raymond M., David V.A., Eizirik E., Roelke M.E., Ghaffari H., O'Brien S.J. (2009) Mapping of the domestic cat 'SILVER' coat color locus identifies a unique genomic location for silver in mammals. Journal of Heredity 100(Suppl 1):S8–S13.

Mercer J.A., Seperack P.K., Strobel M.C., Copeland N.G., Jenkins N.A. (1991) Novel myosin heavy chain encoded by murine dilute coat colour locus. *Nature* 349:709–713.

Mercer J.F. (1998) Menkes syndrome and animal models. *American Journal of Clinical Nutrition* 67(5 Suppl):1022S–1028S.

Mercer J.F., Llanos R.M. (2003) Molecular and cellular aspects of copper transport in developing mammals. Journal of Nutrition 133(5 Suppl 1):1481S–1484S

Mercer J.F., Paynter J.A., Grimes A. (1994) The toxic milk mouse does have elevated hepatic metallothionein mRNA. *Biochemical Journal* 304:317–318.

Michaud E.J., Bultman S.J., Stubbs L.J., Woychik R.P. (1993) The embryonic lethality of homozygous *lethal yellow mice (A^y/A^y)* is associated with the disruption of a novel RNA-binding protein. *Genes & Development* 7:1203–1213.

Michaud E.J., Bultman S.J., Klebig M.L., Van Vugt M.J., Stubbs L.J., Russell L.B., Woychik R.P. (1994) A molecular model for the genetic and phenotypic characteristics of the mouse *lethal yellow* (A^y) mutation. *Proceedings of the National Academy of Sciences USA* 91:2562–2566.

Millar S.E. (2003) WNTS: Multiple genes, multiple functions. *Journal of Investigative Dermatology* 120:7–8.

Millar S.E., Miller M.W., Stevens M.E., Barsh G.S. (1995) Expression and transgenic studies of the mouse agouti gene provide insight into the mechanisms by which mammalian coat color patterns are generated. *Development* 121(10):3223–3232.

Miller M.W., Duhl D.M., Vrieling H., Cordes S.P., Ollmann M.M., Winkes B.M., Barsh G.S. (1993) Cloning of the mouse *agouti* gene predicts a secreted protein ubiquitously expressed in mice carrying the *lethal yellow* mutation. *Genes & Development* 7(3):454–467.

Miller W.J., Hollander W.F. (1986) The sex-linked black cat fallacy: a textbook case. *Journal of Heredity* 77(6):463–464.

Mintz B. (1967) Gene control of mammalian pigmentary differentiation. I. Clonal origin of melanocytes. *Proceedings of the National Academy of Sciences USA* 58(1):344–351.

Mintz B. (1971a) Clonal basis of mammalian differentiation. In: *Control Mechanisms of Growth and Differentiation*. 25th Symposium of the Society of Experimental Biology. Davies D.D., Balls M. (eds), pp. 345–370. Cambridge University Press, Cambridge.

Mintz B. (1971b) Genetic mosaicism in vivo: development and disease in allophenic mice. *Federation Proceedings* 30(3):935–943.

Mintz B., Klein-Szanto A.J. (1992) Malignancy of eye melanomas originating in the retinal pigment epithelium of transgenic mice after genetic ablation of choroidal melanocytes. *Proceedings of the National Academy of Sciences USA* 89(23):11421–11425.

Mitani H., Kamei Y., Fukamachi S., Oda S., Sasaki T., Asakawa S., Todo T., Shimizu N. (2006) The medaka genome: why we need multiple fish models in vertebrate functional genomics. *Genome Dynamics* 2:165–182.

Mizoguchi M. (2004) Melanocyte development: with a message of encouragement to young women scientists. *Pigment Cell Research* 17(5):533–544.

Mjaatvedt C.H., Kern C.B., Norris R.A., Fairey S., Cave C.L. (2005) Normal distribution of melanocytes in the mouse heart. *Anatomical Record Part A Discoveries in Molecular Cellular and Evolutionary Biology* 285:748–757.

Mochi M., Ono T., Matsubara Y., Eguchi G. (1998) Spontaneous transdifferentiation of quail pigmented epithelial cell is accompanied by a mutation in the Mitf gene. *Developmental Biology* 196:145–159.

Mollaaghababa R., Pavan W.J. (2003) The importance of having your SOX on: role of SOX10 in the development of neural crest-derived melanocytes and glia. *Oncogene* 22:3024–3034.

Monaco A.P., Chelly J. (1995) Menkes and Wilson diseases. *Advances in Genetics* 33:233–253.

Montagna W., Ellis R.A. (ed.) (1958) *The Biology of Hair Growth*. Academic Press, New York.

Montoliu L., Larue L., Beermann F. (2004) On the use of regulatory regions from pigmentary genes to drive the expression of transgenes in mice. *Pigment Cell Research* 17(2):188–190.

Montoliu L., Oetting W.S., Bennett D.C. (2009) *Color Genes*. European Society for Pigment Cell Research. www.espcr.org/micemut/.

Moore K.A., Lemischka I.R. (2006) Stem cells and their niches. *Science* 311:1880–1885.

Moore K.J. (1995) Insight into the microphthalmia gene. *Trends in Genetics* 11(11):442–448.

Moore K.J., Swing D.A., Copeland N.G., Jenkins N.A. (1994) The murine dilute suppressor gene encodes a cell autonomous suppressor. *Genetics* 138(2):491–497.

Moore K.J., Swing D.A., Rinchik E.M., Mucenski M.L., Buchberg A.M., Copeland N.G., Jenkins N.A. (1998) The murine dilute suppressor gene dsu suppresses the coat-color phenotype of three pigment mutations that alter melanocyte morphology, d, ash and ln. *Genetics* 119(4):933–941.

Moore R., Larue L. (2004) Cell surface molecules and truncal neural crest ontogeny: a perspective. *Birth Defects Research C Embryo Today* 72(2):140–150.

Moore R., Champeval D., Denat L., Tan S.-S., Faure F., Julien-Grille S., Larue L. (2004) Involvement of cadherins 7 and 20 in mouse embryogenesis and melanocyte transformation. Oncogene 23:6726–6735.

Morgan H.D., Sutherland H.G., Martin D.J., Whitelaw E. (1999) Epigenetic inheritance at the agouti locus in the mouse. *Nature Genetics* 23:314–318.

Morgan N.V., Pasha S., Johnson C.A., Ainsworth J.R., Eady R.A.J., Dawood B., McKeown C., Trembath R.C., Wilde J., Watson S.P., Maher E.R. (2006) A germline mutation in BLOC1S3 / Reduced Pigmentation causes a novel variant of Hermansky-Pudlak Syndrome (HPS8). *American Journal of Human Genetics* 78:160–166.

Mori M., Metzger D., Garnier J.M., Chambon P., Mark M. (2002) Site-specific somatic mutagenesis in the retinal pigment epithelium. *Investigative Ophthalmology & Visual Science* 43(5):1384–1388.

Morizot D.C., Slaugenhaupt S.A., Kallman K.D., Chakravarti A. (1991) Genetic linkage map of fishes of the genus *Xiphophorus* (Teleostei: Poeciliidae). *Genetics* 127(2):399–410.

Morii E., Tsujimura T., Jippo T., Hashimoto K., Takebayashi K., Tsujino K., Nomura S., Yamamoto M., Kitamura Y. (1996) Regulation of mouse mast cell protease 6 gene expression by transcription factor encoded by the mi locus. *Blood* 88(7):2488–2494.

Moriyama M., Osawa M., Mak S.S., Ohtsuka T., Yamamoto N., Han H., Delmas V., Kageyama R., Beermann F., Larue L., Nishikawa S. (2006) Notch signaling via Hes1 transcription factor maintains survival of melanoblasts and melanocyte stem cells. *Journal of Cell Biology* 173(3):333–339.

Morrison S.J., Perez S.E., Qiao Z., Verdi J.M., Hicks C., Weinmaster G., Anderson D.J. (2000) Transient Notch activation initiates an irreversible switch from neurogenesis to gliogenesis by neural crest stem cells. *Cell* 101:499–510.

Morrison-Graham K., Weston J.A. (1993) Transient steel factor dependence by neural crest-derived melanocyte precursors. *Developmental Biology* 159:346–352.

Morse H.C. III (ed.) (1978) *Origins of Inbred Mice*. Elsevier. http://www.informatics.jax.org/morsebook/.

Motohashi H., Hozawa K., Oshima T., Takeuchi T., Takasaka T. (1994) Dysgenesis of melanocytes and cochlear dysfunction in mutant microphthalmia (mi) mice. *Hear Research* 80(1):10–20.

Mundy N.I. (2005) A window on the genetics of evolution: MC1R and plumage colouration in birds. *Proceedings Biological Sciences* 272(1573):1633–1640.

Murisier F., Beermann F. (2006) Genetics of pigment cells: lessons from the tyrosinase gene family. *Histology and Histopathology* 21(5):567–578.

Murisier F., Guichard S., Beermann F. (2007) Distinct distal regulatory elements control tyrosinase expression in melanocytes and the retinal pigment epithelium. *Developmental Biology* 303(2):838–847.

Murisier F., Aydin I.T., Guichard S., Brunschwiler S., Beermann F. (2008) Expression from a bacterial artificial chromosome containing the Dct gene locus. *Pigment Cell and Melanoma Research* 21:212–215.

Murphy M., Reid K., Williams D.E., Lyman S.D., Bartlet P.F. (1992) Steel factor is required for maintenance, but not differentiation of melanocyte precursors in the neural crest. *Developmental Biology* 153:396–401.

Murphy W.J., Pevzner P.A., O'Brien S.J. (2004) Mammalian phylogenomics comes of age. *Trends in Genetics* 20(12):631–639.

Murray M., Pawelek J.M., Lamoreux M.L. (1983) New regulatory factors for melanogenesis: developmental changes in neonatal mice of various genotypes. *Developmental Biology* 100(1):120–126.

Nagase T., Ishikawa K., Miyajima N., Tanaka A., Kotani H., Nomura N., Ohara O. (1998) Prediction of the coding sequences of unidentified human genes. IX. The complete sequences of 100 new cDNA clones from brain which can code for large proteins in vitro. *DNA Research* 5:31–39.

Nagao K., Toyoda M., Takeuchi-Inoue K., Fujii K., Yamada M., Miyashita T. (2005) Identification and characterization of multiple isoforms of a murine and human tumor suppressor, patched, having distinct first exons. *Genomics* 85:462–471.

Nagashima K., Torii S., Yi Z., Igarashi M., Okamoto K., Takeuchi T., Izumi T. (2002) Melanophilin directly links Rab27a and myosin Va through its distinct coiled-coil regions. *FEBS Letters* 517(1–3):233–238.

Nagle D.L., Martin-DeLeon P., Hough R.B., Bucan M. (1994) Structural analysis of chromosomal rearrangements associated with the developmental mutations *Ph*, W^{19H} and *Rw* on mouse chromosome 5. *Proceedings of the National Academy of Sciences USA* 91:7237–7241.

Nagle D.L., McGrail S.H., Vitale J., Woolf E.A., Dussault B.J. Jr, DiRocco L., Holmgren L., Montagno J., Bork P., Huszar D. et al. (1999) The mahogany protein is a receptor involved in suppression of obesity. *Nature* 398(6723):148–152.

Nakagawa S., Takeichi M. (1995) Neural crest cell-cell adhesion controlled by sequential and subpopulation-specific expression of novel cadherins. *Development* 121(5):1321–1332.

Nakagawa S., Takeichi M. (1998) Neural crest emigration from the neural tube depends upon regulated cadherin expression. *Development* 125:2963–2971.

Nakayama A., Nguyen M.-T.T., Chen C.C., Opdecamp K., Hodgkinson C.A., Arnheiter H. (1998) Mutations in microphthalmia, the mouse homolog of the human deafness gene MITF, affect neuroepithelial and neural crest-derived melanocytes differently. *Mechanisms of Development* 70:155–166.

Nakayama K., Nakayama K., Negishi I., Kuida K., Sawa H., Loh D.Y. (1994) Targeted disruption of Bcl-2 alpha beta in mice: occurrence of gray hair, polycystic kidney disease, and lymphocytopenia. *Proceedings of the National Academy of Sciences USA* 91(9):3700–3704.

Napolitano A., Vincensi M.R., Di Donato P., Monfrecola G., Prota G. (2000) Microanalysis of melanins in mammalian hair by alkaline hydrogen peroxide degradation: Identification of a new structural marker of pheomelanins. *Journal of Investigative Dermatology* 114:1141–1147.

Nascimento A.A., Roland J.T., Gelfand V.I. (2003) Pigment cells: a model for the study of organelle transport. *Annual Review of Cell and Developmental Biology* 19:469–491.

Nataf V., Lecoin L., Eichmann A., Le Douarin N.M. (1996) Endothelin-B receptor is expressed by neural crest cells in the avian embryo. *Proceedings of the National Academy of Sciences USA* 93:9645–9650.

Nazarian R., Falcon-Perez J.M., Dell'Angelica E.C. (2003) Biogenesis of lysosome-related organelles complex 3 (BLOC-3): a complex containing the Hermansky-Pudlak syndrome (HPS) proteins HPS1 and HPS4. *Proceedings of the National Academy of Sciences USA* 100(15):8770–8775.

Nelson W.J., Nusse R. (2004) Convergence of Wnt, beta-catenin, and cadherin pathways. *Science* 303(5663):1483–1487.

Neveu M.M., Holder G.E., Ragge N.K., Sloper J.J., Collin J.R., Jeffery G. (2006) Early midline interactions are important in mouse optic chiasm formation but are not critical in man: a significant distinction between man and mouse. *European Journal of Neurochemistry* 23(11):3034–3042.

Newton J.M., Wilkie A.L., He L., Jordan S.A., Metallinos D.L., Holmes N.G., Jackson I.J., Barsh G.S. (2000) Melanocortin 1 receptor variation in the domestic dog. *Mammalian Genome* 11(1):24–30.

Newton J.M., Orlow S.J., Barsh G.S. (1996) Isolation and characterization of a mouse homolog of the X-linked ocular albinism (OA1) gene. *Genomics* 37(2):219–225.

Newton J.M., Wilkie A.L., He L., Jordan S.A., Metallinos D.L., Holmes N.G., Jackson I.J., Barsh G.S. (2000) Melanocortin 1 receptor variation in the domestic dog. *Mammalian Genome* 11(1):24–30.

Newton J.M., Cohen-Barak O., Hagiwara N., Gardner J.M., Davisson M.T., King R.A., Brilliant M.H. (2001) Mutations in the human orthologue of the mouse *underwhite* gene *(uw)* underlie a new form of oculocutaneous albinism, OCA4. *American Journal of Human Genetics* 69(5):981–988.

Nguyen T., Novak E.K., Kermani M., Fluhr J., Peters L.L., Swank R.T., Wei M.L. (2002) Melanosome morphologies in murine models of hermansky-pudlak syndrome reflect blocks in organelle development. *Journal of Investigative Dermatology* 119:1156–1164.

Nguyen T., Wei M.L. (2007) Hermansky-Pudlak HPS1/pale ear gene regulates epidermal and dermal melanocyte development. *Journal of Investigative Dermatology* 127(2):421–428.

Nichols D.H. (1981) Neural crest formation in the head of the mouse embryo as observed using a new histological technique. *Journal of Embryology and Experimental Morphology* 64:105–120.

Nishikawa S.-I., Osawa M. (2007) Generating quiescent stem cells. *Pigment Cell Research* 20:263–270.

Nishikawa S., Kusakabe M., Yoshinaga K., Ogawa M., Hayashi S., Kunisada T., Era T., Sakakura T. (1991) In utero manipulations of coat color formation by a monoclonal anti-c-kit antibody: two distinct waves of c-kit-dependency during melanocyte development. *EMBO Journal* 10:2111–2118.

Nishimura E., Yoshida H., Kunisada T., Nishikawa S. (1999) Regulation of E- and P-cadherin expression correlated with melanocyte migration and diversification. *Developmental Biology* 215:155–166.

Nishimura E.K., Jordan S.A., Oshima H., Yoshida H., Osawa M., Moriyama M., Jackson I.J., Barrandon Y., Miyachi Y., Nishikawa S.-I. (2002) Dominant role of the niche in melanocyte stem-cell fate determination. *Nature* 416:854–860.

Nishimura E.K., Granter S.R., Fisher D.E. (2005) Mechanisms of hair graying: Incomplete melanocyte stem cell maintenance in the niche. *Science* 307(5710):720–724.

Nordlund J.J., Boissy R.E., Hearing V.J., King R.A., Oetting W.S., Ortonne J.-P. (eds) (2006) *The Pigmentary System: Physiology and Pathophysiology*, 2nd edn. Blackwell Publishing, London.

Norris B.J. (2008) Riding 'white' horses reveals new insights into pigmentation and melanoma. *Pigment Cell and Melanoma Research* 21(6):596–597.

Norris B.J., Whan V.A. (2008) A gene duplication affecting expression of the ovine ASIP gene is responsible for white and black sheep. *Genetic Research* 18:1282–1293.

Novak E.K., Reddington M., Zhen L., Stenberg P.E., Jackson C.W., McGarry M.P., Swank R.T. (1995) Inherited thrombocytopenia caused by reduced platelet production in mice with the gunmetal pigment gene mutation. *Blood* 85:1781–1789.

Nusse R. (2003) Wnts and Hedgehogs: lipid-modified proteins and similarities in signaling mechanisms at the cell surface. *Development* 130:5297–5305.

O'Brien S.J., Wienberg J., Lyons L.A. (1997) Comparative genomics: lessons from cats. *Trends in Genetics* 13(10):393–399.

O'Brien S.J., Menotti-Raymond M., Murphy W.J., Nash W.G., Wienberg J., Stanyon R., Copeland N.G., Jenkins N.A., Womack J.E., Marshall Graves J.A. (1999) The promise of comparative genomics in mammals. *Science* 286(5439):458–462:479–481.

O'Brien S.J., Johnson W., Driscoll C., Pontius J., Pecon-Slattery J., Menotti-Raymond M. (2008) State of cat genomics. *Trends in Genetics* 24(6):268–279.

Oetting W.S. (2002) New insights into ocular albinism type 1 (OA1): mutations and polymorphisms of the OA1 gene. *Human Mutation* 19:85–92.

Oetting W.S., Setaluri V. (2006) In: Nordlund J.J. et al. (eds) *The Pigmentary System*, 2nd edn, pp. 213–229. Blackwell Publishing, Oxford.

Ogden S.K., Fei D.L., Schilling N.S., Ahmed Y.F., Hwa J., Robbins D.J. (2008) G protein $G\alpha_i$ functions immediately downstream of Smoothened in Hedgehog signalling. *Nature* 456:967–970.

Oh J., Ho L., Ala-Mello S., Amato D., Armstrong L., Bellucci S., Carakushansky G., Ellis J.P., Fong C.T., Green J.S. et al. (1998) Mutation analysis of patients with Hermansky-Pudlak syndrome: a frameshift hot spot in the HPS gene and apparent locus heterogeneity. *American Journal of Human Genetics* 62(3):593–598.

Oiso N., Riddle S.R., Serikawa T., Kuramoto T., Spritz R.A. (2004) The rat ruby (R) locus is Rab38: Identical mutations in Fawn-hooded and Tester-Moriyama rats derived from an ancestral Long Evans rat substrain. *Mammalian Genome* 15(4):307–314.

Olivares C., Solano F., Garcia-Borron J.C. (2003) Conformation-dependent post-translational glycosylation of tyrosinase. *Journal of Biological Chemistry* 278:15735–15743.

Ollmann M.M., Wilson B.D., Yang Y.K., Kerns J.A., Chen Y., Gantz I., Barsh G.S. (1997) Antagonism of central melanocortin receptors *in vitro* and *in vivo* by agouti-related protein. *Science* 278(5335):135–138.

Ollmann M.M., Lamoreux M.L., Wilson B.D., Barsh G.S. (1998) Interaction of Agouti protein with the melanocortin 1 receptor in vitro and in vivo. *Genes & Development* 12(3):316–330.

Olson T.A. (1982) *Inheritance of Coat Colour in Cattle: a Review*. Iowa State University and Agricultural Experiment Station Bulletin, Ames, IA.

Olson T.A. (1999) *The Genetics of Cattle*. CAB International, Wallingford.

Ono H., Kawa Y., Asano M., Ito M., Takano A., Kubota Y., Matsumoto J., Mizoguchi M. (1998) Development of melanocyte progenitors in murine Steel mutant neural crest explants cultured with stem cell factor, endothelin-3, or TPA. *Pigment Cell Research* 11(5):291–298.

Opdecamp K., Nakayama A., Nguyen M.-T.T., Hodgkinson C.A., Pavan W.J., Arnheiter H. (1997) Melanocyte development *in vivo* and in neural crest cell cultures: crucial dependence on the Mitf basic-helix-loop-helix-zipper transcription factor. *Development* 124:2377–2386.

Orlow S.J. (1995) Melanosomes are specialized members of the lysosomal lineage of organelles. *Journal of Investigative Dermatology* 105(1):3–7.

Orlow S.J., Lamoreux M.L., Pifko-Hirst S., Zhou B.K. (1993) Pathogenesis of the *platinum* (c^p) mutation, a model for oculocutaneous albinism. *Journal of Investigative Dermatology* (2):137–140.

Orlow S.J., Zhou B.K., Chakraborty A.K., Drucker M., Pifko-Hirst S., Pawelek J.M. (1994) High-molecular weight forms of tyrosinase and the tyrosinase-related proteins: evidence for a melanogenic complex. *Journal of Investigative Dermatology* 103:196–201.

Osawa M., Egawa G., Mak S.S., Moriyama M., Freter R., Yonetani S., Beermann F., Nishikawa S. (2005) Molecular characterization of melanocyte stem cells in their niche. *Development* 132(24):5589–5599.

Oshima H., Rochat A., Kedzia C., Kobayashi K., Barrandon Y. (2001) Morphogenesis and renewal of hair follicles from adult multipotent stem cells. *Cell* 104:233–245.

Ostrander E.A., Galibert F., Patterson D.F. (2000) Canine genetics comes of age. *Trends in Genetics* 16:(3)117–123.

Ostrander E.A., Giger U., Lindblad-Toh K., Ostrander E.A. (eds) (2005) *The Dog and Its Genome*. Cold Spring Harbor Press, Cold Spring Harbor, NY.

O'Sullivan T.N., Wu X.S., Rachel R.A., Huang J.D., Swing D.A., Matesic L.E., Hammer J.A. III., Copeland N.G., Jenkins N.A. (2004) Dsu functions in a MYO5A-independent pathway to suppress the coat color of dilute mice. *Proceedings of the National Academy of Sciences USA* 101(48):16831–16736.

Ota M., Ito K. (2006) BMP and FGF2 regulate neurogenin-2 expression and the differentiation of sensory neurons and glia. *Developmental Dynamics* 235(3):646–655.

Ozeki H., Ito S., Wakamatsu K., Hirobe T. (1995) Chemical characterization of hair melanins in various coat-color mutants of mice. *Journal of Investigative Dermatology* 105:361–366.

Ozeki H., Ito S., Wakamatsu K., Ishiguro I. (1997a) Chemical characterization of pheomelanogenesis starting from dihydroxyphenylalanine or tyrosine and cysteine. Effects of tyrosinase and cysteine concentrations and reaction time. *Biochimica et Biophysica Acta* 1336(3):539–548.

Ozeki H., Wakamatsu K., Ito S., Ishiguro I. (1997b) Chemical characterization of eumelanins with special emphasis on 5,6-dihydroxyindole-2-carboxylic acid content and molecular size. *Analytical Biochemistry* 248(1):149–157.

Paladini R.D., Saleh J., Qian C., Xu G.X., Rubin L.L. (2005) Modulation of hair growth with small molecule agonists of the hedgehog signaling pathway. *Journal of Investigative Dermatology* 125(4):638–646.

Palmisano I., Bagnato P., Palmigiano A., Innamorati G., Rotondo G., Altimare D., Venturi C., Sviderskaya E.V., Piccirillo R., Coppola M. et al. (2008) The ocular albinism type 1 protein, an intracellular G protein-coupled receptor, regulates melanosome transport in pigment cells. *Human Molecular Genetics* 17(22):3487–3501.

Palmiter R.D., Brinster R.L. (1986) Germ-line transformation of mice. *Annual Review of Genetics* 20:465–499.

Palmiter R.D., Brinster R.L., Hammer R.E., Trumbauer M.E., Rosenfeld M.G., Birnberg N.C., Evans R.M. (1982) Dramatic growth of mice that develop from eggs microinjected with metallothionein-growth hormone fusion genes. *Nature* 300(5893):611–615.

Parichy D.M., Reedy M.V., Erickson C.A. (2006) Regulation of melanoblast migration and differentiation. In: Nordlund J.J. et al. (eds) *The Pigmentary System*, 2nd edn, pp. 108–139. Blackwell Publishing, Oxford.

Pastural E., Barrat F.J., Dufourcq-Lagelouse R., Certain S., Sanal O., Jabado N., Seger R., Griscelli C., Fischer A., de Saint Basile G. (1997) Griscelli disease maps to chromosome 15q21 and is associated with mutations in the myosin-Va gene. *Nature Genetics* 16(3):289–292.

Pavan W.J., Tilghman S.M. (1994) *Piebald lethal* (s^l) acts early to disrupt the development of neural crest-derived melanocytes. *Proceedings of the National Academy of Sciences USA* 91:7159–7163.

Pawson T., Nash P. (2000) Protein–protein interactions define specificity in signal transduction. *Genes & Development* 14:1027–1047.

Pérez-Losada J., Sanchez-Martin M., Rodriguez-Garcia A., Sanchez M.L., Orfao A., Flores T., Sanchez-Garcia I. (2002) Zinc-finger transcription factor Slug contributes to the function of the stem cell factor c-kit signaling pathway. *Blood* 100:1274–1286.

Pérez-Losada J., Sánchez-Martín M., Pérez-Caro M., Pérez-Mancera P.A., Sánchez-Garcia I. (2003) The radio resistance biological function of the SCF/kit signaling pathway is mediated by the zinc-finger transcription factor Slug. *Oncogene* 22:4205–4211.

Pérez-Oliva A.B., Fernéndez L.P., Detorre C., Herráiz C., Martínez-Escribano J.A., Benítez J., Lozano Teruel J.A., García-Borrón J.C., Jiménez-Cervantes C., Ribas G. (2009) Identification and functional analysis of novel variants of the human melanocortin 1 receptor found in melanoma patients. *Human Mutation* 30(5):811–822.

Perou C.M., Leslie J.D., Green W., Li L., Ward D.M., Kaplan J. (1997) The Beige/Chediak-Higashi syndrome gene encodes a widely expressed cytosolic protein. *Journal of Biological Chemistry* 272(47):29790–29794.

Perron M., Boy S., Amato M.A., Viczian A., Koebernick K., Pieler T., Harris W.A. (2003) A novel function for Hedgehogsignalling in retinal pigment epithelium differentiation. *Development* 130:1565–1577.

Perry W.L., Hustad C.M., Swing D.A., Jenkins N.A., Copeland N.G. (1995) A transgenic mouse assay for agouti protein activity. *Genetics* 140(1):267–274.

Phan L.K., Lin F., LeDuc C.A., Chung W.K., Leibel R.L. (2002) The mouse *mahoganoid* coat color mutation disrupts a novel C3HC4 RING domain protein. *Journal of Clinical Investigation* 110(10):1449–1459.

Philipp U., Hamann H., Mecklenburg L., Nishino S., Mignot E., Günzel-Apel A.-R., Schmutz S.M., Leeb T. (2005) Polymorphisms within the canine MLPH gene are associated with dilute coat color in dogs. *BMC Genetics* 6:34.

Phillips R.J.S. (1961) 'Dappled,' a new allele at the mottled locus in the house mouse. *Genetic Research (Cambridge)* 2:290–295.

Pielberg R.G., Golovko A., Sundström E., Curik I., Lennartsson J., Seltenhammer M.H., Druml T., Binns M., Fitzsimmons C., Lindgren G. et al. (2008) A cis-acting regulatory mutation causes premature hair graying and susceptibility to melanoma in the horse. *Nature Genetics* 40(8):1004–1009.

Pla P., Moore R., Morali O.G., Grille S., Martinozzi S., Delmas V., Larue L. (2001) Cadherins in neural crest cell development and transformation. J. Cell Phys. 189:121–132.

Pla P., Larue L. (2003) Involvement of endothelin receptors in normal and pathological development of neural crest cells. *International Journal of Developmental Biology* 47(5):315–325.

Pontius J.U., Mullikin J.C., Smith D.R., Agencourt Sequencing Team, Lindblad-Toh K., Gnerre S., Clamp M., Chang J., Stephens R., Neelam B. et al., NISC Comparative Sequencing Program, O'Brien S.J. (2007) Initial sequence and comparative analysis of the cat genome. *Genome Research* 17(11):1675–1689.

Poole T.W., Silvers W.K. (1976) An experimental analysis of the recessive yellow coat color mutant in the mouse. *Developmental Biology* 48(2):377–381.

Porter S., Larue L., Mintz B. (1991) Mosaicism of tyrosinase-locus transcription and chromatin structure in dark vs. light melanocyte clones of homozygous chinchilla-mottled mice. *Developmental Genetics* 12(6):393–402.

Porter S.D., Meyer C.J. (1994) A distal tyrosinase upstream element stimulates gene expression in neural-crest-derived melanocytes of transgenic mice: position-independent and mosaic expression. *Development* 120(8):2103–2111.

Postlethwait J.H. (2006) The Zebrafish genome: a review and msx gene case study. *Genome Dynamics* 2:183–197.

Postlethwait J.H., Yan Y.L., Gates M.A., Horne S., Amores A., Brownlie A., Donovan A., Egan E.S., Force A., Gong Z. et al. (1998) Vertebrate genome evolution and the zebrafish gene map. *Nature Genetics* 18:345–349.

Potterf S.B., Furumura M., Sviderskaya E.V., Santis C., Bennett D.C., Hearing V.J. (1998) Normal tyrosine transport and abnormal tyrosinase routing in pink-eyed dilution melanocytes. *Experimental Cell Research* 244(1):319–326.

Potterf S.B., Furumura M., Dunn K.J., Arnheiter H., Pavan W.J. (2000) Transcription factor hierarchy in Waardenburg syndrome: regulation of MITF expression by SOX10 and PAX3. *Human Genetics* 107(1):1–6.

Potterf S.B., Mollaaghababa R., Hou L., Southard-Smith E.M., Hornyak T.J., Arnheiter H., Pavan W.J. (2001) Analysis of SOX10 function in neural crest-derived melanocyte development: SOX10-dependent transcriptional control of dopachrome tautomerase. *Developmental Biology* 237:245–257.

Pratt B.M. (1982) Site of gene action of the white allele (Mi^{wh}) of the microphthalmia locus: a dermal-epidermal recombination study. *Journal of Experimental Zoology* 220:93–101.

Price E.R., Ding H.-F., Badalian T., Bhattacharya S., Takemoto C., Yao T.-P., Hemesath T.J., Fisher D.E. (1998) Lineage specific signaling in melanocytes: c-Kit stimulation recruits p300/CBP to Microphthalmia. *Journal of Biological Chemistry* 273(29):17983–17986.

Prosser H., Rastan S. (2003) Manipulation of the mouse genome: a multiple impact resource for drug discovery and development. *Trends in Biotechnology* 21(5):224–232.

Prota G. (2000) Melanins, melanogenesis and melanocytes: looking at their functional significance from the chemist's viewpoint. *Pigment Cell Research* 13(4):283–293.

Provance D.W. Jr, James T.L., Mercer J.A. (2002) Melanophilin, the product of the leaden locus is required for targeting of myosin-Va to melanosomes. *Traffic* 3(2):124–132.

Puig I., Yajima I., Bonaventure J., Delmas V., Larue L. (2009) The tyrosinase promoter is active in a subset of vagal neural crest cells during early development in mice. *Pigment Cell and Melanoma Research* 22(3):331–334.

Puri N., Gardner J.M., Brilliant M.H. (2000) Aberrant pH of melanosomes in *Pink-Eyed Dilution (p)* mutant melanocytes. *Journal of Investigative Dermatology* 115:607–613.

Quevedo W.C., Holstein T.J. (2006) General biology in mammalian pigmentation. In: Nordlund J.J. et al. (eds) *The Pigmentary System*, 2nd edn. Blackwell Publishing, Oxford.

Rajewsky K., Gu H., Kuhn R., Betz U.A., Muller W., Roes J., Schwenk F. (1996) Conditional gene targeting. *Journal of Clinical Investigation* 98(3):600–603.

Rao C., Foernzler D., Loftus S.K., Liu S., McPherson J.D., Jungers K.A., Apte S.S., Pavan W.J., Beier D.R. (2003) A defect in a novel ADAMTS family member is the cause of the *belted* white-spotting mutation. *Development* 130(19):4665–4672.

Raposo G., Marks M.S. (2007) Melanosomes – dark organelles enlighten endosomal membrane transport. *Nature Reviews/Molecular Cell Biology* 8:786–797.

Rawles M.E. (1940a) The development of melanophores from embryonic mouse tissues grown in the coelom of chick embryos. *Proceedings of the National Academy of Sciences USA* 26(12):673–680.

Rawles M.E. (1940b) The pigment-forming potency of early chick blastoderms. *Proceedings of the National Academy of Sciences USA* 26(1):86–94.

Rawles M.E. (1947) Origin of pigment cells from the neural crest in the mouse embryo. *Physiological Zoology* 20:248–266.

Rawles M.E. (1948) Origin of melanophores and their role in development of color patterns in vertebrates. *Physiological Review* 28(4):383–408.

Rawls J.F., Frieda M.R., McAdow A.R., Gross J.P., Clayton C.M., Heyen C.K., Johnson S.L. (2003) Coupled mutagenesis screens and genetic mapping in zebrafish. *Genetics* 163(3):997–1009.

Raymond S.M., Jackson I.J. (1995) The RPE is required for the development and maintenance of the neural retina. *Current Biology* 5(11):1286–1295.

Read A.P., Newton V.E. (1997) Waardenburg syndrome. *Journal of Medical Genetics* 34(8):656–665.

Rees J. (2003) Genetics of hair and skin color. *Annual Review of Genetics* 37:67–90.

Reid K., Nishikawa S., Bartlett P.F., Murphy M. (1995) Steel factor directs melanocyte development *in vitro* through selective regulation of the number of c-kit+ progenitors. *Developmental Biology* 169:568–579.

Relaix F., Polimeni M., Rocancourt D., Ponzetto C., Schafer B.W., Buckingham M. (2003) The transcriptional activator PAX3-FKHR rescues the defects of Pax3 mutant mice but induces a myogenic gain-of-function phenotype with ligand-independent activation of Met signaling in vivo. *Genes & Development* 17(23):2950–2965.

Relaix F., Rocancourt D., Mansouri A., Buckingham M. (2004) Divergent functions of murine Pax3 and Pax7 in limb muscle development. *Genes & Development* 18(9):1088–1105.

Rhim H., Dunn K.J., Aronzon A., Mac S., Cheng M., Lamoreux M.L., Tilghman S.M., Pavan W.J. (2000) Spatially restricted hypopigmentation associated with an Ednrbs-modifying locus on mouse chromosome 10. *Genome Research* 10(1):17–29.

Rinchik E.M., Stoye J.P., Frankel W.N., Coffin J., Kwon B.S., Russell L.B. (1993a) Molecular analysis of viable spontaneous and radiation-induced albino (c)-locus mutations in the mouse. *Mutation Research* 286(2):199–207.

Rinchik E.M., Bultman S.J., Horsthemke B., Lee S.T., Strunk K.M., Spritz R.A., Avidano K.M., Jong M.T., Nicholls R.D. (1993b) A gene for the mouse pink-eyed dilution locus and for human type II oculocutaneous albinism. *Nature* 361(6407):72–76.

Robbins L.S., Nadeau J.H., Johnson K.R., Kelly M.A., Roselli-Rehfuss L., Baack E., Mountjoy K.G., Cone R.D. (1993) Pigmentation phenotypes of variant extension locus alleles result from point mutations that alter MSH receptor function. *Cell* 72(6):827–834.

Robinson R. (1971) *Genetics for Cat Breeders*. Pergamon Press, Oxford.

Robinson M.S., Bonifacino J.S. (2001) Adaptor-related proteins. *Current Opinion in Cell Biology* 13(4):444–453.

Robinson S.J., Healy E. (2002) Human melanocortin 1 receptor (MC1R) gene variants alter melanoma cell growth and adhesion to extracellular matrix. *Oncogene* 21(52):8037–8046.

Röder I.V., Petersen Y., Choi K.R., Witzemann V., Hammer J.A.III, Rudolf R. (2008) Role of Myosin Va in the plasticity of the vertebrate neuromuscular junction in vivo. *PLoS ONE* 3(12):e3871.

Romerdahl C.A., Stephens L.C., Bucana C., Kripke M.L. (1989) The role of ultraviolet radiation in the induction of melanocytic skin tumors in inbred mice. *Cancer Communications* 1(4):209–216.

Römpler H., Rohland N., Lalueza-Fox C., Willerslev E., Kuznetsova T., Rabeder G., Bertranpetit J., Schöneberg T., Hofreiter M. (2006) Nuclear gene indicates coat-color polymorphism in mammoths. *Science* 313:362.

Rosemblat S., Durham-Pierre D., Gardner J.M., Nakatsu Y., Brilliant M.H., Orlow S.J. (1994) Identification of a melanosomal membrane protein encoded by the pink-eyed dilution (type II oculocutaneous albinism) gene. *Proceedings of the National Academy of Sciences USA* 91(25):12071–12075.

Rosenblum E.B., Hoekstra H.E., Nachman M.W. (2004) Adaptive reptile color variation and the evolution of MC1R gene. *Evolution* 58(80):1794–1808.

Rosenquist T.H., Finnell R.H. (2007) Another key role for the cardiac neural crest in heart development. *American Journal of Physiology Heart and Circulatory Physiology* 292(3):H1225–H1226.

Rottapel R., Reedijk I.M., Williams D.E., Lyman S.D., Anderson D.M., Pawson T., Bernstein A. (1991) The Steel/W transduction pathway: Kit autophosphorylation and its association with a unique subset of cytoplasmic signaling proteins is induced by the Steel Factor. *Molecular and Cellular Biology* 11(6):3043–3051.

Rouzaud F., Kadekaro A.L., Abdel-Malek Z.A., Hearing V.J. (2005) MC1R and the response of melanocytes to ultraviolet radiation. *Mutation Research* 571(1–2):133–512.

Rovasio R.A., Delouvee A., Yamada K.M., Timpl R., Thiery J.P. (1983) Neural crest cell migration: requirements for exogenous fibronectin and high cell density. *Cell Biology* 96:462–473.

Rowe D.W., McGoodwin E.B., Martin G.R., Sussman M.D., Grahn D., Faris B., Franzblau C. (1974) A sex-linked defect in the cross-linking of collagen and elastin associated with the mottled locus in mice. *Journal of Experimental Medicine* 139(1):180–192.

Rubin B.P., Antonescu C.R., Scott-Browne J.P., Comstock M.L., Gu Y., Tanas M.R., Ware C.B., Woodell J. (2005) A knock-in mouse model of gastrointestinal stromal tumor harboring kit K641E. *Cancer Research* 65(15):6631–6639.

Ruiz i Altaba A., Mas C., Stecca B. (2007) The Gli code: an information nexus regulating cell fate, stemness and cancer. *Trends in Cell Biology* 17:438–447.

Russell E.S. (1946) A quantitative histological study of the pigment found in the coat-color mutants of the house mouse. I. Variable attributes of the pigment granules. *Genetics* 31:327–346.

Russell E.S. (1949) A quantitative histological study of the pigment found in the coat-color mutants of the house mouse. IV. The nature of the effects of genic substitution in five major allelic series. *Genetics* 34:146–166.

Russell L.B. (ed.) (1978) *Genetic Mosaics and Chimeras in Mammals. Basic Life Sciences*, vol. 12. Plenum Press.

Russell L.B. (1979) Analysis of the albino-locus region of the mouse. II. Mosaic mutants. *Genetics* 91(1):141–147.

Russell L.B., Raymer G.D. (1979) Analysis of the albino-locus region of the mouse. III. Time of death of prenatal lethals. *Genetics* 92(1):205–213.

Russell L.B., Russell W.L., Kelly E.M. (1979) Analysis of the albino-locus region of the mouse. I. Origin and viability. *Genetics* 91(1):127–139.

Russell L.B., Montgomery C.S., Raymer G.D. (1982) Analysis of the albino-locus region of the mouse: IV. Characterization of 34 deficiencies. *Genetics* 100(3):427–453.

Russell L.B., Montgomery C.S., Cacheiro N.L., Johnson D.K. (1995) Complementation analyses for 45 mutations encompassing the *pink-eyed dilution (p)* locus of the mouse. *Genetics* 141(4):1547–1562.

Ruvinsky A., Sampson J. (eds) (2001) *The Genetics of the Dog*. CAB International, Wallingford.

Saito H., Yasumoto K.-I., Takeda K., Takahashi K., Yamamoto H., Shibahara S. (2003) Microphthalmia-associated transcription factor in the Wnt signaling pathway. *Pigment Cell Research* 16:261–265.

Sakai D., Suzuki T., Osumi N., Wakamatsu Y. (2006) Cooperative action of Sox9, Snail2 and PKA signaling in early neural crest development. *Development* 133:1323–1333.

Sakakura Y., Sato H., Shiiya A., Tamba M., Sagara J., Matsuda M., Okamura N., Makino N., Bannai S. (2007) Expression and function of cystine/glutamate transporter in neutrophils. *Journal of Leukocyte Biology* 4:974–982.

Salazar G., Craige B., Styers M.L., Newell-Litwa K.A., Doucette M.M., Wainer B.H., Falcon-Perez J.M., Dell'Angelica E.C., Peden A.A., Werner E., Faundez V. (2006) BLOC-1 complex deficiency alters the targeting of adaptor protein complex-3 cargoes. *Molecular Biology of the Cell* 17(9):4014–4026.

Sanchez-Martin M., Rodriguez-Garcia A., Perez-Losada J., Sagrera A., Read A.P., Sanchez-Garcia I. (2002) SLUG (SNAI2) deletions in patients with Waardenburg disease. *Human Molecular Genetics* 11:3231–3236.

Sanchez-Martin M., Perez-Losada J., Rodriguez-Garcia A., Gonzales-Sanchez B., Korf B.R., Kuster W., Moss C., Spritz R.A., Sanchez-Garcia I. (2003) Deletion of the SLUG (SNAI2) gene results in human piebaldism. *American Journal of Medical Genetics* 122A:125–132.

Sánchez-Más J., Hahmann C., Gerritsen I., García-Borrón J.C., Jiménez-Cervantes C. (2004) Agonist-independent, high constitutive activity of the human melanocortin 1 receptor. *Pigment Cell Research* 17(4):386–395.

Sarangarajan R., Boissy R.E. (2001) Tyrp1 and oculocutaneous albinism type 3. *Pigment Cell Research* 14(6):437–444.

Sarna T., Swartz H.A. (2006) The physical properties of melanins. In: Nordlund J.J. et al. (eds) *The Pigmentary System*, 2nd edn, pp. 310–341. Blackwell Publishing, Oxford.

Sato H., Shiiya A., Kimata M., Maebara K., Tamba M., Sakakura Y., Makino N., Sugiyama F., Yagami K., Moriguchi T., Takahashi S., Bannai S. (2005) Redox imbalance in cystine/glutamate transporter-deficient mice. *Journal of Biological Chemistry* 280(45):37423–37429.

Sato N., Leopold P.L., Crystal R.G. (1999) Induction of the hair growth phase in postnatal mice by localized transient expression of Sonic hedgehog. *Journal of Clinical Investigation* 104:855–864.

Sauer B. (1998) Inducible gene targeting in mice using the Cre/lox system. *Methods (San Diego, Calif.)* 14(4):381–392.

Sauer B., Henderson N. (1988) Site-specific DNA recombination in mammalian cells by the Cre recombinase of bacteriophage P1. *Proceedings of the National Academy of Sciences USA* 85(14):5166–5170.

Savagner P., Yamada K.M., Thiery J.P. (1997) The zinc-finger protein Slug causes desmosome dissociation, and initial and necessary step for growth factor-induced epithelial-mesenchymal transition. *Journal of Cell Biology* 137:1403–1419.

Schaible R.H. (1969) Clonal distribution of melanocytes in piebald-spotted and variegated mice. *Journal of Experimental Zoology* 172:181–199.

Schepsky A., Bruser K., Gunnarsson G.J., Goodall J., Hallsson J.H., Goding C.R., Steingrimsson E., Hecht A. (2006) The microphthalmia-associated transcription factor Mitf interacts with B-Catenin to determine target gene expression. *Molecular and Cellular Biology* 26(23):8914–8927.

Schiaffino M.V., Tacchetti C. (2005) The ocular albinism type 1 (OA1) protein and the evidence for an intracellular signal transduction system involved in melanosome biogenesis. *Pigment Cell Research* 18(4):227–233.

Schmidt A., Beermann F. (1994) Molecular basis of dark-eyed albinism in the mouse. *Proceedings of the National Academy of Sciences USA* 91:4756–4760.

Schmidt-Küntzel A., Eizirik E., O'Brien S.J., Menotti-Raymond M. (2005) Tyrosinase and tyrosinase related protein 1 alleles specify domestic cat coat color phenotypes of the albino and brown loci. *Journal of Heredity* 96(4):289–301.

Schmidt-Küntzel A., Nelson G., David V.A., Schäffer A.A., Eizirik E., Roelke M.E., Kehler J.S., Hannah S.S., O'Brien S.J., Menotti-Raymond M. (2009) A domestic cat X chromosome linkage map and the sex-linked orange locus: mapping of orange, multiple origins and epistasis over nonagouti. *Genetics* 181(4):1415–1425.

Schmutz S.M., Berryere T.G. (2007) Genes affecting coat colour and pattern in domestic dogs: a review. *Animal Genetics* 38(6):539–549.

Schmutz S.M., Berryere T.G., Goldfinch A.D. (2002) TYRP1 and MC1R genotypes and their effects on coat color in dogs. *Mammalian Genome* 13:380–387.

Schmutz S.M., Berryere T.G., Ellinwood N.M., Kerns J.A., Barsh G.S. (2003) MClR studies in dogs with melanistic mask or brindle patterns. *Journal of Heredity* 94(1):69–73.

Schmutz S.M., Berryere T.G., Ciobanu D.C., Mileham A.J., Schmidtz B.H., Fredholm M. (2004) A form of albinism in cattle is caused by a tyrosinase frameshift mutation. *Mammalian Genome* 15:62–67.

Schonthaler H.B., Lampert J.M., von Lintig J., Schwarz H., Geisler R., Neuhauss S.C. (2005) A mutation in the *silver* gene leads to defects in melanosome biogenesis and alterations in the visual system in the zebrafish mutant *fading vision*. *Developmental Biology* 284(2):421–436.

Schouwey K., Delmas V., Larue L., Zimber-Strobl U., Strobl L.J., Radtke F., Beermann F. (2007) Notch1 and Notch2 receptors influence progressive hair graying in a dose-dependent manner. *Developmental Dynamics* 236(1):282–289.

Schwahn D.J., Timchenko N.A., Shibahara S., Medrano E.E. (2005) Dynamic regulation of the human dopachrome tautomerase promoter by MITF, ER-a and chromatin remodelers during proliferation and senescence of human melanocytes. *Pigment Cell Research* 18:203–213.

Scott G.A. (2007) Melanosome trafficking and transfer. In: Nordlund J.J. et al. (eds) *The Pigmentary System*, 2nd edn, pp. 171–180. Blackwell Publishing, Oxford.

Searle A.G. (1968) *Comparative Genetics of Coat Colour in Mammals*. Logos Press, London.

Sefton M., Sanchez S., Nieto M.A. (1998) Conserved and divergent roles for members of the Snail family of transcription factors in the chick and mouse embryo. *Development* 125:3111–3121.

Seitz J.J., Schmutz S.M., Thue T.D., Buchanan F.C. (1999) A missense mutation in the bovine MGF gene is associated with the roan phenotype in Belgian Blue and Shorthorn cattle. *Mammalian Genome* 10:710–712.

Seong E., Saunders T.L., Stewart C.L., Burmeister M. (2004) To knockout in 129 or in C57BL/6: that is the question. *Trends in Genetics* 20(2):59–62.

Seperack P.K., Mercer J.A., Strobel M.C., Copeland N.G., Jenkins N.A. (1995) Retroviral sequences located within an intron of the dilute gene alter dilute expression in a tissue-specific manner. *EMBO Journal* 14:2326–2332.

Setty S.R., Tenza D., Truschel S.T., Chou E., Sviderskaya E.V., Theos A.C., Lamoreux M.L., Di Pietro S.M., Starcevic M., Bennett D.C. et al. (2007) 'BLOC-1 is required for cargo-specific sorting from vacuolar early endosomes toward lysosome-related organelles.' *Molecular Biology of the Cell* 18(3):768–780.

Setty S.R., Tenza D., Sviderskaya E.V., Bennett D.C., Raposo G., Marks M.S. (2008) Cell-specific ATP7A transport sustains copper-dependent tyrosinase activity in melanosomes. *Nature* 454:1142–1146.

Sextier-Sainte-Claire Deville F., Ziller C., Le Douarin N. (1992) Developmental potentialities of cells derived from the truncal neural crest in clonal cultures. *Developmental Brain Reseach* 20; 66:1–10.

Shibahara S., Takeda K., Yasumoto K., Udono T., Watanabe K., Saito H., Takahashi K. (2001) Microphthalmia-associated transcription factor (MITF): multiplicity in structure, function, and regulation. *Journal of Investigative Dermatology Symposium Proceedings* 6(1):99–104.

Shima A., Himmelbauer H., Mitani H., Furutani-Seiki M., Wittbrodt J., Schartl M. (2003) Fish genomes flying. Symposium on Medaka Genomics. *EMBO Reports* 4(2):121–125.

Shin M.K., Russell L.B., Tilghman S.M. (1997) Molecular characterization of four induced alleles at the Ednrb locus. *Proceedings of the National Academy of Sciences USA* 94(24):13105–13110.

Shin M.K., Levorse J.M., Ingram R.S., Tilghman S.M. (1999) The temporal requirement for endothelin receptor-B signalling during neural crest development. *Nature* 402(6761):496–501.

Shutter J.R., Graham M., Kinsey A.C., Scully S., Lüthy R., Stark K.L. (1997) Hypothalamic expression of ART, a novel gene related to *agouti*, is up-regulated in obese and diabetic mutant mice. *Genes & Development* 11(5):593–602.

Sidman R.L., Appel S.H., Fullier J.F. (1965) Neurological mutants of the mouse. *Science* 150(3695):513–516.

Sieber-Blum M. (1989) Commitment of neural crest cells to the sensory neuron lineage. *Science* 243:1608–1611.

Sieber-Blum M., Cohen A. (1980) Clonal analysis of quail neural crest cells: They are pluripotent and differentiate *in vitro* in the absence of non-neuronal crest cells. *Developmental Biology* 80:96–106.

Silver D.L., Hou L., Somerville R., Young M.E., Apte S.S., Pavan W.J. (2008) The secreted metalloprotease ADAMTS20 is required for melanoblast survival. *PLoS Genetics* 4(2):e1000003.

Silvers W.K. (1979) *The Coat Colors of Mice*. Springer-Verlag, New York; www.informatics.jax.org/wksilvers/.

Silvers W.K., Russell E.S. (1955) An experimental approach to action of genes at the *agouti* locus in the mouse. *Journal of Experimental Zoology* 130:199–220.

Siracusa L.D. (1994) The agouti gene: turned on to yellow. *Trends in Genetics* 10(12):423–428.

Sitaram A., Piccirillo R., Palmisano I., Harper D.C., Dell'Angelica E.C., Schiaffino M.V., Marks M.S. (2009) Localization to mature melanosomes by virtue of cytoplasmic dileucine motifs is required for human OCA2 function. *Molecular Biology of the Cell* 20(5):1464–1477.

Slominski A., Paus R. (1993) Melanogenesis is coupled to murine anagen: toward new concepts for the role of melanocytes and the regulation of melanogenesis in hair growth. *Journal of Investigative Dermatology* 101:90s–97s.

Slominski A., Plonka P.M., Pisarchik A., Smart J.L., Tolle V., Wortsman J., Low M.J. (2005) Preservation of eumelanin hair pigmentation in proopiomelanocortin-deficient mice on a nonagouti (*a/a*) genetic background. *Endocrinology* 146(3):1245–1253.

Smeyne R.J., Klein R., Schnapp A., Long L.K., Bryant S., Lewin A., Lira S.A., Barbacid M. (1994) Severe sensory and sympathetic neuropathies in mice carrying a disrupted Trk/NGF receptor gene. *Nature* 368(6468):246–249.

Smit D.J., Smith A.G., Parsons P.G., Muscat G.E., Sturm R.A. (2000) Domains of Brn-2 that mediate homodimerization and interaction with general and melanocytic transcription factors. *European Journal of Biochemistry* 267:6413–6422.

Smith A.G., Luk N., Newton R.A., Roberts D.W., Sturm R.A., Muscat G.E. (2008) Melanocortin-1 receptor signaling markedly induces the expression of the NR4A nuclear receptor subgroup in melanocytic cells. *Journal of Biological Chemistry* 283(18):12564–12570.

Smith D.E., del Amo F.F., Gridley T. (1992) Isolation of Sna, a mouse gene homologous to the *Drosophila* genes snail and escargot: its expression pattern suggests multiple roles during post-implantation development. *Development* 116:1033–1039.

Smits B.M., Cuppen E. (2006a) Rats go genomic. *Genome Biology* 7(2):306.

Smits B.M.G., Cuppen E. (2006b) Rat genetics: the next episode. *Trends in Genetics* 22(4):232–240.

Solano F., García-Borrón J.C. (2006) In: Nordlund J.J. et al. (eds) *The Pigmentary System*, 2nd edn, pp. 261–281. Blackwell Publishing, Oxford.

Solano F., Martínez-Esparza M., Jiménez-Cervantes C., Hill S.P., Lozano J.A., García-Borrón J.C. (2000) New insights on the structure of the mouse *silver* locus and on the function of the silver protein. *Pigment Cell Research* 13(Suppl 8):118–124.

Sone M., Orlow S.J. (2007) The ocular albinism type 1 gene product, OA1, spans intracellular membranes 7 times. *Experimental Eye Research* 85(6):806–816.

Soriano P. (1997) The PDGFα receptor is required for neural crest cell development and for normal patterning of the somites. *Development* 124:2691–2700.

Soriano P. (1999) Generalized lacZ expression with the ROSA26 Cre reporter strain. *Nature Genetics* 21(1):70–71.

Southard-Smith E.M., Kos L., Pavan W.J. (1998) Sox10 mutation disrupts neural crest development in Dom Hirschsprung mouse model. *Nature Genetics* 18:60–64.

Somerville R.P.T., Longpre J.-M., Jungers K.A., Engle J.M., Ross M., Evanko S., Wight T.N., Leduc R., Apte S.S. (1994) Characterization of ADAMTS-9 and ADAMTS-20 as a distinct ADAMTS subfamily related to Caenorhabditis elegans GON-1. *Journal of Biological Chemistry* 278:9503–9513.

Somerville R.P., Longpre J.M., Jungers K.A., Engle J.M., Ross M., Evanko S., Wight T.N., Leduc R., Apte S.S. (2003) Characterization of ADAMTS-9 and ADAMTS-20 as a distinct ADAMTS subfamily related to *Caenorhabditis elegans* GON-1. *Journal of Biological Chemistry* 278(11):9503–9513.

Spanakis E., Lamina P., Bennett D.C. (1992) Effects of the developmental colour mutations silver and recessive spotting on proliferation of diploid and immortal mouse melanocytes in culture. *Development* 114(3):675–680.

Sponenberg D.P. (2009) *Equine Color Genetics*, 3nd edn. Wiley-Blackwell, Oxford.

Sponenberg D.P., Lamoreux M.L. (1985) Inheritance of tweed, a modification of merle, in Australian shepherd dogs. *Journal of Heredity* 76(4):303–304.

Sponenberg D.P., Bigelow B. (1987) An extension locus mosaic Labrador retriever dog. *Journal of Heredity* 78(6):406.

Sponenberg D.P., Ito S., Wakamatsu K., Eng L.A. (1988a) Pigment types in sheep, goats, and llamas. *Pigment Cell Research* 1(6):414–8.

Sponenberg D.P., Ito S., Eng L.A., Schwink K. (1988b) Pigment types of various color genotypes of horses. *Pigment Cell Research* 1(6):410–413.

Spritz R.A. (2007) The genetics of generalized vitiligo and associated autoimmune diseases. *Pigment Cell Research* 20(4):271–278.

Spritz R.A. (2008) The genetics of generalized vitiligo. *Current Directions in Autoimmunity* 10:244–257.

Spritz R.A., Hearing V.J. Jr (1994) Genetic disorders of pigmentation. *Advances in Human Genetics* 22:1–45.

Staleva L., Manga P., Orlow S.J. (2002) Pink-eyed dilution protein modulates arsenic sensitivity and intracellular glutathione metabolism. *Molecular Biology of the Cell* 13(12):4206–4220.

Starcevic M., Dell'Angelica E.C. (2004) Identification of snapin and three novel proteins (BLOS1, BLOS2, and BLOS3/*reduced pigmentation*) as subunits of biogenesis of lysosome-related organelles complex-1 (BLOC-1). *Journal of Biological Chemistry* 279(27):28393–283401.

Steel K.P. (2002) Varitint-waddler: a double whammy for hearing. *Proceedings of the National Academy of Sciences USA* 99:14613–14615.

Steel K.P., Brown D.M. (1994) Genes & Deafness. *Trends in Genetics* 10:428–435.

Steel K.P., Bussoli T.J. (1999) Deafness genes. *Trends in Genetics* 15(6):207–211.

Steel K.P., Davidson D.R., Jackson I.J. (1992) TRP-2/DT, a new early melanoblast marker, shows that steel growth factor (c-kit ligand) is a survival factor. *Development* 115:1111–1119.

Steingrímsson E. (2008) All for one, one for all: alternative promoters and Mitf. *Pigment Cell and Melanoma Research* 21:412–414.

Steingrímsson E., Moore K.J., Lamoreux M.L., Ferré-D'Amaré A.R., Burley S.K., Zimring D.C., Skow L.C., Hodgkinson C.A., Arnheiter H., Copeland N.G. et al. (1994) Molecular basis of mouse *microphthalmia (mi)* mutations helps explain their developmental and phenotypic consequences. *Nature Genetics* 8(3):256–263.

Steingrímsson E., Arnheiter H., Hallsson J.H., Lamoreux M.L., Copeland N.G., Jenkins N.A. (2003) Interallelic complementation at the mouse Mitf locus. *Genetics* 163(1):267–276.

Steingrímsson E., Copeland N.G., Jenkins N.A. (2004) Melanocytes and the microphthalmia transcription factor network. *Annual Review of Genetics* 38:365–411.

Stemple D.L., Anderson D.J. (1992) Isolation of a stem cell for neurons and glia from the mammalian neural crest. *Cell* 71:973–985.

Stephenson D.A., Novak E.K., Chapman V.M. (1998) Analysis of the *Kit* and *Pdgfra* genes in the *patch-extended* (*Ph^e*) mutation. *Genetic Research (Cambridge)* 72:205–210.

Steward T.A., Wagner E.F., Mintz B. (1982) Human beta-globin gene sequences injected into mouse eggs, retained in adults, and transmitted to progeny. *Science* 217(4564):1046–1048.

Strickland F.M., Pathak S., Multani A.S., Pelley R.P., Donawho C.K. (2003) Molecular characterization of new melanoma cell lines from C3H mice induced by ethanol plus ultraviolet radiation. *Cancer Research* 63(13):3503–3510.

Strom M., Hume A.N., Tarafder A.K., Barkagianni E., Seabra M.C. (2002) A family of Rab27-binding proteins. Melanophilin links Rab27a and myosin Va function in melanosome transport. *Journal of Biological Chemistry* 277:25423–25430.

Sturm R.A. (2002) Skin colour and skin cancer – MC1R, the genetic link. *Melanoma Research* 12(5):405–416.

Sun K., Johnson B.S., Gunn T.M. (2007) Mitochondrial dysfunction precedes neurodegeneration in mahogunin (Mgrn1) mutant mice. *Neurobiology of Aging* 28(12):1840–1852.

Sundberg J.P. (ed.) (1994) *Handbook of Mouse Mutations with Skin and Hair Abnormalities: Animal Models and Biomedical Tools*. CRC Press, Boca Raton, FL.

Suto J., Matsuura S., Imamura K., Yamanaka H., Sekikawa K. (1998) Genetics of obesity in KK mouse and effects of A(y) allele on quantitative regulation. *Mammalian Genome* 9(7):506–510.

Suto J., Wakamatsu K., Yamanaka H., Ito S., Sekikawa K. (2000) Quantitative trait loci that modify the sootiness of yellow pigmentation in KK-A(y)/a mice. *Mammalian Genome* 11(8):639–644.

Suzuki I., Cone R., Im S., Nordlund J., Abdel-Malek Z. (1996) Binding of melanotropic hormones to the melanocortin receptor MC1R on human melanocytes stimulates proliferation and melanogenesis. *Endocrinology* 137:1627–1633.

Suzuki N., Hirata M., Kondo S. (2003a) Traveling stripes on the skin of a mutant mouse. *Proceedings of the National Academy of Sciences USA* 100(17):9680–9685.

Suzuki T., Li W., Zhang Q., Novak E.K., Sviderskaya E.V., Wilson A., Bennett D.C., Roe B.A., Swank R.T., Spritz R.A. (2001) The gene mutated in cocoa mice, carrying a defect of organelle biogenesis, is a homologue of the human Hermansky-Pudlak syndrome-3 gene. *Genomics* 78(1/2):30–37.

Suzuki T., Li W., Zhang Q., Karim A., Novak E.K., Sviderskaya E.V., Hill S.P., Bennett D.C., Levin A.V., Nieuwenhuis H.K., Fong C.T., Castellan C., Miterski B., Swank R.T., Spritz R.A. (2002) Hermansky-Pudlak syndrome is caused by mutations in HPS4, the human homolog of the mouse light-ear gene. *Nature Genetics* 30:321–324.

Suzuki T., Oiso N., Gautam R., Novak E.K., Panthier J.J., Suprabha P.G., Vida T., Swank R.T,. Spritz R.A. (2003b) The mouse organellar biogenesis mutant *buff* results from a mutation in Vps33a, a homologue of yeast vps33 and Drosophila carnation. *Proceedings of the National Academy of Sciences USA* 100(3):1146–1150.

Svärd J., Heby-Henricson K., Persson-Lek M., Rozell B., Lauth M., Bergström A., Ericson J., Toftgård R., Teglund S. (2006) Genetic elimination of Suppressor of fused reveals an essential repressor function in the mammalian Hedgehog signaling pathway. *Developmental Cell* 10:187–197.

Svedine S., Wang T., Halaban R., Hebert D.N. (2004) Carbohydrates act as sorting determinants in ER associated degradation of tyrosinase. *Journal of Cell Science* 117:2937–2949.

Sviderskaya E.V., Wakeling W.F., Bennett D.C. (1995) A cloned, immortal line of murine melanoblasts inducible to differentiate to melanocytes. *Development* 121(5):1547–1557.

Sviderskaya E.V., Bennett D.C., Ho L., Bailin T., Lee S.T., Spritz R.A. (1997) Complementation of hypopigmentation in p-mutant (pink-eyed dilution) mouse melanocytes by normal human P cDNA, and defective complementation by OCA2 mutant sequences. *Journal of Investigative Dermatology* 108(1):30–34.

Sviderskaya E.V., Hill S.P., Evans-Whipp T.J., Chin L., Orlow S.J., Easty D.J., Cheong S.C., Beach D., DePinho R.A., Bennett D.C. (2002) p16(Ink4a) in melanocyte senescence and differentiation. *Journal of the National Cancer Institute* 94(6):446–454.

Swank R.T., Novak E.K., McGarry M.P., Rusiniak M.E., Feng L. (1998) Mouse models of Hermansky Pudlak syndrome. *Pigment Cell Research* 11(2):60–80.

Swank R.T., Jiang S.Y., Reddington M., Conway J., Stephenson D., McGarry M.P., Novak E.K. (1993) Inherited abnormalities in platelet organelles and platelet formation and associated altered expression of low molecular weight guanosine triphosphate-binding proteins in the mouse pigment mutant gunmetal. *Blood* 81(10):2626–2635.

Sweet H.O. (1983) Dilute suppressor, a new suppressor gene in the house mouse. *Journal of Heredity* 74(4):305–306.

Sweet H.O., Brilliant M.H., Cook S.A., Johnson K.R., Davisson M.T. (1998) A new allelic series for the underwhite gene on mouse chromosome 15. *Journal of Heredity* 89(6):546–551.

Switonski M., Szczerbal I., Nowacka J. (2004) *Journal of Applied Genetics* 45(2):195–214.

Tachibana M. (1999) Sound needs sound melanocytes to be heard. *Pigment Cell Research* 12:344–354.

Tachibana M., Hara Y., Vyas D., Hodgkinson C., Fex J., Grundfast K., Arnheiter H. (1992) Cochlear disorder associated with melanocyte anomaly in mice with a transgenic insertional mutation. *Molecular and Cellular Neuroscience* 3:433–445.

Tachibana M., Kobayashi Y., Matsushima Y. (2003) Mouse models for four types of Waardenburg Syndrome. *Pigment Cell Research* 16:448–454.

Takayama H., LaRochelle W.J., Anver M., Bockman D.E., Merlino G. (1996) Scatter factor/hepatocyte growth factor as a regulator of skeletal muscle and neural crest development. *Proceedings of the National Academy of Sciences USA* 93:5866–5871.

Takayama H., LaRochelle W.J., Sharp R., Otsuka T., Kriebel P., Anver M., Aaronson A., Merlino G. (1997) Diverse tumorigenesis associated with aberrant development in mice overexpressing hepatocyte growth factor/scatter factor. *Proceedings of the National Academy of Sciences USA* 94:701–706.

Takeda K., Yasumoto K., Takada R., Takada S., Watanabe K., Udono T., Saito H., Takahashi K., Shibahara S. (2000) Induction of melanocyte-specific microphthalmia-associated transcription factor by Wnt-3a. *Journal of Biological Chemistry* 275:14013–6.

Tamate H.B., Takeuchi T. (1984) Action of the *e* locus of mice in the response of phaeomelanic hair follicles to α-melanocyte-stimulating hormone *in vitro*. *Science* 224:1241–1242.

Taraviras S., Marcos-Gutierrez C.V., Durbec P., Jani H., Grigoriou M., Sukumaran M., Wang L.-C., Hynes M., Raisman G., Pachnis V. (1999) Signaling by the RET receptor tyrosine kinase and its role in the development of the mammalian enteric nervous system. *Development* 126:2785–2797.

Terry R.B., Archer S., Brooks S., Bernoco D., Bailey E. (2004) Assignment of the appaloosa coat colour gene (LP) to equine chromosome 1. *Animal Genetics* 35(2):134–137.

Terry R.W., Kwee L., Baldwin H.S., Labow M.A. (1997) Cre-mediated generation of a VCAM-1 null allele in transgenic mice. *Transgenic Research* 6(5):349–356.

Theos A.C., Truschel1 S.T., Raposo G., Marks M.S. (2005a) The silver locus product Pmel17/gp100/SI/ME20: controversial in name and in function. *Pigment Cell Research* 18(5):322–336.

Theos A.C., Tenza D., Martina J.A., Hurbain U., Peden A.A., Sviderskaya E.V., Stewart A., Robinson M.S., Bennett D.C., Cutler D.F., Bonifacino J.S., Marks M.S., Raposo G. (2005b) Functions of adaptor protein (AP-3 and AP-1) in tyrosinase sorting from endosomes to melanosomes. *Molecular Biology of the Cell* 16:5356–5372.

Theos A.C., Berson J.F., Theos S.C., Herman K.E., Harper D.C., Tenza D., Sviderskaya E.V., Lamoreux M.L., Bennett D.C., Raposo G., Marks M.S. (2006) Dual loss of ER export and endocytic signals with altered melanosome morphology in the silver mutation of Pmel17. *Molecular Biology of the Cell* 17(8):3598–3612.

Thiery J.-P. (2003) Epithelial-mesenchymal transitions in development and pathologies. *Current Opinion in Cell Biology* 15:740–746.

Thomas A.J., Erickson C.A. (2008) The making of a melanocyte: the specification of melanoblasts from the neural crest. *Pigment Cell and Melanoma Research* 21(6):598–610.

Tissot R.G., Beattie C.W., Amoss M.S. Jr, Williams J.D., Schumacher J. (1993) Common swine leucocyte antigen (SLA) haplotypes in NIH and Sinclair miniature swine have similar effects on the expression of an inherited melanoma. *Animal Genetics* 24(3):191–193.

Tobin D.J., Hagen E., Botchkarev V.A., Paus R. (1998) Do hair bulb melanocytes undergo apoptosis during hair follicle regression (catagen)? *Journal of Investigative Dermatology* 111:941–947.

Tolmachova T., Anders R., Stinchcombe J., Bossi G., Griffiths G.M., Huxley C., Seabra M.C. (2004) A general role for Rab27a in secretory cells. *Molecular Biology of the Cell* 15:332–344.

Tonks I.D., Nurcombe V., Paterson C., Zournazi A., Prather C., Mould A.W., Kay G.F. (2003) Tyrosinase-Cre mice for tissue-specific gene ablation in neural crest and neuroepithelial-derived tissues. *Genesis* 37(3):131–138.

Townsend D., Witkop C.J., Mattson J. (1981) Tyrosinase subcellular distribution and kinetic parameters in wild type and *c*-locus mutant C57BL/6J. *Journal of Experimental Zoology* 216:113–119.

Toyoda R., Kasai A., Sato S., Wada S., Saiga H., Ikeo K., Gojobori T., Numakunai T., Yamamoto H. (2004) Pigment cell lineage-specific expression activity of the ascidian tyrosinase-related gene. *Gene* 12:61–69.

Toyofuku K., Wada I., Valencia J.C., Kushimoto T., Ferrans V.J., Hearing V.J. (2001) Oculocutaneous albinism types 1 and 3 are ER retention diseases: mutation of tyrosinase or Tyrp1 can affect the processing of both mutant and wild-type proteins. *FASEB Journal* 15(12):2149–2161.

Toyofuku K., Valencia J.C., Kushimoto T., Costin G.-E., Virador V.M., Vieira W.D., Ferrans V.J., Hearing V.J. (2002) The etiology of oculocutaneous albinism (OCA) Type II: The pink protein modulates the processing and transport of tyrosinase. *Pigment Cell Research* 15:217–224.

Trainor P.A., Krumlauf R. (2001) Hox genes, neural crest cells and branchial arch patterning. *Current Opinion in Cell Biology* 13:698–705.

Tshori S., Gilon D., Beeri R., Nechushtan H., Kaluzhny D., Pikarsky E., Razin E. (2006) Transcription factor MITF regulates cardiac growth and hypertrophy. *Journal of Clinical Investigation* 116(10):2673–2681.

Tsukamoto K., Jackson I.J., Urabe K., Montague P.M., Hearing V.J. (1992) A second tyrosinase-related protein, TRP-2, is a melanogenic enzyme termed DOPAchrome tautomerase. *EMBO Journal* 11(2):519–526.

Turner P., Robinson R. (1980) Melanin inhibitor: a dominant gene in the domestic cat. *Journal of Heredity* 71(6):427–428.

Udono T., Yasumoto K., Takeda K., Amae S., Watanabe K., Saito H., Fuse N., Tachibana M., Takahashi K., Tamai M., Shibahara S. (2000) Structural organization of the human microphthalmia-associated transcription factor gene containing four alternative promoters. *Biochimica et Biophysica Acta* 1491(1–3):205–219.

Uehara S., Izumi Y., Kubo Y., Wang C.C., Mineta K., Ikeo K., Gojobori T., Tachibana M., Kikuchi T., Kobayashi T. et al. (2009) Specific expression of Gsta4 in mouse cochlear melanocytes: a novel role for hearing and melanocyte differentiation. *Pigment Cell and Melanoma Research* 22(1):111–119.

Vance K.W., Goding C.R. (2004) The transcription network regulating melanocyte development and melanoma. *Pigment Cell Research* 17:318–325.

Van Raamsdonk C.D., Fitch K.R., Fuchs H., de Angelis M.H., Barsh G.S. (2004) Effects of G-protein mutations on skin color. *Nature Genetics* 36(9):961–968.

Van Steensel M.A.M., Happle R., Steijlen P.M. (2000) Molecular Genetics of the hair follicle. The State of the Art. *Proceedings of the Society Experimental Biology and Medicine* 223:1–7.

Vega S., Morales A.V., Ocana O.H., Valdés F., Fabregat I., Nieto M.A. (2007) Snail blocks the cell cycle and confers resistance to cell death. *Genes & Development* 18:1131–1143.

Veis D.J., Sorenson C.M., Shutter J.R., Korsmeyer S.J. (1993) Bcl-2-deficient mice demonstrate fulminant lymphoid apoptosis, polycystic kidneys, and hypopigmented hair. *Cell* 75(2):229–240.

Vella C.M., Shelton L.M., McGonagle J.J., Stanglein T.W. (2003) *Robinson's Genetics for Cat Breeders and Veterinarians*. Butterworth-Heinemann, Oxford.

Vijayasaradhi S., Xu Y., Bouchard B., Houghton A.N. (1995) Intracellular sorting and targeting of melanosomal membrane proteins: identification of signals for sorting of the human brown locus protein, GP75. *Journal of Cell Biology* 130(4):807–820.

Vleminck K., Kemler R. (1999) Cadherins and tissue formation: integrating adhesion and signaling. *Bioessays* 21(3):211–220.

Vogel P., Read R.W., Vance R.B., Platt K.A., Troughton K., Rice D.S. (2008) Ocular albinism and hypopigmentation defects in *Slc2424a5* -/- mice. Veterinary Pathology 45:264–279.

Voisey J., Van Daal A. (2002) Agouti: from mouse to man, from skin to fat. *Pigment Cell Research* 15:10–18.

Vrieling H., Duhl D.M., Millar S.E., Miller K.A., Barsh G.S. (1994) Differences in dorsal and ventral pigmentation result from regional expression of the mouse agouti gene. *Proceedings of the National Academy of Sciences USA* 91(12):5667–5671.

Wakamatsu K., Hirobe T., Ito S. (2007a) High levels of melanin-related metabolites in plasma from *pink-eyed dilution* mice. *Pigment Cell Research* 20;222–224.

Wakamatsu K., Hu D.-N., McCormick S.A., Ito S. (2007b) Characterization of melanin in human iridal and choroidal melanocytes from eyes with various colored irides. *Pigment Cell and Melanoma Research* 20:1–9.

Walker W.P., Aradhya S., Hu C.L., Shen S., Zhang W., Azarani A., Lu X., Barsh G.S., Gunn T.M. (2007) Genetic analysis of attractin homologs. *Genesis* 45(12):744–756.

Wang R., Tang P., Wang P., Boissy R.E., Zheng H. (2006) Regulation of tyrosinase trafficking and processing by presenilins: Partial loss of function by familial Alzheimer's disease mutation. *Proceedings of the National Academy of Sciences USA* 103(2):353–358.

Wang X., Erf G.F. (2003) Melanocyte-specific cell mediated immune response in vitiliginous Smyth line chickens. *Journal of Autoimmunity* 21(2):149–160.

Wang X., Erf G.F. (2004) Apoptosis in feathers of Smyth line chickens with autoimmune vitiligo. *Journal of Autoimmunity* 22(1):21–30.

Wang Y., Huso D., Cahill H., Ryugo D., Nathans J. (2001) Progressive cerebellar, auditory, and esophageal dysfunction caused by targeted disruption of the frizzled-4 gene. *Journal of Neurochemistry* 21(13):4761–4771.

Wasmeier C., Romao M., Plowright L., Bennett D.C., Raposo G., Seabra M.C. (2006) Rab38 and Rab32 control post-Golgi trafficking of melanogenic enzymes. *Journal of Cell Biology* 175(2):271–281.

Watabe H., Valencia J.C., Yasumoto K.-I., Kushimoto T., Ando H., Muller J., Vieira W.D., Mizoguchi M., Ettore A., Hearing V.J. (2004) Regulation of tyrosinase processing and trafficking by organellar pH and by proteasome activity. *Journal of Biological Chemistry* 279(9):7971–7981.

Watanabe A., Takeda K., Ploplis., Tachibana M. (1998) Epistatic relationship between Waardenburg syndrome genes MITF and PAX3. *Nature Genetics* 18:283–286.

Watanabe K., Takeda K., Yasumoto K., Udono T., Saito H., Ikeda K., Takasaka T., Takahashi K., Kobayashi T., Tachibana M., Shibahara S. (2002) Identification of a distal enhancer for the melanocyte-specific promoter of the MITF gene. *Pigment Cell Research* 15(3):201–211.

Waters J.M., Richardson G.D., Colin A.B. Jahoda C.A.B. (2007) Hair follicle stem cells. *Seminars in Cell & Developmental Biology* 18:245–254.

Wayne R.K. (1993) Molecular evolution of the dog family. *Trends in Genetics* 9(6):218–224.

Wayne R.K., Ostrander E.A. (2007) Lessons learned from the dog genome. *Trends in Genetics* 23(11):557–567.

Wegner M. (2005) Secrets to a healthy Sox life: lessons for melanocytes. *Pigment Cell Research* 18(2):74–85.

Wehrle-Haller B. (2003) The role of kit-ligand in melanocyte development and epidermal homeostasis. *Pigment Cell Research* 16:287–296.

Wehrle-Haller B., Weston J.A. (1995) Soluble and cell-bound forms of steel factor activity play distinct roles in melanocyte precursor dispersal and survival on the lateral neural crest migration pathway. *Development* 121:731–742.

Wehrle-Haller B., Weston J.A. (1997) Receptor tyrosine kinase-dependent neural crest migration in response to differentially localized growth factors. *Bioessays* 19:337–345.

Wehrle-Haller B., Morrison-Graham K., Weston J.A. (1996) Ectopic c-kit expression affects the fate of melanocyte precursors in Patch mutant embryos. *Developmental Biology* 177:463–474.

Wehrle-Haller B., Meller M., Weston J.A. (2001) Analysis of melanocyte precursors in Nf1 mutants reveals that MGF/KIT signaling promotes directed cell migration independent of its function in cell survival. *Developmental Biology* 232:471–483.

Wei K., Chen J., Akrami K., Galbraith G.C., Lopez I.A., Chen F. (2007) Neural crest cell deficiency of c-myc causes skull and hearing defects. *Genesis* 45(6):382–390.

Weiner L., Han R., Scicchitano B.M., Li J., Hasegawa K., Grossi M., Lee D., Brissette J.L. (2007) Dedicated epithelial recipient cells determine pigmentation patterns. *Cell* 130(5):932–942.

Weston J.A. (1991) Sequential segregation and fate of developmentally restricted intermediate cell populations in the neural crest lineage. *Current Topics in Developmental Biology* 25:133–153.

Whitney J.B. III, Lamoreux M.L. (1982) Transposable elements controlling genetic instabilities in mammals. *Journal of Heredity* 73:12–18.

Wick G., Andersson L., Hala K., Gershwin M.E., Selmi C., Erf G.F., Lamont S.J., Sgonc R. (2006) Avian models with spontaneous autoimmune diseases. *Advances in Immunology* 92:71–117.

Wilkie A.L., Jordan S.A., Jackson I.J. (2002) Neural crest progenitors of the melanocyte lineage: coat colour patterns revisited. *Development* 129(14):3349–3357.

Wilkinson M.G., Millar J.B.A. (2000) Control of the eukaryotic cell cycle by MAP kinase signaling pathways. *FASEB Journal* 14:2147–2157.

Wilson S.M., Yip R., Swing D.A., O'Sullivan T.N., Zhang Y., Novak E.K., Swank R.T., Russell L.B., Copeland N.G., Jenkins N.A. (2000) A mutation in Rab27a causes the vesicle transport defects observed in ashen mice. *Proceedings of the National Academy of Sciences USA* 97:7933–7938.

Wilson Y.M., Richards K.L., Ford-Perriss M.L., Panthier J.-J., Murphy M. (2004) Neural crest cell lineage segregation in the mouse neural tube. *Development* 131:6153–6162.

Winder A.J., Wittbjer A., Rosengren E., Rorsman H. (1993) The mouse *brown (b)* locus protein has dopachrome tautomerase activity and is located in lysosomes in transfected fibroblasts. *Journal of Cell Science* 106(Pt 1):153–166.

Winder A., Kobayashi T., Tsukamoto K., Urabe K., Aroca P., Kameyama K., Hearing V.J. (1994) The tyrosinase gene family – interactions of melanogenic proteins to regulate melanogenesis. *Cellular and Molecular Biology Research* 40(7–8):613–626.

Wolfe H.G., Coleman D.L. (1966) Pigmentation. In: *Biology of the Laboratory Mouse*, pp. 405–425. McGraw Hill, New York.

Wolff G.L. (2003) Regulation of yellow pigment formation in mice: a historical perspective. *Pigment Cell Research* 16(1):2–15.

Wolnicka-Glubisz A., Noonan F.P. (2006) Neonatal susceptibility to UV induced cutaneous malignant melanoma in a mouse model. *Photochemistry and Photobiology Sciences* 5(2):254–260.

Womack J.E. (1987) Comparative gene mapping: a valuable new tool for mammalian developmental studies. *Developmental Genetics* 8(4):281–293.

Womack J.E. (2006) The bovine genome. *Genome Dynamics* 2:69–78.

Wong C.E., Paratore C., Dours-Zimmermann M.T., Rochat A., Pietri T., Suter U., Zimmermann D.R., Dufour S., Thiery J.P., Meijer D. et al. (2006) Neural crest-derived cells with stem cell features can be traced back to multiple lineages in the adult skin. *Journal of Cell Biology* 175:1005–1015.

Wright S. (1917) Color inheritance in mammals. II. The mouse. *Journal of Heredity* 8:373–378.

Wright S. (1925) The factors of the albino series of guinea-pigs and their effects on black and yellow pigmentation. *Genetics* 10(3):223–260.

Wright S. (1928) An eight-factor cross in the Guinea Pig. *Genetics* 13(6):508–531.

Wright S., Chase H.B. (1936) On the genetics of the spotted pattern of the Guinea Pig. *Genetics* 21(6):758–787.

Wu J.J., Chen J.X., Rothman T.P., Gerson M.D. (1999a) Inhibition of in vitro enteric neuronal development by endothelin-3: mediation by endothelin B receptors. *Development* 126:1161–1173.

Wu M., Hemesath T.J., Takemoto C.M., Horstmann M.A., Wells A.G., Price E.R., Fisher D.Z., Fisher D.E. (1999b) c-Kit triggers dual phosphorylations, which couple activation and degradation of the essential melanocyte factor Mi. *Genes & Development* 14:301–312.

Wu X., Bowers B., Rao K., Wei Q., Hammer J.A. 3rd (1998) Visualization of melanosome dynamics within wild type and dilute melanocytes suggests a paradigm for myosin V function in vivo. *Journal of Cell Biology* 143:1899–1918.

Wu X., Rao K., Bowers M.B., Copeland N.G., Jenkins N.A., Hammer J.A. III. (2001) Rab27a enables myosin Va-dependent melanosome capture by recruiting the myosin to the organelle. *Journal of Cell Science* 114:1091–1100.

Wu X.S., Rao K., Zhang H., Wang F., Sellers J.R., Matesic L.E., Copeland N.G., Jenkins N.A., Hammer J.A. III. (2002) Identification of an organelle receptor for myosin-Va. *Nature Cell Biology* 4:271–278.

Yajima I., Larue L. (2008) The location of heart melanocytes is specified and the level of pigmentation in the heart may correlate with coat color. *Pigment Cell and Melanoma Research* 21:471–476.

Yajima I., Sato S., Kimura T., Yasumoto K., Shibahara S., Goding C.R., Yamamoto H. (1998) An L1 element intronic insertion in the black-eyed white (Mitf[mi-bw]) gene: the loss of a single Mitf isoform responsible for the pigmentary defect and inner ear deafness. *Human Molecular Genetics* 8(8):1431–1441.

Yajima I., Endo K., Sato S., Toyoda R., Wada H., Shibahara S., Numakunai T., Ikeo K., Gojobori T., Goding C.R., Yamamoto H. (2003) Cloning and functional analysis of ascidian Mitf in vivo: insights into the origin of vertebrate pigment cells. *Mechanisms of Development* 120(12):1489–1504.

Yajima I., Belloir E., Bourgeois Y., Kumasaka M., Delmas V., Larue L. (2006) Spatiotemporal gene control by the Cre-ERT2 system in melanocytes. *Genesis* 44(1):34–43.

Yamaguchi H., Aiba A., Nakamura K., Nakao K., Sakagami H., Goto K., Kondo H., Katsuki M. (1996) Dopamine D2 receptor plays a critical role in cell proliferation and proopiomelanocortin expression in the pituitary. *Genes Cells* 1(2):253–268.

Yamaguchi Y., Hearing V.J. (2009) Physiological factors that regulate skin pigmentation. *Biofactors* 35(2):193–199.

Yanfeng W., Saint-Jeannet J.-P., Klein P.S. (2003) Wnt-frizzled signaling in the induction and differentiation of the neural crest. *Bioessays* 25:317–325.

Yang Y.K., Thompson D.A., Dickinson C.J., Wilken J., Barsh G.S., Kent S.B., Gantz I. (1999) Characterization of Agouti-related protein binding to melanocortin receptors. *Molecular Endocrinology* 13(1):148–155.

Yang W., Li C., Ward D.M., Kaplan J., Mansour S.L. (2000) Defective organellar membrane protein trafficking in Ap3b1-deficient cells. *Journal of Cell Science* 113(Pt 22):4077–4086.

Yasuda H., Galli S.J., Geissler E.N. (1993) Cloning and functional analysis of the mouse c-kit promoter. *Biochemical and Biophysical Research Communications* 191(3):893–901.

Yasumoto K., Takeda K., Saito H., Watanabe K., Takahashi K., Shibahara S. (2002) Microphthalmia-associated transcription factor interacts with LEF-1, a mediator of Wnt signaling. *EMBO Journal* 21:2703–2714.

Yasumoto K.-I., Watabe H., Valencia J.C., Kushimoto T., Kobayashi T., Appella E., Hearing V.J. et al. (2004) Epitope mapping of the melanosomal matrix protein gp100 (PMEL17). Rapid processing in the endoplasmic reticulum and glycosylation in the early golgi compartment. *Journal of Biological Chemistry* 279(27):28330–28338.

Yaswen L., Diehl N., Brennan M.B., Hochgeschwender U. (1999) Obesity in the mouse model of pro-opiomelanocortin deficiency responds to peripheral melanocortin. *Nature Medicine* 5(9):1066–1070.

Yonetani S., Moriyama M., Nishigori C., Osawa M., Nishikawa S.-I. (2007) In vitro expansion of immature melanoblasts and their ability to repopulate melanocyte stem cells in the hair follicle. *Journal of Investigative Dermatology* 128(2):408–420.

Yoshida H., Kunisada T., Kusakabe M., Nishikawa S., Nishikawa S.-I. (1996) Distinct stages of melanocyte differentiation revealed by analysis of nonuniform pigmentation patterns. *Development* 122:1207–1214.

Young A., Powelson E.B., Whitney I.E., Raven M.A., Nusinowitz S., Jiang M., Birnbaumer L., Reese B.E., Farber D.B. (2008) Involvement of OA1, an intracellular GPCR, and G alpha i3, its binding protein, in melanosomal biogenesis and optic pathway formation. *Investigative Ophthalmology & Visual Science* 49(7):3245–3252.

Yu J.-K., Meulemans D., McKeown S.J. et al. (2008) Insights from the amphioxus genome on the origin of vertebrate neural crest. *Genome Research* 18:1127–1132.

Zdarsky E., Favor J., Jackson I.J. (1990) The molecular basis of *brown*, an old mouse mutation, and of an induced revertant to wild type. *Genetics* 126(2):443–449.

Zhang Q., Zhao B., Li W., Oiso N., Novak E.K., Rusiniak M.E., Gautam R., Chintala S., O'Brien E.P., Zhang Y. et al. (2003) Ru2 and Ru encode mouse orthologs of the genes mutated in human Hermansky-Pudlak syndrome types 5 and 6. *Nature Genetics* 33(2):145–153.

Zhang Y., Andl T., Yang S.H., Teta M., Liu F., Seykora J.T., Tobias J.W., Piccolo S., Schmidt-Ullrich R., Nagy A. et al. (2008) Activation of beta-catenin signaling programs embryonic epidermis to hair follicle fate. *Development* 135(12):2161–2172.

Zhao S., Hung F.-C., Colvin J.S., White A., Dai W., Lovicul F.J., Ornitz D.M., Overbeek P.A. (2001) Patterning the optic neuroepithelium by FGF signaling and Ras activation. *Development* 130:1565–1577.

Zhen L., Jiang S., Feng L., Bright N.A., Peden A.A., Seymour A.B., Novak E.K., Elliott R., Gorin M.B., Robinson M.S., Swank R.T. (1999) Abnormal expression and subcellular distribution of subunit proteins of the AP-3 adaptor complex lead to platelet storage pool deficiency in the pearl mouse. *Blood* 94(1):146–155.

Zhou J., Callapina M., Goodall G.J., Brüne B. (2004) Functional integrity of nuclear factor kappaB, phosphatidylinositol 3'-kinase, and mitogen-activated protein kinase signaling allows tumor necrosis factor alpha-evoked Bcl-2 expression to provoke internal ribosome entry site-dependent translation of hypoxia-inducible factor 1alpha. *Cancer Research* 64(24):9041–9048.

Zhou B.P., Hung M.C. (2005) Wnt, hedgehog and snail: sister pathways that control by GSK-3beta and beta-Trcp in the regulation of metastasis. *Cell Cycle* 4(6):772–776.

Zhu L., Lee H.O., Jordan C.S., Cantrell V.A., Southard-Smith E.M., Shin M.K. (2004) Spatiotemporal regulation of endothelin receptor-B by SOX10 in neural crest-derived enteric neuron precursors. *Nature Genetics* 36(7):732–737.

Ziegler I. (2003) The pteridine pathway in zebrafish: regulation and specification during the determination of neural crest cell-fate. *Pigment Cell Research* 16:172–182.

Zsebo K.M., Williams D.A., Geissler E.N., Broudy V.C., Martin F.H., Atkins H.L., Hsu R.Y., Birkett N.C., Okino K.H., Murdock D.C. et al. (1990) Stem cell factor is encoded at the Sl locus of the mouse and is the ligand for the c-kit tyrosine kinase receptor. *Cell* 63(1):213–224.

Index

Note: page numbers in *italics* refer to figures; those in **bold** refer to tables. Genetic locus terms have been integrated under the current locus name, except where very long, in which case the current locus symbol is used.

Printed and bound by CPI Group (UK) Ltd, Croydon, CR0 4YY
15/09/2026
14962396-0001